FRAMEMAKER - CREATING AND PUBLISHING CONTENT

About the Authors

Matt Sullivan

Matt is the founder of Tech Comm Tools (you can subscribe to his free newsletter at www.techcommtools.com). Along with FrameMaker consulting, Matt is most interested in helping organizations improve their use of video and interactive media within their documentation, and delivering that content to users via mobile devices.

Along with training and consulting services, you can access free and paid courseware at http://training.techcommtools.com.

Connect with Matt on social platforms using his personal social id, mattrsullivan, or his Tech Comm Tools id, tc2ls.

Rick Quatro

Rick is president of Carmen Publishing Inc. and has been working with FrameMaker since 1993. He has written over 6,600 scripts since 1997 for clients worldwide. He is recognized as one of the top FrameMaker automation experts in the world.

FrameMaker - Creating and Publishing Content

Matt R. Sullivan

FrameMaker - Creating and Publishing Content

Copyright © 2015, Matt R. Sullivan

Tech Comm Tools, and their logos are trademarks of Matt R. Sullivan. Adobe and FrameMaker are trademarks of Adobe Systems Incorporated. All other trademarks used herein are the properties of their respective owners and are used for identification purposes only.

No part of this book may be reproduced or transmitted in any form or by any means, graphic, electronic, or mechanical, including photocopying, recording, taping, or by any information storage or retrieval system, without written permission from Matt R. Sullivan.

Every effort was made to ensure this book is accurate. However, Matt R. Sullivan assumes no responsibility for errors or omissions, or for the use of information in this book.

ISBN-13: 978-0-9967157-9-9

Written and composed in Adobe® FrameMaker®, this book contains material adapted from *Publishing Fundamentals: Unstructured FrameMaker 11* (ISBN 978-0-9828118-5-6).

Cover design: Sean Byrne, updated by Matt R. Sullivan

Table of Contents

Part I, Front Matter

Acknowledgments xix
Preface xx
Introduction xxi
New features by version xxii
New since FrameMaker 8 xxii
Who should read this book xxiv
What's in this book xxiv
Conventions used in this book xxviii
Mistakes were made xxviii

Part I, Getting started with Framemaker

Chapter 1: FrameMaker strengths 3
FrameMaker strengths 4
Uses for FrameMaker 8

Chapter 2: Creating your first document 9
Starting FrameMaker 10
Setting up a new document 11
    Exploring the standard templates 12
    Using a file as a template 13
    Creating a blank document 13
    Using an existing file 15
Opening, saving, closing, and printing documents 15
Understanding the document window 16
    Changing the zoom settings 16
    Displaying nonprinting items in the document window 17
    Changing view options 19
FrameMaker interface options 19
    Workspaces 19
    Toolbars 20
    Icons 20
    Document tabs 21
    Pods 21
    Managing open files 22
    Restore last session 23

Understanding file configuration and preferences 23
    Storing format definitions 23
    Importing and embedding graphics 23
    Document control 23
    Using the Fonts pod 25
    Handling missing fonts 25
    Organizing directories and files 26
    Saving file information with XMP 27
    Setting automatic backup and save features 28
Switching between structured and unstructured interfaces 29
    File naming conventions 29

Part II, Creating and Manipulating Text

Chapter 3: Word-Processing Features 33
Standard Word-Processing Features 33
    Undoing and redoing changes 34
    Inserting special characters 35
    Searching and replacing 38
    Spell-checking 43
    Thesaurus 51
Configuring Default Text Options 51
    Smart quotes 52
    Smart spaces 53
    Restricting line breaks 53
    Formatting superscripts, subscripts, and small caps 54
Tracking Changes 55
    Using the Track Text Edits tool 55
    PDF review 57
    Displaying the word count 58
    Using change bars 58
    Displaying line numbers 59
    Comparing documents and books 60
Working with Footnotes 62
    Inserting footnotes 64
    Using the same number for several footnotes 65
    Customizing footnote properties 66
Importing Text from Other Applications 69
    MS Word (copy into document) 69
    Copying, pasting, and converting text 71

Chapter 4: Establishing a workflow in FrameMaker 73
Workflow considerations 74
Planning 75
Creating and editing content 76
File organization 76
    Location of files 77
    Folder structure 77
    Book structure 78

Illustration	78
Editing/reviewing	78
Production editing	79
Indexing	79
Printing	79
Single sourcing and digital publishing	80
Evaluating methods	80
Digital publishing workflows	85
Language support	87
Unicode support	88
Dictionary support	88

Chapter 5: Templates — 89

User roles	89
Definition of a FrameMaker template	90
Why you should care about templates	90
Getting started with templates	91
Importing settings from a template file	91
What's imported?	92
What makes a "good" template?	92
Understanding template interactions	93
Tips and tricks	98
Paragraph tags	98
Character tags	99
Table tags	100
Reference pages	100
Entering Special Characters in dialog boxes	100
Naming conventions	104
Capitalization	105
Special characters	105
Taking keyboard shortcuts into account	105
Separating out "housekeeping" tags	106
Documenting your template	107
Using single-purpose templates	107

Chapter 6: Formatting text with paragraph tags — 109

The purpose of paragraph tags	109
Applying paragraph tags	110
Using the paragraph catalog to apply a tag	110
Using the formatting bar	111
Selecting a menu choice	112
Using Smart Insert for paragraph formats	112
Avoiding formatting overrides	112
Modifying paragraph tags	113
Paragraph Designer settings	114
Creating paragraph tags	126
Renaming paragraph tags	127
Setting properties across the entire paragraph catalog	128
Updating selected paragraphs globally	128

  Deleting paragraph tags 129
  Autonumbering details 130
    Basic autonumbering 131
    Numbered steps (1, 2, 3 and a, b, c) 132
    Incorporating chapter and volume numbers 134
    Creating numbered headings 136
    Marking the end of a story 137

Chapter 7: Formatting text with character tags 139
  Maintaining consistency with character tags 139
  Applying Character Tags 140
    Selecting character format with the Character Catalog 140
    Selecting character format with a menu choice 140
    Using Smart Insert for character formats 140
  Modifying character tags 141
    As Is character properties 142
    Avoiding character tag overrides 143
  Creating character tags 143
    Renaming character tags 145
    Updating character properties globally 145
  Removing character tag formatting 146
  Deleting character tags 147
  Additional character tags tips 148

Chapter 8: Understanding table design 149
  Table tag advantages 149
  Inserting tables 150
    Selecting cells 151
    Moving and deleting tables 151
  Changing the assigned table tag 152
  Modifying a table tag 152
    Changing table designer settings 154
    Globally updating table tags 158
  Creating a table tag 158
  Customizing tables outside the Table Designer 158
    Adding and deleting rows and columns 159
    Resizing columns 160
    Applying paragraph tags 162
  Customizing cell ruling and shading 163
  Merging table cells 165
  Rotating table cells 167
  Sorting table data 168
  Deleting table tags 169
  Creating tables from text 170
  Converting a table to text 172

Chapter 9: Cross-references 175
  Advantages of cross-references 175

Inserting Cross-References	177
Creating paragraph-based cross-references	177
Creating spot cross-references	178
Setting Up Cross-Reference Formats	179
Formatting cross-references	180
Cross-reference building blocks	181
Cross-reference examples	184
Deleting cross-reference formats	184
Updating cross-references	185
Updating Cross-References in a Book	185
Forcing a cross-reference update	185
Automatic updating problems	186
Preventing automatic updates	186
Renaming files without breaking cross-references	187
Correcting unresolved (broken) cross-references	187
Changing the referenced file	188
Replacing broken cross-references	189
Converting cross-references to text	189

Chapter 10: Storing content in variables — 191

Advantages of variables	191
Inserting variables	192
Using the Variables pod	192
Using Smart Insert for variables	193
System variable definitions	193
Modifying system variables	193
Valid locations for system variables	193
Date and time variables	194
Numerical variables	195
Filename variables	196
Table variables	197
Running header/footer variables	197
Updating System Variables	201
Creating user variables	201
Modifying user variable definitions	203
Converting user variables to text	204
Deleting user variables	205

Part III, Controlling Page Layout

Chapter 11: Understanding master pages — 209

Advantages of master pages	209
Displaying master pages	210
Assigning master pages to body pages	210
Manually assign master pages	210
Mapping paragraph tags to master pages	211

Creating and managing master pages 214
    Creating default master pages 215
    Creating custom master pages 215
    Renaming master pages 216
    Rearranging master pages 217
Modifying master pages 217
    Adding Text Flows 218
    Setting up headers and footers 218
    Setting up a watermark effect 219
    Creating a landscape master page 221
    Creating bleeding tabs 225
Removing master page overrides 229
    Updating master pages and page layouts 229
Deleting master pages 231

Chapter 12: Text flows 233
Text flow considerations 233
Understanding text flows 234
    Text frame overflows 235
    Understanding text frame properties 235
    Adding text flows to master pages 236
    Drawing text frames on the body page 237
    Selecting text frames 238
    Resizing text frames 238
    Moving and copying text frames 239
    Deleting text frames 240
Customizing text flows 240
    Connecting text flows 241
    Creating room for side heads 242
    Aligning text across columns 246
    Splitting text frames 251
    Disconnecting text frames 251
    Changing the flow tag 252
    Switching from single- to double-sided pages 253

Chapter 13: Importing graphic content 255
Advantages of using external graphics 255
Anchoring Graphics 256
    Inserting an anchored frame 256
    Positioning anchored frames 259
    Importing a graphic 259
    Shrink-wrapping an anchored frame 265
    Relinking missing imported graphics 266
    Setting anchored frame object properties 267
Graphic Formats 270
    Choosing the best graphics format 270
    Image facets 273
    Cross-platform images 273
    Transparency 274

　　　　Placing graphics on the reference pages 275
　　　　　　Placing graphics above and below the paragraph 275
　　　　　　Placing graphics "beside" a paragraph 277
　　　　　　Changing the name of a graphic frame 277
　　　　Importing graphics on a master page 278

Chapter 14: Object styles 279
　　　　A case for using object styles 280
　　　　Create an object style 280
　　　　　　Create an object style from an object 281
　　　　　　Create an object style from scratch 281
　　　　Display the Object Style pod with a shortcut 282
　　　　Using object styles in a regular workflow 282

Chapter 15: Placing rich media 283
　　　　Supported formats 283
　　　　Placing a multimedia file 284
　　　　Full-motion video 285
　　　　　　Set movie poster 285
　　　　　　Set graphic name 286
　　　　　　Create video player controls 286
　　　　　　Create cue points 287
　　　　　　Activate SWF and FLV by default 288
　　　　Interactive simulations 288
　　　　3D objects 288
　　　　　　Inserting a multimedia links table 289
　　　　　　Link 3D part to text 289
　　　　　　Activate 3D By default 290
　　　　　　Display 3D and multimedia in pop-up windows 290
　　　　　　Other 3D control options 290
　　　　　　3D object support for JavaScript 291
　　　　QR codes 292

Chapter 16: FrameMaker's graphics tools 293
　　　　Advantages of FrameMaker graphics tools 294
　　　　Drawing basic shapes 294
　　　　　　Working with grids 297
　　　　　　Selecting objects 298
　　　　　　Deleting objects 299
　　　　Modifying objects 299
　　　　　　Resizing 303
　　　　　　Reshaping 304
　　　　　　Smoothing corners 306
　　　　　　Changing the number of sides 306
　　　　　　Joining lines 307
　　　　　　Cropping and Masking Graphics 309

Rearranging objects 310
    Aligning 310
    Distributing 312
    Grouping 313
    Running text around an object 313
    Layering objects 314
Changing the orientation 315
    Rotating 315
    Flipping horizontally or vertically 317
Transparency in FrameMaker graphics 318
Image hotspots 318

Part IV, Building Books

Chapter 17: Setting up book files 321
Advantages of using book files 321
Creating a book file 322
What the book window tells you 322
Managing files in a book 323
    Adding files 323
    Adding special book structures 325
    Opening, closing, and saving all files 327
    Removing files 327
    Rearranging files 327
    Renaming files 328
Updating a book 329
    Troubleshooting book updates 330
Managing numbering 332
Printing a book 341
Modifying files from the book 342
    Spell-checking and finding/changing items in book files 342
    Book-level features 343
    Choosing files for other book-level operations 343
    Book features available inside files 344
    Paging through files in a book 344

Chapter 18: Creating tables of contents 345
Setting up table of contents file 345
    The initial table of contents file 347
Customizing the table of contents 348
    Locating the TOC reference flow 349
    Understanding the TOC flow entries 350
    Formatting the table of contents 352
    Using character-level formatting 352
    Formatting examples 352
Mini TOC 356

Chapter 19: Creating indexes 357
The mechanics of a generated index 357

Creating the index file	358
Creating index entries	359
Basic entries	361
Inserting Unicode in index entries	362
Editing and deleting index entries	362
Creating subentries	363
Stacking multiple entries in a single index marker	363
Creating ranges	363
Creating references to synonyms ("See")	364
Changing sorting order for a single entry	365
Formatting the index	365
Ignoring characters while sorting	367
Modifying page separators	367
Changing the sort order	368
Changing the group titles	369
Formatting the page number	370
Creating ranges automatically	370
Eliminating unwanted chapter numbers	371

Chapter 20: Creating glossaries 373

Generated versus static glossary files	374
Marking glossary definitions for digital publishing	374
Marking a glossary definitions for print or PDF	375
Marking glossary terms for digital publishing	375
Create the generated glossary for print or PDF	376
Modifying the reference pages	376

Chapter 21: Creating Other Generated Files 377

Examples of other generated files	377
Creating paragraph-based lists	380
List of figures	381
List of tables	381
List of paragraphs	381
Alphabetical list of paragraphs	382
Creating lists of other items	382
List of markers	382
Alphabetical list of markers	382
List of references	383
Generated indexes	385
Index of authors	385
Index of subjects	386
Index of markers	386
Index of references	386

Part V, Creating Output

Chapter 22: Print, PDF output, and package 389

A comparison of print and electronic formats	390
Setting up a FrameMaker PDF review	391

Printing your documents 391
    Printing an individual document 392
    Printing from the book file 392
    Specifying pages 392
    Printing several copies 393
    Changing the paper size 393
    Printing double-sided documents 394
    Skipping blank pages 394
Advanced printing options 395
    Printing thumbnails 395
    Printing spot color in black and white 395
    Printing low-resolution images 396
    Printing registration marks 396
    Changing the printer 397
    About color separations 397
Setting Adobe PDF document properties 398
    Adobe PDF Settings sheet 398
    Configuring PDF job options 399
Creating PDF Files 401
    PDF Setup dialog options 401
    Generating PDF bookmarks 403
    Generating tagged PDF files 405
    Optimizing PDF files 407
    Creating hyperlinks to other PDF files 409
Adobe PDF printer setup 410
Package 412
    Choosing Package options 412
    Settings 412

Chapter 23: Digital Publishing 413

Accessing the Publish pod 413
Exploring the Publish pod 414
    Available formats 414
    Pod controls 415
    Style Mapping tab 415
    Notes on style mapping 416
    Outputs tab 416
    Responsive HTML5 notes 417
    Mobile App notes 418
    WebHelp notes 420
    ePub notes 421
    Kindle notes 423
    Microsoft HTML Help notes 424
    Exporting XML output 424

Chapter 24: Color output ... 425
- Advantages of defining custom colors ... 426
- Understanding types of color ... 427
  - Understanding process color ... 428
  - Understanding spot color ... 429
- Applying color to text and objects ... 430
- Managing color definitions ... 432
  - Modifying existing colors ... 432
  - Adding colors from a color library ... 434
  - Creating custom colors ... 437
  - Renaming colors ... 437
- Viewing colors ... 438
  - Selecting a color view ... 438
  - Setting up color views ... 439
- Deleting colors ... 439
- Controlling colors for output ... 440

Part VI, Advanced Techniques

Chapter 25: Setting up conditional text ... 443
- How conditional text works ... 444
  - Notes regarding conditional text ... 444
  - Strategies for using conditional text ... 445
- Conditional text examples ... 445
- Using conditional text in digital publishing ... 446
  - Use conditions for multiple outputs ... 446
  - Use conditions for personalized dynamic content filtering ... 446
- Applying condition tags ... 446
- Removing a single condition tag ... 448
- Removing all condition tags from text ... 449
- Using condition indicators ... 449
- Planning conditional text ... 450
  - Common condition tags ... 451
  - Alternatives to conditional text ... 451
- Creating condition tags ... 452
- Modifying condition tags ... 453
- Deleting a condition tag ... 453
- Showing and hiding conditional text ... 454
  - Showing All Conditional Text ... 454
  - Choosing the method for showing conditional text ... 454
  - Showing or hiding conditional text as per condition ... 457
  - Using conditional text expressions ... 458
  - Modifying expressions ... 460
  - Deleting expressions ... 460
  - Examples of expressions ... 460

Chapter 26: Automation with ExtendScript ... 465
Advantages of using ExtendScript ... 465
A brief history of FrameMaker automation ... 465
Getting started with scripting ... 466
Interacting with FrameMaker ExtendScript ... 466
Using the Script Library pod ... 467
Creating and editing scripts ... 469
Creating an alert box ... 470
Automating text entry ... 470
Objects, Properties, and Methods ... 472
Finding the correct objects, properties, and methods ... 473
Principles of successful script writing ... 476
Start small ... 477
Use functions ... 477
Working with selections ... 478

Chapter 27: Creating Interactive Content with Hypertext ... 485
Advantages of hypertext interactivity ... 486
Setting up a basic hypertext link ... 487
Creating Link Destinations ... 488
Creating a link to a destination ... 489
Creating the active area in text ... 490
Activating links ... 490
Creating hotspots ... 490
Using hotspot properties to create links ... 491
Using hotspot mode to create links ... 491
Using a text frame as a hotspot ... 491
Creating a text link to a web address ... 492
Creating an email link ... 492
Using markers to create notes in PDF files ... 493
Locking and unlocking view-only documents ... 494
Hypertext Command Reference ... 495

Chapter 28: Writing equations ... 499
Using native FrameMaker equation editor ... 499
Understanding the Equations pod ... 500
Inserting equations ... 502
Using the Equations Pod ... 503
Typing an equation ... 504
Selecting equations and math elements ... 504
Navigating through equations ... 505
Moving equations ... 506
Modifying equations ... 506
Moving text using keyboard shortcuts ... 506
Moving text using the equations palette ... 507
Deleting equations ... 508

Formatting equations	509
Changing equation fonts	509
Changing equation font sizes	509
Inserting automatic line breaks	510
Changing the equation size	510
Applying character tags	511
Evaluating equations	511
Working with MathML	511

Chapter 29: Content reuse with text insets ... 513
Advantages of using text insets ... 513
A text inset example ... 514
Working with text in insets ... 517
Considering text insets and other reuse options ... 518
Planning modular text ... 518
    Breaking text down into modules ... 518
    File storage ... 519
    Information retrieval ... 519
    Controlling formatting in text insets ... 519
Creating a text inset ... 521
Managing text insets ... 522
    Opening the source file ... 524
    Converting text insets to text ... 524

Chapter 30: Dropbox and Cloud Collaboration ... 525
Configure Dropbox ... 525
    Share for review ... 525
    Open files ... 526
    Save files locally and access content offline ... 526

Chapter 31: Using a Content Management System (CMS) ... 527
Default connectors ... 527
Creating a custom connection to a CMS ... 528
Working with WebDAV ... 528
Managing workgroups ... 528
    Setting up a CMS connection ... 529
    Using the Repository Manager ... 530
    Uploading the current document ... 530

Part VII, Appendixes

Appendix A: Resources ... 533
FrameMaker and Technical Communication Web Resources ... 533
Mailing Lists/User Groups ... 533
Third-Party Tools and Plug-Ins ... 534
Database Publishing from FrameMaker ... 536
Online Manuals ... 536
Reporting Bugs ... 536

Appendix B: Shortcuts 537
Appendix C: Building blocks 541
    Building block usage 541
Appendix D: Customizing maker.ini 547
    Editing maker.ini 547
    Setting clipboard pasting order 548
    Updating graphics from web addresses 548
    Changing the substitution fonts 548
Appendix E: Preference settings 551
    Global 551
        General 551
        Interface 552
        Alerts 553
        Pods 554
        Launch 555
        DropBox 556
    CMS 557
        Documentum 557
        SharePoint 557
        DITA Exchange 558
    Spelling 559
        Dictionary 559
        Spelling options 559
    Simplified XML 560
    MathML 560
Appendix F: Maker Interchange Format 561
    Sample MIF code 561
    Creating a MIF file 565
    Opening a MIF file in FrameMaker 566
    Viewing a MIF file 566
    Cool stuff you can do with MIF 566
        Eliminating file corruption problems 567
        Making a file available to an older version of FrameMaker ... 567
        Creating a character tag for a vertical baseline shift 567
        Using MIF fragments to update catalog settings 568
        Performing global search-and-replace operations 569
        Writing Your Own Conversion Tools 570

Index 571

Front Matter

Acknowledgments

This much-asked-for update to the *Publishing Fundamentals: FrameMaker 11* work would not be possible without the contributions of the original Scriptorium team, in particular, Sarah O'Keefe and Alan Pringle.

A huge thanks goes out to Rick Quatro for contributing the excellent chapter on ExtendScript.

In addition, I'd like to thank the Adobe FrameMaker product team, especially engineering. Access to Adobe's teams has without question made this a better book.

Thanks also to the FrameMaker prerelease team; participation in the prerelease program gave us insights to the feature sets that make up this most current release of FrameMaker.

Matt Sullivan
September, 2015

Preface

Much has changed in FrameMaker since the FrameMaker 11 version of this book, yet much remains the same. My hope is that this book will give you insight into the workflow that has made FrameMaker the standard in technical book publishing for the last 20 years.

The original text, with content authored and directed by Sarah O'Keefe, has been completely revamped to address the features, user interface, and workflows available in FrameMaker up to and through FrameMaker 2015.

This version continues to move more to my perspective as a user of FrameMaker since version 1 on the Mac.

While I have always updated to the latest version of FrameMaker, there are many who do not. For this reason, I've left in many references to earlier versions of FrameMaker. Most significantly, there was

- a change to book functionality with version 5.5
- a blending of the structured and unstructured version in version 7
- an interface overhaul in version 9,
- significant changes to electronic publishing in versions 12 onward.

If what you see in the FrameMaker 2015 release does not match what you see in the book, consider updating to the latest patches for your release (very important!) or downloading a trial version of FrameMaker to compare functionality between the versions. Make sure you update to the latest patch for the trial of FrameMaker as well.

Specifically, this print version references the FrameMaker 13.0.1 release, though a few screen captures may be from the 13.0.0 release.

Of course, if you do download the trial, *use test copy, or copies of your content, not the originals!*

I have been a FrameMaker implementation specialist and trainer since FrameMaker 5, and have seen fantastic changes in the product, especially as it continues to expand support for video and interactive media. The latest version of FrameMaker also greatly enhances support for mobile content delivery, improving the Responsive HTML5 controls and experience, and providing support for mobile app creation.

This book is intended as user reference, not as a training guide. If you are quite new to FrameMaker, I suggest you visit training.techcommtools.com for free and paid FrameMaker courseware, as well as look at my schedule of FrameMaker training classes available online or in southern California.

The book is available in both print (amazon.com) and EPUB (techcommtools.com) formats.

Introduction

Welcome to *FrameMaker - Creating and Publishing Content*. Adobe's FrameMaker software is the industry leader in technical publishing, and the release of FrameMaker 2015 continues that tradition. FrameMaker specializes in long-document formatting and automates mundane but essential tasks, such as maintaining running headers and footers, and updating tables of contents. It offers an easy path to producing multiple output formats, including print, PDF, and mobile formats.

Just as a truck with a lift gate is better than an SUV for moving furniture, FrameMaker is better than other options for moving heavy, messy content. Its referencing, and digital publishing features are simply unmatched. Use this book to look up specific features, manage content via templates, and produce beautiful HTML5 and mobile apps.

This book shows new users how to use unstructured FrameMaker, the most popular of the FrameMaker user interface options, to streamline their publishing workflows for greater efficiency and productivity.

I've provided new screenshots throughout, as well as coverage of the new features in FrameMaker 12 and FrameMaker 2015. There's even something for the most knowledgeable Frame folks. There's a separate list of the features added by version, and lots of best-practice details transposed from my own "shop-worn" copy.

One of the most significant changes since the FrameMaker 11 version of this book is the emergence of the mobile web. FrameMaker has kept pace with new technologies like HTML5, EPUB, and Kindle. This book gives you the detail you need for many of the available FrameMaker options for multichannel output.

In addition to writing about FrameMaker, I practice what we preach. I've trained, consulted and written about FrameMaker since the early versions. In fact, the book you are holding was itself produced in FrameMaker 2015.

If you would like to express interest in additional FrameMaker or Technical Communication Suite subjects, please visit www.framemaker11book.com.

New features by version

New since FrameMaker 8

FrameMaker 9 New Features
- New user interface (UI)
- Added groups and folders to books
- PDF comment workflow
- Content management improvements
- RoboScreenCapture included
- Character Palette including Unicode characters
- History pod

FrameMaker 10 New Features
- Drag-and-drop editing
- Filter by attribute (structured feature)
- Auto spell check
- Find/Change overrides
- Table catalog
- Repeat last operation
- ExtendScript
- Improved placement of multimedia
- Set poster for multimedia
- Set background color
- Suppress alerts
- Improved linking with RoboHelp (Technical Communication Suite version)

FrameMaker 11 New Features
- Interactive multimedia links for 3D objects
- Object styles
- Identify and refer to content using line numbers
- Hotspots
- Smart Insert
- 3D object part links
- Preferences dialog enhancements
- New publishing formats in FrameMaker Publishing Server (requires separate purchase of FrameMaker Publishing Server)
- More markers
- APIs to automate the CMS connectors' functionality

FrameMaker 12 New Features

- Background color behind paragraphs
- Direct digital publishing (HTML5, WebHelp, EPUB, Kindle, MS HTML Help)
- QR code generation
- Table usability improvements
- Ability to resize pods
- Package utility
- Smart filters in pods
- Restore Session capability
- Currently Opened Files pod
- Regular Expression support
- Integration with Adobe Experience Manager
- Mobile PDF review
- Simultaneous review and authoring workflow
- Dropbox (and cloud based filesharing) integration

FrameMaker 2015 New Features

- Right to left language support for editing and publishing
- Conditional text extended to book components
- Use custom fonts in EPUB
- Publish mobile apps (iOS and Android)
- Customize WebHelp skins
- Dynamic content filtering in HTML5 and mobile apps
- Improvements to direct digital publishing
- Auto-insert table continuation variables
- Conditional table columns
- Mini TOC
- Improvements to visual conditional indicators
- Improvements in table fill display
- Totally new MS Word import wizard
- In-app messaging

Who should read this book

The information in this book will be useful for both new and long-time FrameMaker users. The book explains unstructured FrameMaker from top to bottom. It details basics, such as creating a document, importing formats from one file to another, and applying paragraph tags. In more advanced sections, this reference describes how to create templates, build books, use modular text, and insert hypertext commands.

In short, there's something here for every user. Beginners will find a wealth of information, organized in order of increasing complexity. Advanced users may want to skip the first few parts and focus on the second half of the book.

What's in this book

This rather large book is divided into several parts to help you find your way around. You'll notice that each part has a handy thumb tab to help find the part you need.

Part I, Getting started with Framemaker

Part I is intended mainly for new users. It provides an overview of FrameMaker's features and interface. All users should consider reading Chapter 4, which describes several different workflows that include FrameMaker.

Part I includes the following chapters:

- *Chapter 1, FrameMaker strengths* offers an overview of FrameMaker's features. It describes some of the features that make FrameMaker unique and explains how you can use them to automate common publishing tasks. This chapter is intended for users who are new to FrameMaker and need an explanation of why FrameMaker makes sense for technical publishing projects.

- *Chapter 2, Creating your first document* explains how to create a new document and open existing documents. It also describes the document window, toolbars, and status bar. Finally, it explains how to configure preferences to suit your workflow. Read this chapter if you are not familiar with the FrameMaker interface.

- *Chapter 3, Word-Processing Features* describes how to type text into FrameMaker. It also explains how to import text from other applications. Generally, FrameMaker users write directly in FrameMaker. This chapter will be helpful to new users who are not yet familiar with basic text manipulation in FrameMaker. For more advanced users, the importing section provides detailed information about how to import content from Microsoft Word files successfully.

- *Chapter 4, Establishing a workflow in FrameMaker* describes the typical publishing workflow and examines how FrameMaker fits into that workflow. It also includes an overview of how to set up a single-sourcing workflow that lets you publish to multiple output formats. If you are starting with a clean slate and can configure your workflow however you want, read this chapter to get an idea of the possibilities.

- *Chapter 5, Templates* pulls together information about many different features. It describes interactions among different components and provides some tips for designing useful templates.

Part II, Creating and Manipulating Text

Part II explains the word-processing features of FrameMaker and how to add and manipulate text, tables, and other items in your documents. This part includes basic information, such as how to apply a paragraph tag, but it also covers more advanced topics, such as how to create a paragraph tag and set up autonumbering. New users will want to read at least the first half of each chapter to learn about each feature; more advanced users will probably focus on the latter half of each chapter.

Part II includes the following chapters:

- *Chapter 6, Formatting text with paragraph tags* begins with an explanation of paragraph-level style sheets and how to apply them. It also provides a detailed explanation of how to create and modify paragraph tags.
- *Chapter 7, Formatting text with character tags* shows you how to apply and create character-level style sheets. In FrameMaker, character-level and paragraph-level style sheets are stored in separate locations. Characters tags are used in several other features, such as cross-references and variables, to provide formatting; this chapter also explains those relationships.
- *Chapter 8, Understanding table design* explains how to create tables and modify their formatting. Like paragraphs, tables have style sheets, and this chapter describes how to set up table styles.
- *Chapter 9, Cross-references* describes how to create pointers from one section of a document to another. FrameMaker automatically maintains and updates these references as pagination in the document changes.
- *Chapter 10, Storing content in variables* details how to use variables to store and format bits of reused text, such as the title of a book or a product name. It also describes how to use system variables to automate running headers and footers, page numbers, and "continued" labels in tables.

Part III, Controlling Page Layout

The chapters in Part III describe managing blocks of text and positioning them on your pages. This part also includes discussions about importing graphics created in other applications and creating graphics in FrameMaker.

Part III includes the following chapters:

- *Chapter 11, Understanding master pages* explains how to set up master pages to determine the page layout in a file, and it shows how to apply and import page layout definitions.
- *Chapter 12, Text flows* describes how to create and connect blocks of text, or text flows. This information is especially helpful for template designers, who need to understand how to position text blocks on a page and how text blocks are connected and disconnected.
- *Chapter 13, Importing graphic content* explains anchored frames, which serve as containers for most graphics. It also describes how to import graphics and rich media files from other applications.
- *Chapter 14, Object styles* introduces object-level style sheets and how to apply them. It also provides a detailed explanation of how to create and modify object styles.
- *Chapter 15, Placing rich media* describes the types of multimedia that may be placed within a FrameMaker file. It also explains the controls available for each media type.
- *Chapter 16, FrameMaker's graphics tools* describes how to use FrameMaker's built-in graphics tools to create art.

Part IV, Building Books

You can create a book file to hold a collection of files that make up a larger document. Among these files, you can include automated tables of contents and index files, which are explained in this part.

Part IV includes the following chapters:

- *Chapter 17, Setting up book files* describes how to set up and modify book files. It explains how to control chapter, page, and other numbering from the book and how to perform global (bookwide) updates on your content.
- *Chapter 18, Creating tables of contents* explains how to choose items for inclusion in a table of contents and how to format those items.
- *Chapter 19, Creating indexes* explains how to create the markers that become index entries and how to format the generated index.
- *Chapter 21, Creating Other Generated Files* describes some of the less well-known generated files. You can, for example, create a list of fonts used in a document, or a list of imported graphics. Several variations on the standard index are also available.

Part V, Creating Output

In Part V, you learn how to create print, PDF, and online formats, and you learn about managing color.

Part V includes the following chapters:

- *Chapter 22, Print, PDF output, and package* provides detailed information about printing FrameMaker files and creating PDF files. It explains how to configure your system for successful PostScript printing and provides information about PDF conversion settings.
- *Chapter 23, Digital Publishing* describes how to use the Publish pod to create content for mobile and online formats like HTML5 and EPUB. The Publish pod is available in FrameMaker 12 or later.
- *Chapter 24, Color output* explains how FrameMaker handles color output and describes how to set up colors in your document.

Part VI, Advanced Techniques

Part VI covers a variety of advanced topics, including template design, conditional text, hypertext, and WebDAV use with FrameMaker.

Part VI includes the following chapters:

- *Chapter 25, Setting up conditional text* explains how to create two (or more) versions of a document in a single file. The conditional text feature is an important part of most single-sourcing environments.
- *Chapter 26, Automation with ExtendScript* provides an overview of FrameMaker automation using Adobe's ExtendScript technology. Scripting tasks can save countless hours of repetitive manual formatting or processing.
- *Chapter 27, Creating Interactive Content with Hypertext* describes how to link documents, create popups, and insert hotspots into graphics with hypertext commands. Many hypertext commands are translated when you convert to HTML or PDF output.
- *Chapter 28, Writing equations* tells you how to create and format equations in your documents.
- *Chapter 29, Content reuse with text insets* provides information about FrameMaker's text importing feature. You can create small text fragments and import them into larger documents to create modular documentation.
- *Chapter 31, Using a Content Management System (CMS)* describes FrameMaker's support for version control and content management through WebDAV. WebDAV software (www.webdav.org) runs on any web server and is accessed via http protocol, so you can store files anywhere and access them over the Internet. This allows you to set up a version control system for authors in many different locations. The "DAV" in WebDAV actually stands for "Distributed Authoring and Versioning."

Part VII, Appendixes

The following appendixes are included:

- *Appendix A, Resources* offers a list of FrameMaker-related resources, including a list of utilities.
- *Appendix C, Building blocks* provides a comprehensive list of the building blocks you use to create tags and formats in FrameMaker.
- *Appendix D, Customizing maker.ini* describes the file where FrameMaker settings are stored and some settings you may want to modify.
- *Appendix E, Preference settings* provides a reference for preference options. Newly expanded options for content management system integration and spelling checker are now found here.
- *Appendix F, Maker Interchange Format* explains Maker Interchange Format (MIF), which is a text-based markup language that describes FrameMaker files. MIF files can be useful for global changes that aren't easily implemented through the FrameMaker interface. It's also widely used as an intermediate format when converting to and from FrameMaker format. FrameMaker files saved ins MIF format can be opened in earlier versions of FrameMaker.

Conventions used in this book

Some of the text in this book uses special formatting to help indicate emphasis or keystrokes. The text conventions are as follows:

Convention	Example	Meaning
Sans Serif Small Caps	<ul><li>press the TAB key</li><li>ESC O P D</li><li>ESC SHIFT+F L K</li></ul>	Indicates keystrokes. If a keyboard shortcut (such as ESC O P D) contains no hyphens, press each key individually. When keys are joined by hyphens, press the joined keys at the same time; for example, SHIFT+F means to type a capital letter F.
Sans Serif Italics	*single-sourcing*	Indicates a defined term, a book title, a placeholder, or text that requires emphasis.
Sans Serif Bold	if you type **2**	Indicates text you type or terms that are being defined in a list.
Sans Serif	<$paratext>	Indicates building block strings or output (for example, Maker Interchange Format output).
Sans Serif Bold with Arrows (>)	**File > Open**	Indicates menu selection and menu choice (select the File menu, then select the Open menu choice).
Sans Serif	Click the Set button.	Indicates a button to be clicked

Mistakes were made

I've tried to make this book as accurate as possible. However, there's the possibility that I was, well, wrong about something. Some features in FrameMaker also may have changed (or bugs may have been fixed) since this book was published.

For changes and errata, visit:

http://www.techcommtools.com/fmbook

Part I

Getting started with Framemaker

Chapter 1: FrameMaker strengths

Adobe FrameMaker is the workhorse that technical communicators depend on year after year. FrameMaker easily creates complex documents, full of technical information, references and numbering. Examples of this are:

- Technical documentation—user guides, installation guides, system administration guides, reference guides, and training materials
- Interactive electronic documents, accessed across all your electronic devices
- Nonfiction books, especially third-party documentation (including this book!)
- Legal and financial documents, such as prospectuses
- Database publishing of directories and catalogs
- Long documents and documents with a long life cycle

Here's a look at what's in this chapter:

FrameMaker strengths 4
Uses for FrameMaker 8

The Technical Communication Suite versions of FrameMaker also enable placement of video, interactive, and rich media into your documents. While this content (other than QR codes) obviously does not translate to print output, the content will work in other electronic output such as PDF, eBook, responsive HTML5, and mobile app output. Of course, the media you place is subject to platform limitations. This means that, for example, Flash Video accessed on an iOS device will not work.

Software requirements for creating technical documentation are different than the requirements for creating short, independent documents such as brochures, newsletters, or annual reports. For technical publishing, stability and reliability are critical, especially as documents get longer. FrameMaker easily handles books of 200 or 2,000 pages. FrameMaker provides a stable, reliable environment in which to create and manage these types of documents.

Most likely, though, you're hoping for something more than an application that doesn't crash often. FrameMaker includes many features that are important for anyone working with long documents in a graphical what-you-see-is-what-you-get (WYSIWYG) environment, shown in the following figure.

Figure 1-1. A typical page in FrameMaker

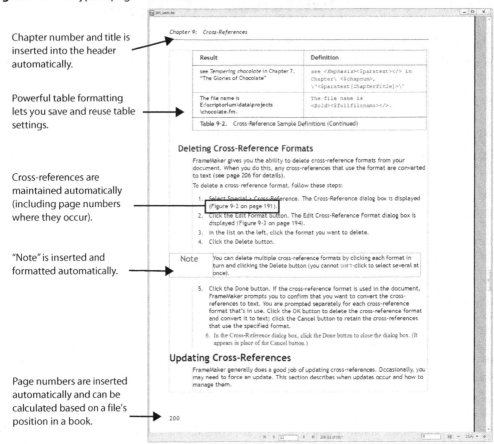

FrameMaker strengths

Highlights of FrameMaker's strengths include the following:

- **Consistent, repeatable, and maintainable formatting.** By using templates (See "Using a file as a template" on page 13.) you can manage many documents with the same look and feel. Templates ensure that formatting is consistent from one document to another, and you can quickly copy updated formatting from file to file. Instead of manual reformatting, you apply the template across affected documents, and FrameMaker changes formatting and repaginates accordingly.
- **Complex document management.** Equations, text variables, conditional text, a staggering array of graphic formats, and many other features help you create and maintain complex documents.
- **Placement of rich media.** You can place video, interactive software simulations, and working 3D models in your docs. You can also provide alternate representations to ensure proper print and electronic output.
- **Stability.** FrameMaker is a stable, reliable application. It maintains your content, references, and numbering so that you can focus on content.

- **Long-document management.** FrameMaker provides strong support for managing pagination, tables of contents, and indexes in books. This lets you process multi-volume books, dozens of chapters, and thousands of pages without having to worry about the accuracy of numbering and references. To rearrange the order of chapters in a book, you just drag and drop the files to their new locations and then update the book. FrameMaker automatically changes the pagination to reflect the new order, changes the chapter numbers as necessary, and updates the table of contents and index.

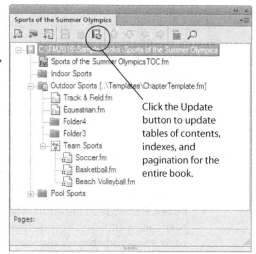

Click the Update button to update tables of contents, indexes, and pagination for the entire book.

- **Cross-media publishing.** FrameMaker can serve as the engine or as a component of a very sophisticated publishing system, in which you create print, Portable Document Format (PDF), Responsive HTML5, Kindle, and EPUB from a single set of documents. This *single-source* publishing makes it possible to write content once and then publish one or more versions of the content across different media.

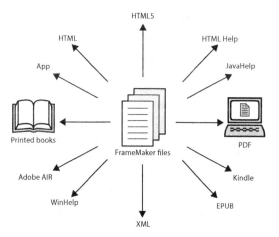

Note   Structured FrameMaker is better for some of these outputs, especially XML.

- **Change management.** FrameMaker natively supports popular content management systems, and also supports WebDAV, which lets you set up your own version control system on a web server. If you work with large books or collaborate with multiple authors, change management can be a critical issue.

Chapter 1: FrameMaker strengths

You'll find many functional similarities between FrameMaker and other desktop publishing applications. Only FrameMaker, however, is *designed* for long-document publishing; its document-management features are robust and reliable. Highlights of FrameMaker's document-management features include the following:

- **Style sheets.** Word processors and desktop publishing packages provide style sheets for paragraph formatting, but FrameMaker takes this concept much farther. A FrameMaker file includes style sheets for paragraphs, characters, tables, cross-references, variables, master pages, object styles and several other items. These style sheets (often referred to as *tags*), are stored in catalogs, which can be imported from one file to another.

- **Books.** A book is a collection of files (often chapters) in a particular sequence. FrameMaker manages the pagination and chapter numbering as information is added to and deleted from the book. A book can also contain grouped chapters, subdirectories of chapters, and other books.

- **Cross-references.** Using cross-references, you create a pointer from one part of the book to another (for example, *see "Widgets" on page 913*). FrameMaker automatically updates the page numbers and displayed text (in our example, Widgets) when content changes. In electronic output, cross-references can contain usable hyperlinks.

- **Tables of contents.** You can automatically generate a list of specific content in the document. After making changes to a document, updating the book automatically updates topic lists and page references.

- **Indexes.** You identify which terms to index by inserting markers in the documents. After that, the index is generated, sorted, and grouped automatically.

- **Tables.** Table support is very strong and includes the ability to save named table formats with settings for lines and shading. Starting in FrameMaker 2015, you can more easily set the color of borders in your table.

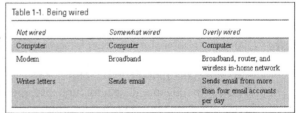

Table 1-1. Being wired

Not wired	Somewhat wired	Overly wired
Computer	Computer	Computer
Modem	Broadband	Broadband, router, and wireless in-home network
Writes letters	Sends email	Sends email from more than four email accounts per day

- **Graphics.** You can import external graphics and embed or link them into the document. Linked graphics are automatically updated in the FrameMaker file when the graphic changes. FrameMaker 11 introduced *object styles*, which allow rapid formatting and management of graphic objects, including lines, shapes and frames. FrameMaker also includes a set of drawing tools for creating and managing graphics.

- **Conditional text.** You can use conditional text to identify information that belongs to a specific version of a document (for example, instructor-only material in a training guide). You can then create document versions that show or hide the tagged material. Conditional text is an important component of any single-sourcing strategy because it lets you identify information that should be excluded from some output (such as "print-only" and "online-only" content).

- **Variables.** These are useful for long text strings, for things that can change frequently, or for phrases that must be formatted a specific way. For example, you can create a variable called ProductName and use it throughout the document wherever the product name is needed. If the product name or formatting changes, you change the value of the variable, changing every occurrence of the variable throughout the document.
- **Modular text.** You can create small files that contain chunks of content and then reuse the chunks (text insets) into larger documents. Modular text lets you reuse chunks in multiple documents and maintain the information from a single file.
- **Equations.** FrameMaker includes an equation editor, which lets you create complex mathematical expressions.
- **Templates.** FrameMaker lets you import named styles from any other FrameMaker document. In most desktop publishing applications, templates are limited to paragraph style sheets and master pages, but in FrameMaker, template items include paragraph tags, character tags (for text range formatting inside paragraphs, such as italics), tables, master pages (page layouts), cross-reference formats, variable names and definitions, conditional text tag names, oject styles, and more.

Chapter 1: FrameMaker strengths

Uses for FrameMaker

It's possible to use FrameMaker to create highly designed documents, such as annual reports and newsletters (where layouts change on almost every page), and glossy full-color publications. However, FrameMaker's document-management features are much more useful in enforcing consistency across hundreds (or thousands) of pages.

FrameMaker is very good at...	FrameMaker isn't ideal for...
• Enforcing consistency throughout documents • Handling long documents • Incorporating video, interactive, and rich media • Publishing for mobile, including mobile apps • Managing books, tables of contents, and indexes • Single-sourcing/multichannel publishing • Formatting tables • Importing and maintaining graphics • Providing cross-platform compatibility • Automated workflows using the built-in ExtendScript scripting language	• Creating simple, unrelated word-processing documents • Working in an environment with users who don't believe in using style sheets or templates in their documents • Creating artistic typography • Creating graphics • Setting up complex, single-use layouts

At first glance, the logic behind FrameMaker's features may seem a bit odd. Keep with it though, because the logic is there. FrameMaker was designed as a tool for long documents, so its strength is in the ability to enforce consistency across your chapter, your book, and all content controlled by a given template (a template is basically a set of naming conventions and formats). In short, if you act in ways that FrameMaker expects, FrameMaker will do all the heavy lifting for you, freeing you up to create or manage content instead of formatting it.

FrameMaker won't solve all of your problems, but if you're looking for a reliable, industrial-strength publishing tool that produces consistent output, it is the software for you.

Chapter 2: Creating your first document

After starting FrameMaker you can either create a new document or open an existing document. After that, you can start learning how to get around in the FrameMaker interface. This chapter explains how to start FrameMaker, create documents, and open them.

Starting FrameMaker	10
Setting up a new document	11
Exploring the standard templates	12
Using a file as a template	13
Creating a blank document	13
Using an existing file	15
Opening, saving, closing, and printing documents	15
Understanding the document window	16
Changing the zoom settings	16
Displaying nonprinting items in the document window	17
Changing view options	19
FrameMaker interface options	19
Workspaces	19
Toolbars	20
Icons	20
Document tabs	21
Pods	21
Managing open files	22
Restore last session	23
Understanding file configuration and preferences	23
Storing format definitions	23
Importing and embedding graphics	23
Document control	23
Using the Fonts pod	25
Handling missing fonts	25
Organizing directories and files	26
Saving file information with XMP	27
Setting automatic backup and save features	28
Switching between structured and unstructured interfaces	29
File naming conventions	29

Starting FrameMaker

You should have a shortcut for FrameMaker installed in the Start menu after installation. Double-click on the desktop shortcut, or select **Start > All Programs** to start the FrameMaker application.

Note   The exact path may vary, depending on installation choices and whether you installed FrameMaker as part of the Adobe Technical Communication Suite.

The first time you start FrameMaker, you are prompted to activate the product, as well as sign in with your Adobe ID. Creating an Adobe ID or signing in is recommended, as it will give you options to collaborate with others via Acrobat.com among other things.

Next, you are prompted to select an interface.

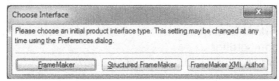

Select one of the following options (you can change this selection later):

- **FrameMaker** (unstructured): Starts the "vanilla," unstructured version of FrameMaker. Lets you work on unstructured documents without the structure features getting in the way. This book explains the unstructured interface.
- **Structured FrameMaker:** Starts the structured version of FrameMaker. Lets you create structured documents, work with enforceable content models like DITA and XML, and see the underlying code.
- **FrameMaker XML Author:** Starting with FrameMaker 12, this option starts FrameMaker in a structured mode intended only for authoring content. This mode removes the controls needed to modify structure models, thus simplifying the content editing process for some users. XML Author is also available as a standalone application.

When you start FrameMaker, you are presented with a welcome screen.

You can choose to create new books or documents from this screen or from the File menu.

Removing the Adobe banner ads

The first few times you start FrameMaker, you might appreciate the resources shown below the Starter Page. When the appreciation fades, go to **Edit > Preferences… > General** to turn them off.

Setting up a new document

When you create a new document, FrameMaker names the file Untitled1.fm (additional documents are named Untitled2.fm, Untitled3.fm, and so on). You can save the document with a more informative name.

There are several ways to create a new document:

- Exploring the standard templates, then choosing one of those templates
- Creating a new document based on a template
- Creating a blank document
- Saving an existing file under a new name

Exploring the standard templates

For each of the default templates, FrameMaker includes a preview and a feature summary. This information helps you determine whether a particular template will meet your needs.

To examine the unstructured templates, follow these steps:

1. Select **File>New>Document**, press CTRL+N, or click the Document button ( ) on the Quick Access Bar to display the New dialog.

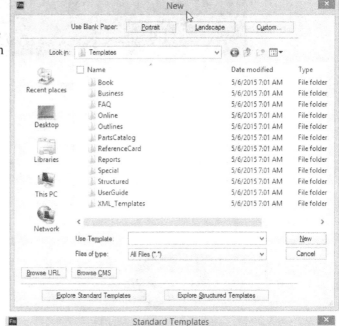

2. Click the Explore Standard Templates button. A list of templates is displayed with previews.

3. Click an item in the template list on the left to display a feature summary and preview for that template on the right. To see a sample document with placeholder text, click the Show Sample button.

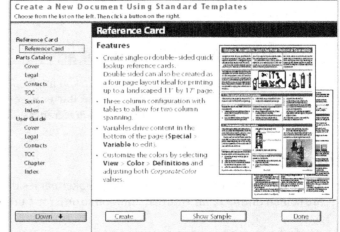

4. To use a particular template, click the Create button.

FrameMaker creates a new, untitled document by making a copy of the template file.

You can now save the document with an appropriate name and begin working.

Using a file as a template

In addition to the standard templates, you can use any FrameMaker file as a template.

Setting up a new document

To use a file as a template, follow these steps:

1. Select **File > New > Document**, press CTRL+N, or click the New File button ( ) on the Quick Access Bar. The New dialog box is displayed (see page 12).

Select the template by doing one of the following:

- To find a standard FrameMaker template, double-click a folder in the Look In window and then double-click the template. You can display a preview of FrameMaker's standard templates by clicking the Explore Standard Templates button.

Tip If you save a file in FrameMaker's template directory, that file will be displayed along with the standard FrameMaker templates. It will not, however, be listed when you explore templates. You can also change the directory used for templates (to display a network location, for example) by customizing the maker.ini settings.

- To use another file as a template, navigate to the directory that contains the file and double-click the file.

2. The new document is displayed.

When you create a new document from a template, FrameMaker makes a complete copy of the file. Any content stored in the original file is included in the new, untitled document (good for contracts or other boilerplate text).

Creating a template file with standard information, perhaps with variables can speed up editing of standard documents.

For more information about variables, see Chapter 10, "Storing content in variables."

Creating a blank document

When you create a blank document, you can use default sizes, choosing portrait or landscape orientation, or you can create a custom page size. Even blank documents contain styles and other information from a default FrameMaker document.

Chapter 2: Creating your first document

To create a blank portrait or landscape document, follow these steps:

1. Select **File > New > Document** or press CTRL+N. The New dialog box is displayed (see page 12).
2. In the Use Blank Paper section, click the Portrait or Landscape button to display the new document.

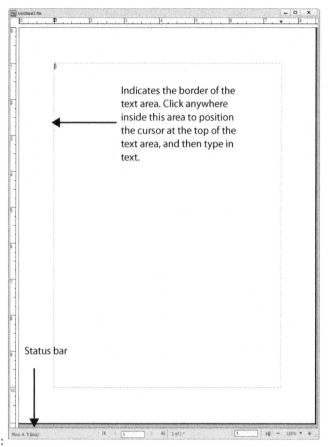

Indicates the border of the text area. Click anywhere inside this area to position the cursor at the top of the text area, and then type in text.

Status bar

To create a document that uses a custom paper size, follow these steps:

1. Select **File > New > Document** or press CTRL+N to display the New dialog.
2. In the Use Blank Paper section, click the Custom button to display the Custom Blank Paper dialog.
3. Click a paper size in the Page Size drop-down list:
    - **Custom:** You specify the page width and height
    - **US Letter:** 8.5 inches wide by 11 inches tall
    - **US Legal:** 8.5 inches wide by 14 inches tall
    - **Tabloid:** 11 inches wide by 17 inches tall
    - **A3 Tabloid:** 29.7 cm wide by 42 cm tall
    - **A4 Letter:** 21 cm wide by 29.7 cm tall
    - **A5 Letter:** 14.8 cm wide by 21 cm tall
    - **B5 Letter:** 17.6 cm wide by 25 cm tall

Opening, saving, closing, and printing documents

Tip To change the measurement units (for example, from inches to picas), select a measurement unit in the Units drop-down list at the bottom right. FrameMaker automatically converts the current measurements into the new units.

4. Modify the margins as appropriate. Page margins are measured from the edge of the page to the edge of the text area, and can be modified later by changing the Master Pages.
5. To create a document with equal-sized columns, specify the number of columns you want in the Columns area. The default is 1. The Gap measurement determines the amount of space between the columns. After you specify the columns and gap value, FrameMaker automatically creates even columns.
6. Set the pagination. Most books use double-sided pagination and start with a right first page. If you specify single-sided pagination, only one master page is created (called Right); if you specify double-sided pagination, two master pages are created (Left and Right). For details, see "Switching from single- to double-sided pages" on page 253.
7. Click the Create button to display your new document.

Using an existing file

Instead of using a document as a template and creating a new file, you can save the template file with a new name to make a copy. To use an existing file, follow these steps:

1. Open a file that has the correct formats.
2. *(optional)* To delete the content in the file, highlight all the information in the file by selecting **Edit > Select All in Flow**, or CTRL+A then press the DELETE key.
3. Select **File > Save As** and save the file with a new name.

You now have a new file and can begin editing it. However, using this method is more prone to operator error than creating a new document from the same file using the **File > New** command.

Opening, saving, closing, and printing documents

FrameMaker includes standard open, save, close, and printing commands:

- To open a file, select **File > Open**, then locate the file you want. Alternately, you can drag files from a Windows Explorer pane onto many parts of your FrameMaker application window to open the file.
- To close a file, select **File > Close**.
- To save a file, select **File > Save**, or select **File > Save As** to save the file with a different name.
- To print a file, select **File > Print**.

Files created in earlier versions of FrameMaker open in newer versions. For example, a file created in FrameMaker 10 opens in FrameMaker 11. However, you cannot open more recent FrameMaker files in earlier versions of FrameMaker. To move a file "back" to an earlier version of FrameMaker, save the file as Maker Interchange Format (or MIF), then open the MIF file in the older version. You will lose features that are not supported in the older versions, but the files usually open cleanly. If you have authored in multiple languages or in languages FrameMaker supports with Unicode, you will need to check your file carefully for changes.

Tip FrameMaker also provides a Save As Document (version - 1) feature, which lets you save files in the previous FrameMaker format directly instead of using MIF as an intermediate format. This feature could be useful if you are working in a mixed version 12/version 2015 environment, but you will have to check your files carefully, especially if you use features unique to FrameMaker 2015.

Understanding the document window

You view and edit your document in the document window. A dotted line indicates the area in which you can insert content. (If you do not see the dotted line, select **View > Borders** to display it.)

At the bottom of the document window, a *status bar* displays information about the current document. By looking at the status bar, you can find out the following information:

- Flow tag, paragraph tag, character tag, and conditional text tag for the currently selected item. (In structured documents, the element name is listed instead of the paragraph and character tags.)
- Current page number and total page count.
- Whether the file has been modified since you last saved it.
- The current percentage zoom at which the file is being displayed.

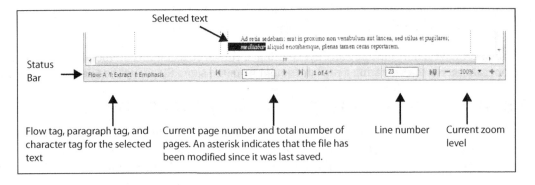

Changing the zoom settings

To change the size at which a document is displayed, you can zoom in or zoom out. The + and - on the status bar let you zoom in or out one step at a time.

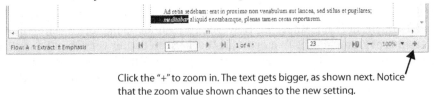

Click the "+" to zoom in. The text gets bigger, as shown next. Notice that the zoom value shown changes to the new setting.

Here is the resulting view:

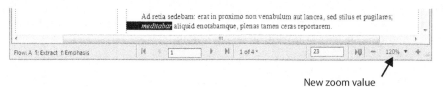

New zoom value

FrameMaker provides ten zoom settings. You cannot add more settings, but you can change the current view to values anywhere between 25% and 1600% by typing a number into the zoom menu. To do this:

1. Click on the zoom percentage to highlight the number.

2. Type in the zoom setting you want and press ENTER or TAB. The document is immediately changed to that zoom level.

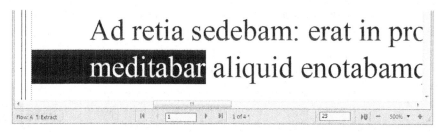

You can also use the Set... option to change the ten preset values to your liking.

Displaying nonprinting items in the document window

You can turn on and off the display of several items in the document window. Listed in the order they appear in the View menu, they are as follows:

- **Borders.** With borders on, the edge of text frames, table cells, and the like are indicated with dotted lines.
- **Text Symbols.** With text symbols on, tabs, end-of-paragraph symbols, markers, and other nonprinting characters are displayed on screen. These are extremely useful when editing or searching for specific items.
- **Rulers.** With rulers on, measurements are shown on the top and right of the document window.
- **Grid Lines.** With grid lines on, a grid pattern is shown to help you align objects on the page.
- **Line Numbers.** These are helpful for reviewers. They help when following irregular spacing and text frames. Line numbers are required for legal pleadings in some jurisdictions
- **Hotspot Indicators.** With hotspot indicators on, you'll get a visual indicator of the graphical objects you've set to act as click-able areas.

FrameMaker also uses many different nonprinting symbols. Though nonprinting, these symbols may cause the document to reflow slightly when displayed in the document. This can slightly affect your pagination, so I recommend turning them off and updating your book prior to output. The following table lists nonprinting text symbols.

Table 2-1.   Nonprinting Text Symbols

Symbol	Description
⟩	**Tab.** Indicates a tab character. Keep in mind that tabs do not move text unless a tab stop is defined for that paragraph. For details, see "Setting tab stops" on page 116.
¶	**End of paragraph symbol.** Indicates the end of a paragraph. When you press ENTER to end a paragraph, FrameMaker inserts this symbol.
§	**End of flow symbol.** Indicates the end of the content in the current flow. As you add more content, the end of flow symbol always stays at the end of the text. You cannot remove this symbol.
⟨	**Forced return.** Indicates a forced line break. Press SHIFT-ENTER to insert a forced return.
␣	**Hard space.** Indicates a nonbreaking space (inserted by pressing CTRL+SPACE), which does not allow a line break to occur at that location.
⊤	**Discretionary hyphen.** Indicates a location at which the word can be hyphenated if necessary, in addition to the hyphenation points defined in the dictionary. Press CTRL+HYPHEN or ESC HYPHEN SHIFT-D to insert a discretionary hyphen. Discretionary hyphens are shown above the text.
—	**Suppress hyphen.** Indicates a location at which the word cannot be hyphenated, even if the dictionary allows hyphenation there. Press ESC N S to suppress hyphenation for a word and insert a suppress hyphen symbol. The suppress hyphenation symbol is displayed under the text.
⊥	**Anchor for table or anchored frame.** Indicates the location at which a table or anchored frame is anchored to a paragraph.
T	**Marker.** Indicates the location of a marker, which contains hidden, nonprinting text. Markers are used for several things, including index entries. FrameMaker automatically inserts markers for cross-references and conditional text, so you should not delete a marker unless you are certain it isn't needed. Markers can occur anywhere in text, including in the middle of a word. They are usually easier to maintain if you put them at the beginning or end of a word, and you might consider applying a a character tag to them to color code them, making it easier to locate and identify them.

Changing view options

The view options let you set measurement units, page options, and more. To change your view options, follow these steps:

1. Select the **View > Options...** command to display the View Options dialog.
2. Set the page scrolling you want. Your choices are as follows:
   - **Vertical:** In the document window, pages scroll vertically. (This is standard behavior in most word-processing and publishing applications.)
   - **Horizontal:** Pages scroll horizontally.
   - **Facing Pages:** Pages are displayed in spreads, with the left and right pages displayed in the document window together.
   - **Variable:** Pages are displayed depending on how the document is zoomed. If facing pages fit in the window, they are displayed; if not, pages scroll vertically.

3. Set the display units with the Display Units and Font Units drop-down lists. The font units are used for font size, leading, and space above and below paragraph. Display units are used for all dialogs and pods.
4. In the Snap section, you can set the spacing for the snap grid, as well as the increment (in degrees) used for snapping a rotated object.
5. In the Display section, Rulers lets you turn display of rulers on and off, as well as set the units and increments displayed.
6. In the Display section, check the items you want to display. Unchecking the Graphics checkbox speeds up scrolling by temporarily displaying your graphics as gray boxes.
7. Click the Set button to save and apply your changes.

FrameMaker interface options

Workspaces

Beginning with FrameMaker 9, FrameMaker now looks much more like Photoshop, Illustrator, and the majority of other Adobe applications.

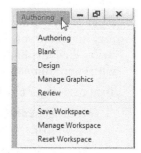

As with these other applications, the visibility and position of toolbars and pods within your workspace is stored, and can be recorded within the Workspaces option in the upper right corner of the application. Use Workspaces to optimize your available pods while performing specic tasks. Workspace files can be shared, and are stored in
C:\Program Files (x86)\Adobe\AdobeFrameMaker11\fminit\WorkSpaces\UnStructured)

Toolbars

FrameMaker provides eight toolbars at the top of the main window, and some have quite a few options. If you like using buttons instead of keyboard shortcuts, check out **View > Toolbars** to show or hide as needed. The buttons each have tooltips to help you navigate the options available.

Toolbars can be docked in one or more rows under the menu bar, or can be dragged into the main document area to create independent floating toolbars.

Figure 2-1. Floating toolbars

Figure 2-2. Docked toolbars

To re-dock toolbars, reset your workspace or drag the handle of the toolbar into the toolbar region below the menu bar until you see a blue bar. The blue bar indicates a legal position for the toolbar.

Toolbar Handle when dragging

A blue bar is displayed when in an allowable position

Developers also provide FrameMaker enhancements for specific workflows and functions. For more information see Appendix A, "Third-Party Tools and Plug-Ins."

Icons

Beginning with FrameMaker 12, you can set your preferences to use larger and/or colored icons. I have tried to make all screen captures in the book using Windows 8.1 and the colored and large options. Of course, the print version won't show the colored icons, but EPUB and other digital publishing output will show color where appropriate.

If you are using a version of FrameMaker before the 2015 release, you may find more variance in the icons within different pods and toolbars. If so, try downloading a trial of the current release to compare functionality with your FrameMaker version.

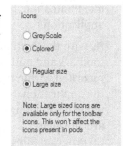

To access larger or colored icons, follow these steps:

1. Select **Edit > Preferences > Global > Interface** to display the icon controls within the Preferences dialog
2. Choose GreyScale (sic) or Colored, as well as Regular or Large as needed.

Document tabs

FrameMaker can display open documents as a series of stacked windows, or as a series of tabs across the top of the document view. When viewing tabs, you can click on a tab to move that file to the front of the working files. When many files are open, you may see a double chevron icon ([>>]) appear to the right of the tabs. Clicking on the chevrons displays a list of open documents, allowing you to quickly navigate to documents whose tabs are currently hidden. To quickly configure display of open documents, select the Arrange Documents ([▦▼]) button.

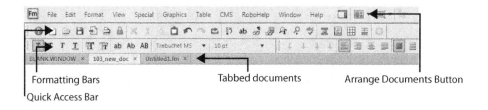

Formatting Bars
Quick Access Bar
Tabbed documents
Arrange Documents Button

Tip The CTRL+CTRL+TAB key will cycle through open documents. This is effective when you have only a handful of documents open in FrameMaker

Pods

Most configuring of text and FrameMaker objects occurs within *pods*. Similar to toolbars, pods can either float or dock to the edges of the application frame. Both floating and docked pods can expand to display contents, or collapse to icons, allowing toggling of pod display, as shown in the following figure. To expand or collapse pods to icons, click the double arrow in the upper right corner of the pod.

Note Adobe has at times referred to *pods* as *pods* and *palettes*. Today, it's pods, but you're likely going to find that I and others use these three terms interchangeably.

Figure 2-3. Floating toolbar, collapsed and expanded

Figure 2-4. Docked toolbar, collapsed and expanded

Managing open files

Beginning with FrameMaker 12, the Currently Opened Files pod allows you to manage, save, and close all of your currently opened files. This pod also shows the complete file path. This is quite useful when working with multiple chapter books. Bonus tip: Double-clicking on a file in the pod displays the document, similar to working with a book file. Select **File > Currently Opened Files** to display the pod.

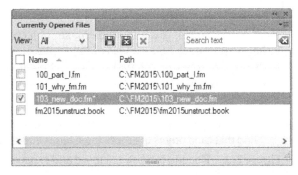

Starting with FrameMaker 12, you'll also see a similar dialog when exiting FrameMaker if you haven't saved all of your files. Earlier versions of FrameMaker will prompt you to save each individually.

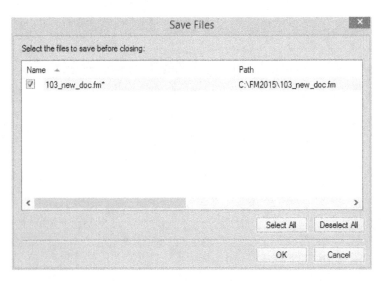

Restore last session

Starting with FrameMaker 12 you can use the restore session feature to reopen FrameMaker to exactly how it was when you closed the program. To do this, select **File > Restore Last Session**. You can also choose to open all recently opened files, even if they were not part of your last FrameMaker session.

This feature is also useful if your machine crashes while using FrameMaker. Upon startup, FrameMaker will prompt you to open the previously opened files.

Understanding file configuration and preferences

FrameMaker files normally use the .fm extension. FrameMaker book files (more on those in Chapter 17, "Setting up book files") normally use .book.

Storing format definitions

A FrameMaker document file is self-contained; it includes all of the formatting or structure information needed to display and print the content. Unlike some other applications, FrameMaker files do not refer to a separate template file; the template information is embedded in each individual document file. For details, see Chapter 5, "Templates" on page 89.

Importing and embedding graphics

When you add graphics to a document, you can either link or embed the graphic. When you link a graphic (referred to as Import by Reference), you insert a pointer from the FrameMaker file to the graphic. This pointer is basically just a file name and path with a small graphic for screen representation, so it's very compact. If you embed a graphic (referred to as Copy into Document), you insert the entire graphic into the FrameMaker file, so the FrameMaker file grows accordingly. It is common for savvy FrameMaker users to link to all graphics within a project, rather than embedding them. However, screenshots, whose FrameMaker representation can be the size of the referenced file itself are a possible exception. As screenshots are likely recaptured and not edited, some authors prefer to copy screenshots directly into their documents.

Tip The vast majority of screen captures in this book were copy/pasted into position, using either RoboScreenCapture or OS screen capture utilities. I find that pasting while at a 160% zoom lets me keep the graphic scaled to 100% of size. This gives me consistency through my documents, and results in quite clean printed output.

Document control

To prevent two people from working on the same file simultaneously, FrameMaker provides a feature called network file locking. If network file locking is active, opening a file creates a .lck file ("lock" file, *not* "ick" file). If a second person attempts to open the locked file, FrameMaker displays a "Document in Use" error message.

You have three choices:

- **Open for Viewing Only:** Opens a read-only version of the file. You cannot make any changes to this file.
- **Open Copy for Editing:** Creates an untitled copy of the file. You can make changes in this file.
- **Reset Lock and Open:** Removes the lock and lets you edit the file. Use this option only when you are certain that nobody else is using the file (for example, if your computer crashed and left incorrect locking information on the server).

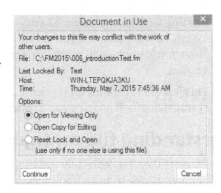

To activate or deactivate network file locking, follow these steps:

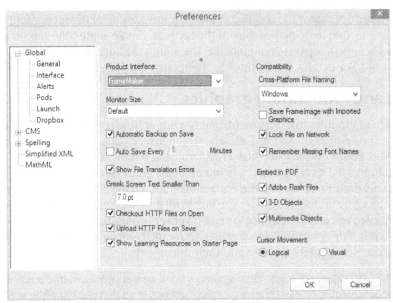

1. Select **Edit > Preferences… > Global > General** to display the Preferences dialog.
2. Check the Lock File on Network checkbox.
3. Click the OK button to save your change.

Remember   Network file locking is turned on by default when you install FrameMaker. Keep in mind, though, that each user must have network file locking activated in their copy of FrameMaker for the feature to function properly and that the .lck file relies heavily on the "honor system" for enforcement.

In a larger group (or if you have people working on the files who are not on the same network), the basic file security provided by file locking may not suffice. If this is the case, you can use a content management system (CMS). FrameMaker also supports file check-in and check-out on a WebDAV server. See Chapter 31, "Using a Content Management System (CMS)," for details.

Understanding file configuration and preferences

Using the Fonts pod

You can track down many font issues using the Fonts pod. To replace a font in your document using the Fonts pod, follow these steps:

1. select **View > pods > Fonts** to display the Fonts pod.
2. Select the Toggle Missing/All Fonts button ([🗛] or [sin]) if needed, and select the font to replace.

3. Select the Replace button ([🗛]) to display the Replace Font pod.
4. Do one of the following:
   - Select a replacement font and click the Apply button.
   - Select an instance of the font and click the Go To Location button ([📄])
   - Select the Refresh button ([🔄]) for an updated view of locations.
5. In the Fonts pod select the Refresh button ([🔄]) to see the results of your font replacement.

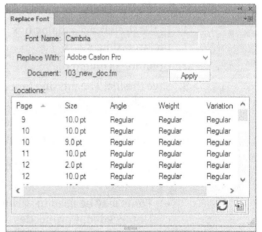

Handling missing fonts

To display text, FrameMaker uses fonts that are installed on your system. When you open a file, any fonts used in the file, but currently unavailable to FrameMaker are identified and listed in the console window. By default, FrameMaker performs only a temporary font replacement. You might have a situation where a consultant's or author's computer lacks fonts required for a document, but the final production computer has the necessary fonts installed. In this scenario, you want FrameMaker to keep track of the missing fonts and replace them only temporarily while the file is being edited. In other cases, you may want a permanent font substitution. You might, for example, receive a file from another company, which uses fonts that you do not have and don't plan to install. In that case, you probably want to get rid of the fonts so that you don't get the missing fonts message every time you open the file.

25

Caution   If some characters in your FrameMaker file are displayed as boxes, question marks, or other strange characters, the missing font may have more characters than the font being substituted. (This is common when a Unicode font is missing and a non-Unicode font is substituted.)
In this case, make sure you have a font that can display all the characters correctly before you allow FrameMaker to permanently substitute a different font.

Font substitution (for both temporary and permanent substitutions) is specified in the maker.ini file. To change the substitution font, see "Changing the substitution fonts" on page 548.

To toggle your missing font setting, follow these steps:

1. Select **Edit > Preferences… > Global > General**. The Preferences dialog box is displayed (see page 24).
2. Check or uncheck the Remember Missing Font Names checkbox.
3. Click the OK button to save your change.

To remove missing fonts from a file, follow these steps:

1. In the Preferences dialog box, turn off Remember Missing Font Names as described in the preceding steps.
2. Open the problem file.
3. Save the file. The missing fonts are replaced with the fonts specified in the maker.ini file. See "Changing the substitution fonts" on page 548.
4. *(optional, but recommended)* Return to the Preferences dialog box and turn on Remember Missing Font Names.

The fonts should now be removed from the file. There may be a few cases where additional measures are needed to remove font calls:

- Fonts may be used in various catalogs, on master pages, or on reference pages. Untagged font properties and text lines may also contain font calls that are difficult to find. Solution: Save file as MIF and search for the font name using a text editor. For details, see Appendix F, "Maker Interchange Format."
- Fonts may be specified in placed meta graphics or vector graphics files. Solution: remove the graphics one at a time and reopen the FrameMaker file to identify the culprit graphic file. If there are many graphics, consider splitting the file and saving as two or more files for testing, then opening each segment to help narrow down the location of the offending graphic file.

Organizing directories and files

Although many companies establish rules about how to store files and set up directories to manage content, a surprising number do not. If you need to establish organization for your documentation library, this section provides some suggestions on how best to manage the files.

The first principle of organizing documents is that a book should have its own directory. Moving the directory (and any associated subdirectories) should move all of the contents of the book. For instance, you can set up a directory organization for a book that includes the book file, chapter files, and graphics in a simple organization.

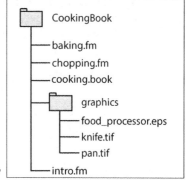

I recommend against naming chapters by their numbers; for example, chap01.fm, chap02.fm, and so on. If you move chapters around, you will either have a big renaming job or have chapters whose file names don't match their numbers. Instead, name chapters by context or subject: production_editing.fm, workflow.fm, and so on.

Putting referenced graphics in a subdirectory makes it easier to keep track of them and to separate the graphics from the FrameMaker content files. Creating subdirectories for each chapter can be appealing when you have lots and lots of graphics, but again, avoid naming these directories by chapter number because of the naming mismatch that will occur if you rearrange chapters. For longer books, consider using a prefix for each chapter so that the graphics sort nicely into chapter groups.

Files that are reused across chapters and especially across books present a problem. Consider creating a special "shared" folder in the book folder or one level up from the book folder if you want to share across different books. Keep in mind that you must remember to send the shared folder in addition to the book folder if you ship the files to someone who cannot access them from your network. Be wary of deep nesting across networks—long path names can cause issues with placed graphics.

If you move files from one root drive to another, it is likely that relative links will change to absolute links. Be aware of this when planning location of working files and network graphics.

To assist in moving or collecting all files referenced by a book, FrameMaker now has an excellent Package function (**File > Package...**). Use it when you need to move, archive, or send files for someone else to work on.

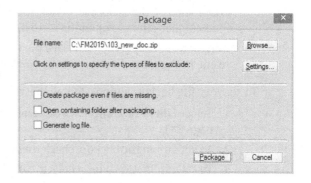

Saving file information with XMP

You can save file information, such as the author, keywords, and title, along with your file. This information is saved as *metadata*, or information about the document. The file information you save is stored in a special format called Extensible Metadata Platform, or XMP. XMP allows you to view metadata without opening the file. You could, for example, define some file information and save your FrameMaker file. An XMP-compatible application or browser can access the XMP data stored for your FrameMaker file, even if the FrameMaker application is not available.

If you make a PDF file from your document, the file information you specify can also be viewed in Adobe Reader.

To set up file information, follow these steps:

1. Open the file or book for which you want to save file information.
2. Select **File > File Info** to display the File Info dialog.
3. Type in the information you want to save. You can fill in any or all of the available fields.
4. Click the Set button to save your changes with this file.

After setting file information, the data you provided is included as XMP metadata when you save the file.

For more details about XMP, consult Adobe's web site (www.adobe.com/products/xmp/).

Implementation of XMP varies, so verify that the XMP data is recognized as you would expect.

Setting automatic backup and save features

FrameMaker provides two complementary features that help you save files automatically:

- **Automatic save.** While you work in FrameMaker, an autosave file (.auto) is created periodically. If you exit FrameMaker normally, the file is deleted. However, if your system crashes (which means you didn't have a chance to close the document properly in FrameMaker), the file is preserved. When you reopen the file you were working on, FrameMaker offers to open the .auto file, which might contain more recent information.
- **Automatic backup.** When you save the working file, a .backup.fm file is created. Backup files reside in the same directory as the working file.

To set your file backup options, follow these steps:

1. Select **Edit > Preferences… > General**. The Preferences dialog box is displayed (see page 24).
2. To set automatic backup, check the Automatic Backup on Save checkbox.
3. To set automatic saving, check the Automatic Save checkbox, then specify the interval at which you want the file saved. (The default is 5 minutes.)
4. Click the Set button to save your changes.

Note FrameMaker saves a crash recovery file (.recover) when a crash occurs. That file might contain more up-to-date information than the .auto or .backup files. I recommend that you inspect the .recover, .auto and .backup files to see which has the most appropriate content.

Switching between structured and unstructured interfaces

To switch between structured and unstructured FrameMaker, you must select the new interface, close FrameMaker, and reopen it.

To change interfaces, follow these steps:

1. Select **Edit > Preferences… > Global > General**. The Preferences dialog box is displayed (see page 24).
2. In the Product Interface drop-down list, select the interface you want to switch to and click OK. You'll be reminded that you must restart FrameMaker for the changes to be applied.
3. Save all your open files.
4. Exit FrameMaker.
5. Restart FrameMaker. The product is started and displays the interface you selected.

File naming conventions

FrameMaker is increasingly used to create online content. Consider avoiding spaces and non-alphanumeric characters for filenames, for imported graphic filenames, and for marker text. Specifically, I have had trouble with EPUB and Responsive HTML5 output when spaces, punctuation, and # characters are present.

Chapter 2: Creating your first document

Part II

Creating and Manipulating Text

Chapter 3: Word-Processing Features

FrameMaker provides full-featured word processing, including standard features such as search and replace. But FrameMaker also goes beyond expected options, giving you more robust options to create and analyze content. FrameMaker extends your capabilities, from the mundane tasks of Undo and Find/Change, to more complex options for change management and text import and reuse.

Standard Word-Processing Features 33
    Undoing and redoing changes 34
    Inserting special characters 35
    Searching and replacing 38
    Spell-checking 43
    Thesaurus 51
Configuring Default Text Options 51
    Smart quotes 52
    Smart spaces 53
    Restricting line breaks 53
    Formatting superscripts, subscripts, and small caps 54
Tracking Changes 55
    Using the Track Text Edits tool 55
    PDF review 57
    Displaying the word count 58
    Using change bars 58
    Displaying line numbers 59
    Comparing documents and books 60
Working with Footnotes 62
    Inserting footnotes 64
    Using the same number for several footnotes 65
    Customizing footnote properties 66
Importing Text from Other Applications 69
    MS Word (copy into document) 69
    Copying, pasting, and converting text 71

Standard Word-Processing Features

FrameMaker has the word-processing features you would expect in typical word-processing programs, but in general, they work much better. Using the Undo and Redo feature, you can remove or re-create multiple changes. But by using the **Edit>History** pod, you can display a list of recent changes and revert to that state of the document. Instead of just one way to insert special characters like nonbreaking spaces, em dashes, or arrows, there are several. My favorite, though, is the Find/Change feature. With Find/Change, you can not only find and replace words, but search for specific styles, FrameMaker components (like tables, or markers), and even identify errors in your document. This is handy for finding inconsistencies like paragraph format overrides and unresolved cross-references. Even the Spelling Checker goes beyond the expected, not only locating misspelled words, extra spaces, and unusual punctuation, , but allowing for a common dictionary across your workgroup.

Undoing and redoing changes

FrameMaker lets you undo or redo changes you have made in a document. In addition, FrameMaker tracks your changes in a History pod (see page 34), from which you can display a list of changes and revert to a specific change. Many actions will clear the History pod for the document. FrameMaker displays a warning prior to performing an action that clears the history by default, but I highly recommend you turn this feature off in the Preferences! The default setting on merely raises your blood pressure about things that you rarely have a choice about.

Undoing or Redoing an Action

You can undo or redo changes in the order in which you made them. For example, if you added text, applied a paragraph tag, and then added more text, you could undo the last text you typed and then undo the paragraph tag.

If you want to undo or redo changes, do the following:

- If you want to undo one or more of your most recent changes, select **Edit > Undo**, click the Undo Typing button ( ), or press CTRL+Z as many times as needed to remove a specific change.
- If you want to redo one or more of your recent changes, select **Edit > Redo**, click the Redo button ( ), or press SHIFT+CTRL+Z as many times as needed to return to the appropriate editing point.

Displaying the History pod

You can display the History pod and select a change to which to revert. The change history is tracked for each open file in its own History pod. If you want to undo multiple changes, using the History is faster and less confusing than performing multiple Undo and Redo commands.

To display the pod and revert to a specific change, follow these steps:

1. Select **Edit > History** or press CTRL+K or ESC E H to display the History pod.
2. Select a change from the list. For example, in the history shown in the illustration, choosing the "Move" entry would remove the "Clear" and the "Anchored Frame" changes.

   To undo every change in the history, select Undo All from the top of the list.

Note Clicking a change in the history reverts that item and all the changes that followed.

Your document reverts to the change you chose. The History pod displays changes that have been removed with asterisks, indicating you can select Redo to restore those changes, or you can select one in the pod to restore that change and any preceding changes. After you perform additional actions that change the document, the Redo history is cleared.

Setting Options for History Warnings

When you perform certain tasks in FrameMaker, such as saving a file, FrameMaker clears the change history, and from that point you can no longer undo or redo changes.

By default, FrameMaker displays a warning whenever you are about to perform an action that clears the change history.

You can specify when FrameMaker displays this warning or you can turn the warning off. To change this setting, either click the "Do not show again" checkbox, or follow these steps:

1. Select **Edit > Preferences > Global > Alerts**.

Note There is a consistent use of *cleaning* instead of *clearing* in the Preferences dialog. On behalf of Adobe, I apologize for this irksome typo.

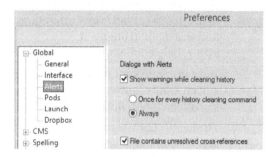

2. Do one of the following:
   - To turn off clear history warnings, uncheck the Show warnings while cleaning history checkbox.
   - To display warnings the first time in each FrameMaker session that you perform an action that clears the history, select the Once for every history cleaning command option under the Show warnings while cleaning history checkbox.
   - To display warnings every time you perform an action that clears the history, select the Always option under the Show warnings while cleaning history checkbox.
3. Click the OK button to save your change.

Tip To help understand which actions clear the change history, consider displaying all warnings when you first start to use FrameMaker. After you are familiar with those actions, consider turning off the warnings to preserve your sanity.

For more on actions that can be undone and ones that clear the change history, see the FrameMaker online help.

Inserting special characters

When you are creating content, you often have to use characters that are not on the keyboard, such as em dashes, special spaces, or dingbats. This section describes working with special characters in FrameMaker.

Using the Symbols Toolbar List

FrameMaker 11 provides an icon in the Formatting toolbar (🔣) to add these commonly used special characters:

Bullet (•)	Register (®)	Ellipsis (…)	Cent (¢)
Plus/minus (±)	Copyright (©)	En dash (–)	Pound (£)
Double dagger (‡)	Trademark (™)	Em dash (—)	Yen (¥)

To insert one of these characters, select it from the drop-down list displayed after you click the Symbols icon. If you need to insert a special character not in this list, use one of the other methods described in the following sections.

You can customize the list to include other symbols by modifying the menus.cfg file, as described in the *Adobe FrameMaker Customizing (Windows)* online manual. At the time of writing, you can find this manual online at
http://help.adobe.com/en_US/FrameMaker/customize_frame_7.pdf.

You can create your own custom list of symbols (and many other items, too) with Auto-Text from Silicon Prairie Software or Shlomo Perets' Express Customization product. For contact information, see Appendix A, "Resources."

The list of characters available in FrameMaker is quite exhaustive. For a more detailed list of special characters and how to insert them, see the following sections in this chapter, as well as Appendix B, "Shortcuts".

Using the FrameMaker Character Palette

To use the Character Palette utility, position your cursor where you need the character in your file, and go to **File > Utilities > Character Palette**. Select the appropriate font family, navigate to the character needed, let the mouse hover the character to get a preview, and click once on the character to insert it into your document.

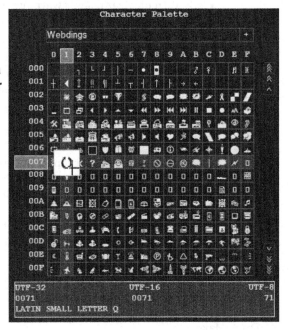

Insert Characters from the Hex Input Palette

The Hex Input palette gives you an alternate way to insert characters from your current font in dialogs and your document based upon either the Unicode character code or its hex value. To enter a character this way, do the following:

1. Place the cursor where you want to insert the character in your document.
2. Select **File>Utilities>Hex Input** to display the Hex Input palette.
3. Type the Unicode number of the character you want to insert. The corresponding character is displayed on the right.

4. To toggle between Unicode character sets, click on the UTF code until you see the desired coding.
5. If you cannot find the character you want to insert, either click the arrow keys to search sequentially or click the character to display the Character palette.
6. Ensure that you have selected the appropriate font from the Character palette. The character for the same hex value changes depending on the font that you select.
7. Press the Enter key when you have located the proper character. If the character is available in your current font, the character is inserted into your document.

Pasting from the Windows Character Map Utility

You can find special characters using the Character Map utility provided with Windows. When you have found a special character in the Windows Character Map utility, you can copy and paste the character into FrameMaker, or you can use the map to find its associated codes.

If you copy and paste the character from the Windows Character Map utility, you must be sure to use the same font in FrameMaker to display the character as you used for choosing the character in the Windows Character Map utility. For example, if you select a special character from the Windows Character Map utility using the Symbol font, the text in FrameMaker must also use the Symbol font, or the character displayed will be incorrect.

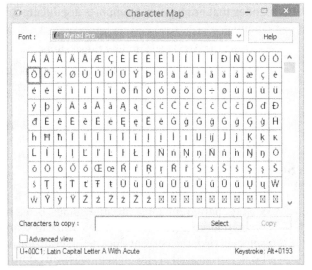

To use the Windows Character Map utility, follow these steps:

1. Locate the Character Map application on your version of windows. If you are unsure of where to find it, press the Windows key on your keyboard to search for the application to display the Character Map dialog.
2. Select the font you are using from the Font drop-down list. This should match the font for your paragraph or character tag in FrameMaker.

3. From the character table, select the character to insert in FrameMaker.
4. Click Select, and then Copy.
5. In FrameMaker, select **Edit > Paste**.

   If question marks or gray boxes are displayed instead of the correct characters in the pasted text, you are copying and pasting Unicode characters, so use paste special. Select **Edit > Paste Special**, select Unicode Text, then click OK.

   To have FrameMaker automatically paste as Unicode text instead of plain text, change the paste order in maker.ini. See "Setting clipboard pasting order" on page 548.

Using Keyboard Sequences

FrameMaker provides keyboard sequences for typing some special characters. Often, the keyboard sequences are complex. For example, to type an em dash, you type CTRL+Q SHIFT+Q. (That is, press and hold CTRL, then press Q, release those keys, then press and hold SHIFT, then press Q.)

Note At the time of writing, various manuals (including FrameMaker Character Sets) from previous FrameMaker versions are available at www.adobe.com/support/documentation/en/framemaker/.

Typing ANSI Codes

You can type a special character by typing its ANSI code using your numeric keypad. The ANSI codes that FrameMaker supports are documented in the *Character Sets (FrameMaker 8)* online manual.

To type the ANSI code, follow these steps:

1. Press the NUM LOCK key on your keyboard to use the numeric keypad, if it is not already turned on. Typically you'll need to use the numeric keypad, not the numbers above your QWERTY keyboard.
2. Press and hold down the ALT key and type the ANSI code on the keyboard for the special character, including the leading zero.

Searching and replacing

The Find/Change feature lets you search for text strings, paragraph tags, variable names, broken cross-references, and other items. You can search selected text, the current document, or an entire book (structured FrameMaker gives you additional options). Suppose that you create a variable for your document title after you typed the title in several places. You can search for the places where you typed the document title and replace the title with the variable (using the Change: By Pasting option).

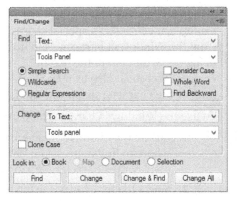

There are several ways to customize your search:

- **Consider Case:** Finds text capitalized like the search item. If you search for *island*, FrameMaker skips the capitalized word *Island*.
- **Whole Word:** Finds the exact words and skips all instances where the text is part of a larger word. If you search for *object*, FrameMaker finds *object* and skips *objective*, *objects*, and *objectification*.
- **Regular Expressions:** Also known as regex, allows searches based on text patterns. Regex is commonly used in markup and programming languages, including HTML. With regex, for example, you could search for US date formats and convert to abbreviations, or European standard formats.
- **Use Wildcards:** Lets you use special characters in place of one or more unspecified characters. Table 3-1 describes how to use each wildcard.

Table 3-1. Using Wildcards in a Search

Wildcard	Description	Example
*	Placeholder for zero or more characters.	s*n finds *sin*, *son*, *soon*, and *sullen*.
?	Placeholder for any single character except for spaces and punctuation.	s?n finds *sin*, *sun*, and *sandal*.
|	Placeholder for one or more spaces and punctuation characters.	gh|the|wo finds *through the woods*.
[ ]	Placeholder for one or more characters you type between the square brackets.	s[ai]ng finds *sang* and *sing*.
[^]	Excludes the characters in brackets from the search.	s[^i]ng finds *sang*, *song*, and *sung*, but not *sing*.
^	Placeholder for the beginning of a line. Not applicable with the Whole Word option.	^si finds *sing* and *single* only at the beginning of a line.
$	Placeholder for the end of a line. This is especially useful for finding text that you do not want to break across lines. Not applicable with the Whole Word option.	Mr.$ finds *Mr.* at the end of a line. If you search for $ed, FrameMaker finds the end of a line followed by *ed* on the next line.

- **Find Backward:** Searches the selection, document, or book starting at the cursor location and moving towards the beginning of the document.
- **Clone Case:** Matches the case of the original item in the replacement. For example, if you type "President" in the search field and "leader" in the replace field and do not check Clone Case, then FrameMaker replaces "President" with "leader." With Clone Case checked, "President" is replaced with "Leader."
- **Look in:** Searches the current document, selected text, or an entire book. The Book option is available only if a book that contains the current file is open in FrameMaker. The Map option is available only when working on DITA content.

FrameMaker can find items other than just text. You pick an item to search for—marker text or an unresolved cross-reference, for example—from a list.

Searching for an Item

To find and change an item in a document, follow these steps:

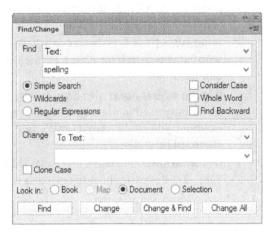

1. You can search a book, a document, or selected text. Do one of the following:
   - To search a document, click anywhere in the document's text.
   - To search a specific section of text, select that text.
   - To search a book, open the book. You can start your search from the book window or inside a file that belongs to that book. To search only specific files in the book, select them first by CTRL+clicking the files you want.
2. Select **Edit > Find/Change** to display the Find/Change pod.
3. In the Find drop-down list, click the item you want to find. Items followed by an ellipsis (…) will display a dialog box in which you select additional options. Items followed by colons (:) require that you type in additional search information in the Find field. Table 3-2 describes the search options.
4. *(optional)* To qualify the search, check any of the Consider Case, Whole Word, Use Wildcards (only if searching for text), or Find Backward checkboxes.
5. To indicate the scope of the search, click the Book, Document, or Selection radio button in the Look in section. (If the book file is displayed, and you want to search only some of its files, then select those files and choose Selection, not Book.) The Map option will be grayed out in unstructured FrameMaker.
6. Click the Find button (or press ENTER) to begin the search. FrameMaker highlights the first match in your document, book, or selection.
7. Click the Find button to search for the next occurrence.

Table 3-2. Search Criteria

Search for...	To find...
Text:	Text you type or paste in the Find field. You can also search for hex, UTF and other character codes. See the FrameMaker Character Sets online manual for details.
Character Format...	Text that uses the specified character formatting. When you select Character Format, the Find Character Format dialog box is displayed, which lets you specify which formatting properties you want to search for. 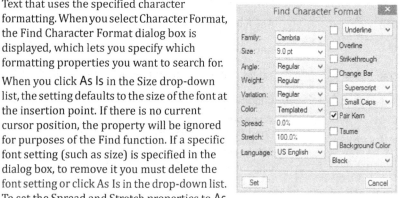 When you click As Is in the Size drop-down list, the setting defaults to the size of the font at the insertion point. If there is no current cursor position, the property will be ignored for purposes of the Find function. If a specific font setting (such as size) is specified in the dialog box, to remove it you must delete the font setting or click As Is in the drop-down list. To set the Spread and Stretch properties to As Is, delete the information shown in the fields. To set the pair kerning to As Is, check the Pair Kern checkbox once. The blue filled checkbox, as shown in the preceding example, indicates the As Is setting.
Paragraph Tag:	The paragraph tag name you specify in the Find field.
Character Tag:	The character tag name you specify in the Find field.
Any Marker	Any type of marker, such as hypertext markers, index markers, cross-reference markers, and so on.
Marker of Type:	Markers of the type you specify in the Find field.
Marker Text:	Markers of any type that contain the text you specify in the Find field.
Any Cross-Reference	Any cross-reference.
Cross-Reference of Format:	Cross-references that use the cross-reference format you specify in the Find field.
Unresolved Cross-Reference	Unresolved (broken) cross-references.
Any Text Inset	Any text inset.
Unresolved Text Inset	Unresolved (broken) text insets.
Any Publisher	Not currently used in FrameMaker (formerly used on Macintosh only).
Any Variable	Any variable.
Variable of Name:	A variable that matches the name you specify in the Find field.

Chapter 3: Word-Processing Features

Table 3-2. Search Criteria (Continued)

Search for…	To find…
Any Rubi	Rubi characters. Rubi are pronunciation cues for Japanese characters.
Anchored Frame	Any anchored frame.
Footnote	Any footnote.
Any Table	Any table.
Table Tag:	Tables that use the table tag you specify in the Find field.
Conditional Text:	Conditional text. When you select this option, the Find Conditional Text dialog box is displayed. Set the condition tags to match the conditional text you want to find, then click the Set button.
Automatic Hyphen	Hyphens inserted by FrameMaker's hyphenation utility.
Text & Character Formats on Clipboard	Text or character formatting you've copied from the FrameMaker document.
Paragraph Format Override	Paragraphs not currently conforming to the paragraph styles as defined in the Paragraph Catalog.
Character Format Override	Text ranges not currently conforming to the character styles as defined in the Character Catalog.
Table Format Override	Tables not currently conforming to the table styles as defined in the Table Catalog.
Object Style Tag	The object tag name you specify in the Find field.
Object Style Format Override	Objects not currently conforming to the object styles as defined in the Object Catalog.

Note FrameMaker currently retains focus on the Find/Change pod after performing a search. This is different from previous versions, which changed focus to the "found" text, potentially resulting in unintended deletion of text.

Replacing an Item

To replace the item you've found, follow these steps:

1. Click a replacement method in the Change drop-down list.
   - **To Text:** Replaces the found item with the text you type in the Change field.
   - **To Character Format:** Displays the Change to Character Format dialog box, which looks similar to the Character Designer. Click the properties you want in the drop-down lists, then click the Set button.
   - **By Pasting:** Replaces the found item with the contents of the clipboard (which can be formatted text or special copied items such as paragraph or conditional text formats).
   - **Remove Override:** Resets properties to the values stored in the appropriate catalog.
   - **Element Tag To:** Used in structured FrameMaker.
   - **Attribute Name To:** Used in structured FrameMaker.
   - **Attribute Value To:** Used in structured FrameMaker.
2. After FrameMaker finds the item you searched for, do one of the following to replace the item:
   - To change only the first match, click the Change button.
   - To change the first match and find the next one, click the Change & Find button.
   - To change all matches, click the Change All button. Click the OK button to replace all matches within your specified scope. If you searched across a book, the item is replaced throughout the book. You may find it helpful (and wise!) to save all your files before using this feature.

Spell-checking

The most significant recent change in FrameMaker's spell check is in the support of Hunspell open source dictionaries in versions 11 and later. This is in addition to the Proximity dictionaries installed in all versions.

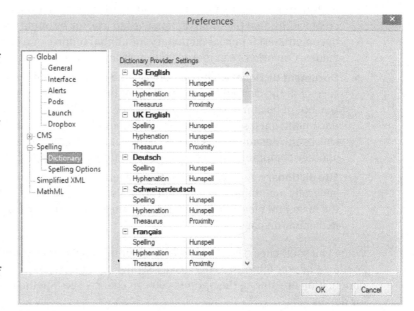

The language setting in paragraph and character tags tells FrameMaker which dictionary to use when spell-checking (and hyphenating) the text. Assign a language of None if you don't want

FrameMaker to spell-check the text. For example, users can set the language in a paragraph or character tag to None to skip spell-checking computer code or foreign language text.

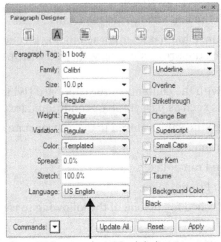 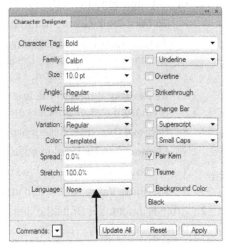

FrameMaker uses the US English dictionary to spell-check text that uses this tag.

The language is None, so FrameMaker doesn't spell-check text that uses this tag.

Regardless of the language chosen, FrameMaker looks in four places while spell-checking:

- **Main dictionary.** The primary dictionary installed with the program. FrameMaker provides support for Proximity and Hunspell dictionaries in several languages. You pick the language in paragraph and character tags. For information on controlling dictionary usage, see "Key dictionary preference features" on page 559.

- **Personal dictionary.** The dictionary of words added during a spell-check. You click the Learn button in the Spelling Checker to add words to this dictionary. FrameMaker also lets you create additional personal dictionaries to use, for example, in different types of documents—a financial dictionary for annual reports or a computer dictionary for user's guides. Coworkers can merge your personal dictionary with their personal dictionaries.

- **Document dictionary.** The dictionary embedded in the document. You add words that are correct in the current document but not in other contexts by clicking the Allow in Document button in the Spelling Checker pod. This dictionary is part of the FrameMaker document, not a separate file like the other dictionaries. As a result, anyone who spell-checks the document has access to the same dictionary, even if the document is opened on another computer. You can write this dictionary to a separate file and clear the dictionary.

- **Site dictionary.** A dictionary containing terms specific to your company, for example, and generally maintained by a system administrator. This file is typically set up on a server, and the maker.ini file of each user must be modified to reflect that location. See "Customizing maker.ini" on page 547.

When you spell-check, FrameMaker looks for misspelled words and items you select in the Spelling Checker options. Extra spaces, straight quotes, repeated words, and unusual punctuation are among the options you can choose. See "Spelling Checker Options" on page 46 for details.

Standard Word-Processing Features

To spell-check, follow these steps:

1. Display the item you want to spell-check by doing one of the following:
   - Position your cursor in the page or document you want to spell-check.
   - Display the book file you want to spell-check. To spell-check specific chapters, CTRL+CLICK each chapter.

2. Select **Edit > Spelling Checker...** to display the Spelling Checker pod.

3. In the Check section, indicate the item you want to spell-check by doing one of the following:
   - In a document, click the Book, Document, or Current Page radio button. (The Book radio button is not available if the book file isn't open.)
   - In a book file, click the Book or Selection radio button. (The Document radio button is not available at the book level.)

4. To start spell-checking, click the Start Checking button, or press ENTER. When FrameMaker finds a misspelled word, possible corrections are displayed. The closest match is displayed in the Correction field. The remaining possibilities are displayed in the list beneath the Correction field.

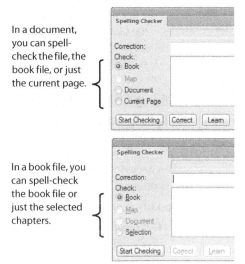

In a document, you can spell-check the file, the book file, or just the current page.

In a book file, you can spell-check the book file or just the selected chapters.

Note   If no misspellings are found, "Spelling OK" is displayed in the upper-left corner of the Spelling Checker pod.

5. Do one of the following:
   - To correct the word, either click the Correct button to accept the first suggested word, double-click another word in the list, or type in the correct word in the Correction field, then click the Correct button. The word on the page is updated.
   - To add the word to the current document dictionary, click the Allow in Document button. The word won't be flagged again in the current document.
   - To add the word to your personal dictionary, click the Learn button. The word won't be flagged again in any document you spell-check using this version of FrameMaker. See "Modifying Your Personal Dictionary" on page 47 for details.
   - To skip the word, click the Start Checking button or press ENTER. You can't add the spell-check options (extra spaces, straight quotes, repeated words, and so on) to the

dictionary, so you should skip a straight quote if you need it in the document. FrameMaker will find skipped items, however, each time you spell-check the document.

Tip When the correct spelling is displayed in the Correction field, select the Automatic Correction checkbox to replace other occurrences of that misspelling automatically as you spell-check. FrameMaker won't prompt you each time the Spelling Checker finds the word. To clear the automatic corrections settings, click the Dictionaries button in the Spelling Checker, deselect the Clear Automatic Corrections checkbox in the Dictionary Functions dialog box (see page 48), then click the OK button.
Automatic corrections are cleared each time you close FrameMaker.

Spelling Checker Options

The Spelling Checker options let you include additional items in a spell-check. You can choose to identify or ignore things like repeated words, two consecutive punctuation marks, straight quotes, extra spaces, and spaces before and after specific characters. You can also choose to spell-check or ignore single character words, all uppercase words, words containing a period, and words with digits.

To modify the Spelling Checker defaults, do the following:

1. Click the Options button in the Spelling Checker to display the Spelling Checker Options dialog.

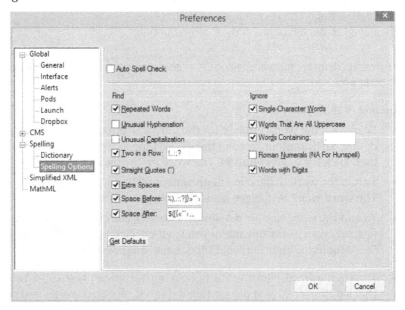

2. Change the options. Table 3-3 describes the options.
3. Click Set.

Table 3-3. Spell-Check Options

Item	Example
Find	
Repeated Words	the the
Unusual Hyphenation	unu-sual
Unusual Capitalization	cRazY
Two in a Row	Why??
Straight Quotes	"happy"
Extra Spaces	party time
Spaces Before and Spaces After	(redacted)
Ignore	
Single-Character Words	a
Words That Are All Uppercase	OLE
Words Containing	window.pane
Roman Numerals	ii
Words with Digits	555-TIME

FrameMaker ordinarily rechecks only paragraphs that have been edited since the prior check. After you modify the Spelling Checker options, you can spell-check the document using the new guidelines. To do so, click the Dictionaries button in the Spelling Checker (see page 35), click the Mark All Paragraphs for Rechecking checkbox in the Dictionary Functions dialog box, and click the OK button.

Note You cannot add items to the dictionary that conflict with Spelling Checker options, such as repeated words or extra spaces. If you want to skip these items, modify the Spelling Checker options and spell-check again.

Modifying Your Personal Dictionary

FrameMaker updates your personal dictionary when you click the Learn and Unlearn buttons. You can also create new personal dictionaries, merge another dictionary into yours, save unknown words to a separate file, and stop using your personal dictionary.

To modify your personal dictionary or create a new one, follow these steps:

1. Click the Dictionaries button in the Spelling Checker pod to display the Dictionary Functions dialog.

2. In the Personal Dictionary drop-down list, do one of the following:

    - To export your personal dictionary to a text file, click Write to File, then click the OK button to display the Write Personal Dictionary to File dialog. Type the name of the dictionary in the File name field, then click the Save button. The text file is created.
    - To switch personal dictionaries, click Change Dictionary, then click the OK button to display the Use File for Personal Dictionary dialog. Click the file you want to use, then click the Use button to display the name of the new personal dictionary.
    - To add entries from another dictionary to your dictionary, click Merge from File, then click the OK button to display the Merge File into Personal Dictionary dialog. Click the file you want to merge into your dictionary, then click the Merge button. The dictionaries are merged.
    - To stop using your personal dictionary, click Set to None, then click the OK button. The name of your current personal dictionary is no longer displayed in the Dictionary Functions dialog. (If you click the Learn or Unlearn button without a personal dictionary selected, FrameMaker warns that you need to select the personal dictionary and try again.)

Modifying the Document Dictionary

FrameMaker updates the document dictionary when you click the Allow in Document button during a spell-check. You can also remove entries, merge another document dictionary into the current one, and save the dictionary to a separate file.

Tip   To edit a file's document dictionary by hand, save your FrameMaker document as a Maker Interchange Format (MIF) file, and then open the MIF file in a text editor. The words you allowed in the document are in the Dictionary section. See Appendix F, "Maker Interchange Format" for details about MIF files.

To perform functions on the document dictionary, follow these steps:

1. Click the Dictionaries button in the Spelling Checker pod. The Dictionary Functions dialog box is displayed (see page 47).

2. In the Document Dictionary drop-down list, do one of the following:
    - To remove all entries from the document dictionary, click Clear, then click the OK button. The words are removed from the document dictionary.
    - To merge another document dictionary with the current document dictionary, click Merge from File, then click the OK button to display the Merge File into Document Dictionary dialog. Click the document dictionary you want to merge with your dictionary, then click the Merge button. The files are merged.

- To export the document dictionary from your FrameMaker document to a separate text file, click Write to File, then click the OK button to display the Write Document Dictionary to File dialog. Type the name of the dictionary in the File name field, then click the Save button. The dictionary file is created.

Tip   You can merge the exported document dictionary with your personal dictionary. See "Modifying Your Personal Dictionary" on page 47 for details.

Customizing Hyphenation

Dictionaries don't just contain lists of words; they also show how words are hyphenated. FrameMaker shows the hyphenation points when you click the Show Hyphenation button in the Spelling Checker. For example, the following dictionary entries suggest hyphenation for three words:

fire-wall
gray-scale
hy-per-link

When you install FrameMaker, dictionaries for different languages and the corresponding hyphenation files are also installed. For example, the *hyphens.itl* file contains hyphenation points for popular Italian words. If you don't like the hyphenation, you correct the hyphenation in the Spelling Checker and click the Learn button. The main dictionary is updated.

Paragraph tags also determine how words are hyphenated:

- **Language:** On the Default Font tab of the Paragraph Designer, you select a language for the paragraph tag, and FrameMaker uses the corresponding hyphenation file, according to the settings in the Dictionary preferences.

- **Hyphenation parameters:** On the Paragraph Designer's Advanced sheet, you turn hyphenation off or on and specify how words are broken across lines. For example, the CellBody paragraph style shown in the following illustration limits the shortest prefix and suffix to three characters each. See "Advanced sheet" on page 124 for details.

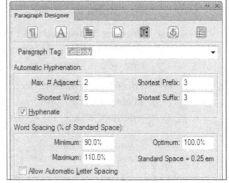

The paragraph tag hyphenation settings take precedence over the hyphenation points. For example, suppose the word *paper* is hyphenated as *pa-per*. According to the predefined hyphenation, the word breaks after the two-letter prefix *pa-*. If the paragraph tag requires at least a three-character prefix, *paper* moves to the next line, and FrameMaker doesn't use the predefined hyphenation.

Paragraph tag allows a two-character prefix at the end of the line. ⟶ We subscribe to the local pa-per, *The News and Observer*.

Paragraph tag requires a three-character prefix, so the word "paper" moves to the next line. Also, the word "Observer" isn't hyphenated due to prefix and suffix rules. ⟶ We subscribe to the local paper, *The News and Observer*.

To display and modify hyphenation, follow these steps:

1. Select **Edit > Spelling Checker** to display the Spelling Checker.
2. Type a word in the Correction field, then click the Show Hyphenation button. The hyphenation points are displayed.
3. To change the hyphenation, modify the word in the Correction field, then click the Learn button. (To remove all hyphenation from a word, type a hyphen before the word, as in -**paper**.) The hyphenation points are updated in your personal dictionary.
4. *(optional)* To update hyphenation throughout the document after you correct hyphenation points, follow these steps:
   a. Click the Dictionaries button in the Spelling Checker. The Dictionary Functions dialog box is displayed (see page 48).
   b. Click the Rehyphenate Document radio button.
   c. Click the OK button. The hyphenation in your document is updated.

Tip To edit hyphenation by hand, open your personal dictionary in a text editor and type the hyphens where you want them.

Hyphenation Shortcuts

In addition to using the Spelling Checker and paragraph tags to control hyphenation, you can insert symbols in specific words.

- **Prevent hyphenation.** To prevent a word from hyphenating at all, insert the suppress hyphenation symbol at the beginning of the word. If you choose, you can insert the symbol where the word breaks. Press ESC N S to insert the symbol. In a variable or cross-reference format, type \+ where you want to prevent hyphenation.
- **Prevent hyphenation at the end of the line.** To prevent a word from hyphenating at the end of a line, insert a nonbreaking hyphen before the hyphen at the end of the line. Press ESC HYPHEN H to insert the symbol.
- **Force hyphenation.** To force a hyphen to display in a specific location, insert a discretionary hyphen where you want the word to break. Press CTRL+D or ESC HYPHEN SHIFT+D.

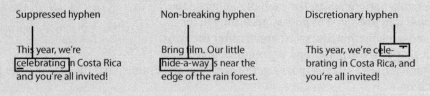

Suppressed hyphen	Non-breaking hyphen	Discretionary hyphen
This year, we're celebrating in Costa Rica and you're all invited!	Bring film. Our little hide-a-way is near the edge of the rain forest.	This year, we're cele- brating in Costa Rica, and you're all invited!

Thesaurus

The FrameMaker Thesaurus identifies the part of speech, synonyms, antonyms, and related words for the selected word or phrase. The FrameMaker Thesaurus isn't as comprehensive as a traditional thesaurus. You also might not find your exact word, but you can try to find the root word instead.

To use the Thesaurus, follow these steps:

1. Do one of the following:
    - If the word doesn't occur in the document or you haven't selected it, select **Edit > Thesaurus** to display the Thesaurus Look Up dialog. Type the word in the Word field, then click the Look Up button. You can also search for the word in a different language by clicking an item in the Language drop-down list. FrameMaker provides a dictionary and associated thesaurus for each of its fully supported languages. See the FrameMaker online help for a list of fully supported languages.
    - To look up a word in the document, select the word, then select **Edit > Thesaurus**. The word, definition, and corresponding entries are displayed in the Thesaurus dialog.

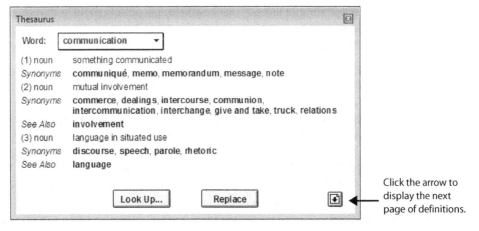

Click the arrow to display the next page of definitions.

2. Click a word in the Synonyms, Antonyms, or See Also sections to explore its meaning. The last ten words you looked up are displayed in the Word drop-down list.
3. Click the Replace button to substitute the word in the Word drop-down list for the highlighted word in your document. (The word must be highlighted, or you'll insert the new word where you inserted the cursor.)

Configuring Default Text Options

Though paragraph tags determine most formatting, FrameMaker lets you specify document defaults to fine-tune the display of text. For example, the Smart Spaces option allows only one consecutive space in the document. When you turn on Smart Quotes, quotes you type are curly. (Quotes you typed before turning on Smart Quotes are not changed.) You can also prevent line breaks after certain characters and specify size and spacing for superscript, subscript, and small caps characters.

You modify text options in an individual file or book file. Because FrameMaker considers the default text options part of the document properties, you can also import the document properties into another FrameMaker file, and the text options are updated. Select **Format>Document>Text Options** to display and change these default text options.

Figure 3-1. Modifying text options

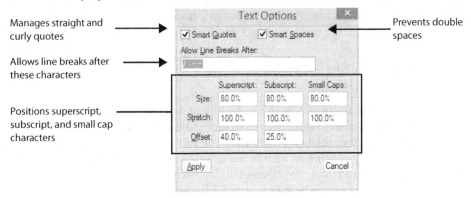

Smart quotes

When you type single and double quotes, the Smart Quotes option ensures that the quotes are curly instead of straight.
For example, instead of typing

' and " you type:

' and "

In some typeface families, curly quotes are more curved than curly. The previous example shows curly quotes in the Cambria typeface. Notice that the quotes look more angled than curved. In the Times New Roman typeface, however, the quotes are curly, as shown in the following sentence:

"Hurry up," cried the impatient customer.

The Spelling Checker can find straight quotes and replace them with curly quotes. This feature is particularly helpful when you import another type of file into FrameMaker. For example, in an imported Microsoft Word document, you need to run the Spelling Checker to find and correct straight quotes. Although FrameMaker catches most of the straight double quotes, it skips over some of the straight single quotes. You need to find and correct the skipped quotes by hand.

Smart Quotes are enabled by default. To turn off smart quotes, follow these steps:

1. In an individual file or book file, select **Format>Document>Text Options**. The Text Options dialog box is displayed (Figure 3-1).
2. Uncheck the Smart Quotes checkbox, then click the Apply button. Your changes are saved in the document.

When Smart Quotes are enabled, you can still type straight quotes, by typing ESC CTRL+' for a single quote, and ESC SHIFT+' for double quotes. Smart Quotes should be language-specific and the glyphs used are controlled by a setting in the maker.ini file.

Configuring Default Text Options

Smart spaces

Smart Spaces prevent you from typing more than one space in a row. This feature provides consistent spacing throughout a document. Although the Smart Spaces text option isn't enabled by default, the spell-checker's Extra Spaces is enabled by default. That means that by default FrameMaker automatically searches for extra spaces during a spell-check, which is especially handy when you import documents of other formats into FrameMaker. For example, an imported Microsoft Word document may have extra spaces after periods. Extra spaces may also have been inserted instead of tabs to position words. When you spell-check the document, FrameMaker finds instances of exactly two spaces and replaces them with a single space.

Note To remove spaces placed in lieu of tabs, search for multiple spaces in the Find/Change pod. Searching for three spaces quickly locates the error, allowing you to quickly select all spaces and replace them with a tab character.

To enable Smart Spaces, follow these steps:

1. In an individual file or book file, select **Format > Document > Text Options**. The Text Options dialog box is displayed (Figure 3-1 on page 52).
2. Check the Smart Spaces checkbox, then click the Apply button. In any new text, you are prevented from typing two spaces in a row. To remove extra spaces from the existing information, you need to spell-check with the Extra Spaces option checked.

Note If you turn off Smart Spaces, you may need to modify the Spelling Checker options to disable checking for extra spaces. See "Spelling Checker Options" on page 46 for more information.

Restricting line breaks

You can force special characters and spaces to move to the next line or prevent them from doing so. By default, line breaks are permitted after the following characters:

- **Forward slash.** /
- **Dash.** -
- **En dash.** –
- **Em dash.** —

These characters are displayed in the Allow Line Breaks After field of the Text Options dialog box (Figure 3-1 on page 52). When you modify characters in the field, FrameMaker instantly updates line breaks in the document.

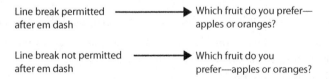

For information about typing special characters in the dialog box, see "Entering Special Characters in dialog boxes" on page 100.

To configure line breaks, follow these steps:

1. In an individual file or book file, select **Format > Document > Text Options**. The Text Options dialog box is displayed (Figure 3-1 on page 52).
2. Modify characters in the Allow Line Breaks After field, then click the Apply button. Your changes are applied to the document.

Formatting superscripts, subscripts, and small caps

The document text options determine the size, offset, and width for superscript, subscript, and small caps characters. When you change the defaults, FrameMaker changes how these characters are displayed across paragraph tags, character tags, footnotes, variables, and other components that include them. These settings apply to the entire document.

You can change the following properties:

- **Size.** The superscript, subscript, and small caps characters are a certain percentage smaller than the default paragraph or character formatting. To change the size of the characters throughout a document, you modify the percentage in the appropriate Size fields of the Text Options dialog.

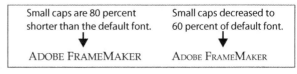

- **Offset.** Superscript and subscript characters are displayed a certain percentage above or below the current line. To increase the offset, you increase the percentages in the Offset fields. To decrease the offset, you decrease the percentage. If you require two different levels of superscript, consider using the subscript with a negative offset value. FrameMaker will complain when doing so, but it'll work!

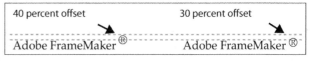

- **Width.** You can change the width of superscript, subscript, and small caps characters by modifying the Stretch fields. In this example, the registration mark is displayed with a 100 percent stretch (the same size as the paragraph tag) and with an 80 percent stretch.

Note In the document text options, the stretch value applies only to superscript, subscript, and small caps characters. You can set the stretch in paragraph and character tags, but those settings are unrelated to the text options.

To change the superscript, subscript, and small caps settings, follow these steps:

1. In an individual file or book file, select **Format > Document > Text Options**. The Text Options dialog box is displayed (Figure 3-1 on page 52).
2. Do any of the following:
   - To change the size, stretch, or offset of superscript and subscript characters, type new percentages in the Size, Stretch, or Offset fields.
   - To change the size or stretch of small caps characters, type new percentages in the Small Caps fields.

3. Click the Apply button. The new settings are applied to the document.

Tracking Changes

You can track changes as they are made using the Track Text Edits tool. After the revisions are made, you can review the edits one at a time, accept all edits, or reject all edits. You can also keep track of changes in other ways in FrameMaker. The basic word count tells you the number of words in the document (see "Displaying the word count" on page 58). You might compare this number to the number of words in a previous version of the document. To show modified text on the page, you can turn on automatic change bars, which are displayed in the margin of the document when any change is made (see "Using change bars" on page 58). For more specific information, you can compare two versions of the same document or book (see "Comparing documents and books" on page 60).

Using the Track Text Edits tool

FrameMaker lets you track text edits as you make them by highlighting inserted and deleted text. You can track revisions, then review the changes, and accept or reject edits. You can also preview the final document with the changes, or preview the original document as it was before changes were made.

You can enable or disable tracking, set the scope of the tracking (Document, Book, or Selected Documents), and track changes by author. The dates of changes saved with the conditional text tag applied.

All of the track text edits options are available through the Special menu, but you are likely to find the Track Text Edits toolbar is much more convenient than a series of third level menu items.

To display the Track Text Edits tool bar, select **View > Toolbars > Track Text Edit Bar**.

Table 3-4 lists the parts of the bar.

Table 3-4. Buttons on the Track Text Edits toolbar

Button	Function
Document	**Scope:** Can be set for Document, Book, or Selected Files.
	Enable/Disable: Turns track text editing on.
	Preview Final/Preview Original/Preview Off: Toggles display of edits.
All Users	**Scope:** Select the author(s) whose edits you want to display.
	Accept/Reject and Show Next: Accepts/rejects the selected edit, removes the track text edit display and moves to the next edit.

55

Table 3-4. Buttons on the Track Text Edits toolbar (Continued)

Button	Function
	Accept/Reject Edit: Accepts/rejects the selected edit, and removes the track text edit display.
	Show Next/Show Previous: Moves to and selects the next or previous text edit.
	Accept All/Reject All: Accepts/rejects all the edits in the current document and removes the track text edit display.

Although tracking uses a form of conditional text, you cannot control these highlights using the Conditional Text pod in FrameMaker.

If you find the visual highlighting of tracking text edits distracting, consider making backup copies of your original files, editing the files, and then using the comparison utility provided with FrameMaker. For instructions, see "Comparing documents and books" on page 60.

For more details about tracking text edits, see "More Information About Track Text Edits" on page 57.

Turning On Tracking

To turn on text edit tracking:

1. Select **Special > Track Text Edits > Enable** or click the Enable/Disable button on the Track Text Edits toolbar.
2. Start working in your document. As you make changes, the changes are highlighted by default as follows:
   - Inserted text is shown as green and underlined. It is also marked with the author's name and date in the condition indicator, which is shown in the status bar.
   - Deleted text is displayed as red and underlined. It is also marked with the author's name and date in the condition indicator, which is shown in the status bar.

Reviewing Edits

You can review each edit in a document and accept or reject it, or you can accept or reject all edits.

To review individual edits in a document:

1. Select **Special > Track Text Edits > Show Next** (or **Show Previous**), or click the Show Next or Show Previous button on the toolbar to select the next or previous text edit.
2. Do one of the following:
   - To accept the edit, select **Special > Track Text Edits > Accept Edit**, or click the Accept Edit button on the toolbar.
   - To reject the edit, select **Special > Track Text Edits > Reject Edit**, or click the Reject Edit button on the toolbar.

To accept or reject all edits in a document:

Tracking Changes

- Select **Special > Track Text Edits > Accept All**, or click the Accept All button on the toolbar to accept all edits.
- Select **Special > Track Text Edits > Reject All**, or click the Reject All button on the toolbar to reject all edits.

Previewing Changes

As you work, you might want to preview the document as it was originally or as it is with all of the text edits. For example, you might want to print both versions for someone to review. You can preview the final document with edits or the original, and then turn the preview off to return to displaying all edits:

- To preview the final document with all edits, select **Special > Track Text Edits > Preview Final**, or click the Preview Final button on the toolbar.
- To preview the original document with no edits, select **Special > Track Text Edits > Preview Original**, or click the Preview Original button on the toolbar.
- To return to viewing all the text edits, select **Special > Track Text Edits > Preview Off**, or click the Preview Off button on the toolbar.

More Information About Track Text Edits

- Tracking is done with conditional text. For best results, choose colors other than red or green for regular conditional text display because these are the default colors used for deleted and inserted text. Also, if you add text marked as conditional while you are tracking changes, the custom text condition is the one displayed. FrameMaker does not display a merged color to show the overlapping conditions.
- If you delete a table row, the text is deleted and not marked.
- When deletions are displayed in your document, the content is still considered part of the document. For example, if you create a table of contents, it will include all headings, including deleted headings. Unresolved cross-references will also still appear. Therefore, when updating books or generating files, you might want to set all files to the Preview Final setting to hide any deleted content.
- For additional options, consider using change bars, a PDF review cycle, or compare document features described in "Using change bars" on page 58 and "Comparing documents and books" on page 60.

PDF review

I've included an entire section on PDF creation and FrameMaker's electronic PDF review feature. Starting with FrameMaker 12, the reviewing features are more robust, and edits made in a PDF review can be applied to the FrameMaker source files. See "Setting up a FrameMaker PDF review" on page 391 for more information.

Displaying the word count

FrameMaker's Word Count feature shows the number of words in a document. You might run a word count along with comparing documents to show the degree to which content has changed. Running a word count may also help you meet a specific word quota.

In the word count, items such as variables and cross-references are included; paragraph autonumbers (step or footnote numbers, for instance) and hidden conditional text are excluded.

To run a word count, follow these steps:

1. Select **File > Utilities > Document Reports**. The Document Reports dialog box is displayed.

2. In the Report list, click Word Count, then click the Run button. The results are displayed.

3. Click the OK button to close the dialog.

> **Note** For Asian language documents, run an Asian Character Count instead of a word count. The number of single- and double-width characters is displayed.

Using change bars

Change bars help readers locate new material, and may be required in aviation documentation and other industries. When you modify a document, change bars are displayed in the margin on the line containing the change, even when you insert only a space. As a result, it's often difficult to decipher exactly what was changed. For instance, the following example shows change bars next to modified lines, but the exact changes are not apparent. You can determine only that information changed in lines 1, 2, and 4; but not in line 3.

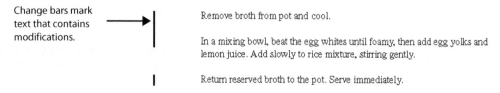

You can customize the color, position, and thickness of automatic change bars and their distance from the text column. These settings apply to the entire document. You're also limited to one change bar style., and cannot define different change bars for individual editors. Character tags have a change bar option, giving you an additional way to mark changes manually.

To toggle and customize automatic change bars, follow these steps:

1. Select **Format > Document > Change Bars** to display the Change Bar Properties dialog.
2. Modify the thickness, position, and color of the change bar as needed.
3. Set the Automatic Change Bars checkbox as needed, then click the Set button.

Caution   Clearing change bars also removes those applied by character tags. You cannot undo this command once change bars are removed. To reapply character tag change bars, import formats back into the document and select Remove Overrides. Save your document before you clear change bars, so you can revert to the saved file if necessary.

Here are the positions available for change bars in the Position drop-down list:

- **Left of Column** *(single-sided document)*: Change bar is displayed in the left margin.
- **Right of Column** *(single-sided document)*: Change bar is displayed in the right margin.
- **Side Closer to Page Edge** *(double-sided document)*: Change bar is displayed in the left margin on a left page or in the right margin on a right page.
- **Side Farther from Page Edge** *(double-sided document)*: Change bar is displayed in the right margin on a left page or in the left margin on a right page.

To change the color, click a color in the Color drop-down list. The colors defined in your document are displayed in the list. You can create new colors if needed. See Chapter 24, "Color output," for details.

Tip   To remove all existing change bars from your document, check the Clear All Change Bars checkbox. All change bars are cleared from the document when you click the Set button.

Displaying line numbers

FrameMaker 11 introduced a line numbering function that works in both the Document View and in the structured FrameMaker Code View. Line numbers print much like Conditional Text indicators (see page 449).

To display line numbers, select the **View > Line Numbers** menu item.

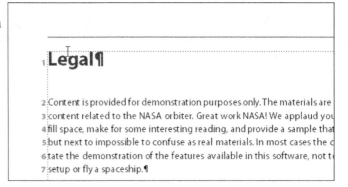

To customize the display of line numbers, select the
Format > Document > Line Numbers menu item. Change options as needed and click the Set button.

Comparing documents and books

The Compare Documents feature lets you compare two versions of the same document or book to see what's changed. In a document comparison, FrameMaker analyzes the contents of matching flows and reports changes to items including the following:

- Text
- Variable definitions
- Marker contents
- Cross-reference formats and the location of cross-reference links
- Anchored frames
- Tables

In a book comparison, FrameMaker lists the modified book components in addition to the document components. For example, the modified items in each file are displayed, along with which files were inserted, modified, or deleted in the book.

There are two types of document comparison reports. In the *composite report,* a marked-up version of the newer document is displayed. The changes are marked with condition tags or special characters—you choose the style before running the report. The *summary report* is only a list of changes. (By default, this report is view-only. Press ESC SHIFT+F L K or the **Toggle View Only** button ( )to make it editable.) FrameMaker lets you run just a summary report or both the composite and summary reports. You can also print and save the reports.

Figure 3-2. The summary and composite document comparison reports

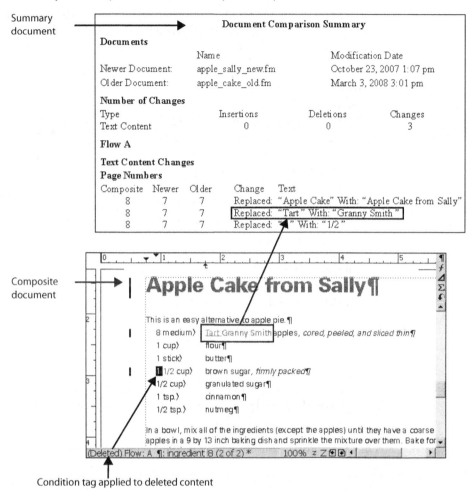

To compare two versions of the same document, follow these steps:

1. Open the old version of the document you want to compare.
2. Open the new version of the document.
3. From the new document, select **File > Utilities > Compare Documents** to display the Compare Documents dialog.
4. If several documents are open, click the name of the document you want to compare in the Older Document drop-down list.

Caution Comparing two completely different documents is possible, but it's usually pointless and might cause unexpected results. You compare documents to pinpoint the changes, and FrameMaker doesn't find changes in unrelated documents!

5. Do one of the following:
   - Click the Summary and Composite Documents radio button to create both a summary of changes and the marked document.
   - Click the Summary Document Only radio button to create a list of changes.
6. *(optional)* Click the Options button to change the mark-up methods to display the Comparison Options dialog.
   - In the Mark Insertions With section, click one of the radio buttons to specify formatting for the inserted characters. The default setting applies a condition tag called Inserted, but you can type a new or existing condition in the Custom Condition Tag field or click the Nothing radio button to avoid marking insertions.

   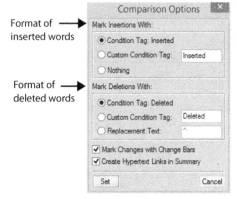

   - In the Mark Deletions With section, click one of the radio buttons to customize formatting for the deleted characters. You can assign the Deleted condition, specify a different condition, or replace the deletion with text, such as "Deleted Text."
   - To display change bars in the margin, check the Mark Changes with Change Bars checkbox.
   - Check the Create Hypertext Links in Summary checkbox to link the changes in the summary document to the original document. This will allow you to CTRL+ALT+click the hyperlinked changes to display the change location in the referenced document.

   Click Set in the Comparison Options dialog.
7. Click the Compare button. The document reports you chose are displayed.

To compare two versions of the same book file, follow these steps:

1. Open the old and new versions of the book file. The two book files must have identical names, and only chapter files with the same names are compared.
2. From the new book file, select **File > Utilities > Compare Books** to display the Compare Books dialog.
3. Click the name of the old book file in the Older Book drop-down list.
4. Click the Compare button. The documents you chose are displayed.

Working with Footnotes

Footnotes provide ancillary information at the bottom of the text column or below a table. You can access the footnote properties for a document or book at **Format > Document > Footnote Properties**. If you need to place endnotes, these are typically placed using custom markers, which are then collected with a custom-generated file. For more information, see Chapter 21, "Creating Other Generated Files" on page 377.

The following items control footnote formatting:

- **Formatting of footnote reference.** FrameMaker formats the main text footnote reference using the current paragraph's formatting.

- **Formatting of footnote.** In the footnote properties of a book or stand-alone document, you set up the footnote number position, prefix or suffix, paragraph tag applied to the footnote, and maximum height. FrameMaker applies the paragraph tag you specify in Footnote Properties to the footnote itself. If Footnote and TableFootnote frames exist on the reference page named Reference, FrameMaker will place the corresponding frame above the appropriate footnote area. You can change the height of the frame to adjust spacing. You can also change the properties of the art within the footnote frame—for example, modify the line width or color. See "Placing graphics above and below the paragraph" on page 275 for more information on reference page graphics. Table footnotes and main footnotes can use different paragraph tags.

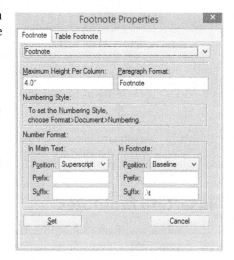

- **Numbering properties.** There are separate properties for main footnotes and table footnotes. If the document is in a book, you select the two properties in the book setup—the footnote format and the numbering options. The footnote format can be numeric, uppercase or lowercase alphabetic, uppercase or lowercase Roman numerals, or a custom format. Numbering options let you specify whether you want the footnote numbering to start over on each page, continue from a previous number, or read the footnote number from the file. You can also specify the starting footnote number for each document in the book. See "Managing numbering" on page 332 for more information.

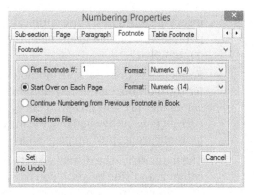

Figure 3-3. Factors affecting footnote formatting

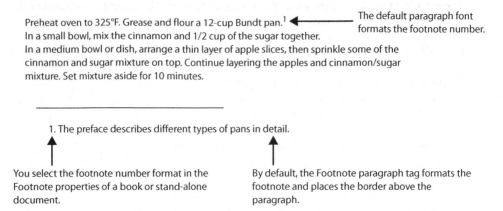

Inserting footnotes

When you insert a footnote, the number is displayed at the insertion point. Your cursor is displayed in a new footnote paragraph, which is located at the bottom of the text column (if your cursor is in the body text) or below the table (if your cursor is in a table).

Footnotes are displayed by their position in the body text or table, not by the order in which you insert the footnotes. For example, if you insert a footnote in the first paragraph and the third paragraph, and then insert one in the second paragraph, FrameMaker renumbers the footnote paragraphs.

FrameMaker does not split footnotes between two pages. For example, if there isn't enough room for the footnote on the current page, the entire footnote moves to the next page. You can sometimes prevent this problem by manually adjusting the text frame so the footnote fits on the current page or by changing the maximum height of the footnote paragraph. See "Customizing footnote properties" on page 66 for details on increasing the maximum height.

The footnote can be displayed in the column or side head space, across all columns, or across all columns and side heads. If you're working in a multicolumn document, you must adjust the pagination settings in the footnote paragraph tag accordingly. See "Modifying paragraph tags" on page 113 for details.

To insert a footnote, follow these steps:

1. Position the cursor where you want the footnote number to be displayed, then select **Special > Footnote**. The footnote number is displayed in the main text, and your cursor is displayed in the new footnote paragraph at the bottom of the column or below the table, depending on the type of footnote.
2. Type your footnote. The footnote is displayed in the document.

Using the same number for several footnotes

When you need to refer to the same footnote number in several places, you insert a cross-reference to the existing footnote. The following example shows the initial footnote in the first line and the cross-referenced footnote in the second line. The two references look the same because the footnote number in the second line uses a cross-reference format that applies the superscript automatically.

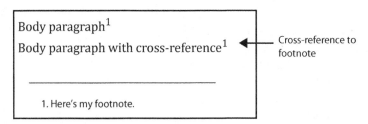

The font properties for the footnote number and cross-reference number are determined by the paragraph tag. For instance, if the footnote paragraph has a serif font and the cross-reference number is in a sans-serif paragraph, the two numbers won't match.

The following example shows the serif footnote number and a sans-serif number for the cross-reference:

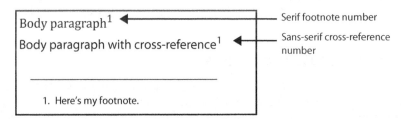

You can prevent this problem by creating a character tag for the cross-reference that specifies the font you want to use for the "imitation" footnotes. The character tag should also include a subscript or superscript property to match the style of the existing footnote. In the cross-reference format, you insert the character tag before the <$paranumonly> building block. The cross-reference format looks something like this:

<footnote><$paranumonly></>

To create a cross-reference to an existing footnote, follow these steps:

1. Create a character tag that matches the footnote paragraph tag's font properties and the style of the footnote number (superscript or subscript).
2. Create a cross-reference format that includes the character tag and <$paranumonly> building block.

3. Position the cursor where you want the cross-reference, then select **Special > Cross-Reference** to display the Cross-Reference pod.

4. Click Paragraphs in the Source Type pop-up menu to display the paragraph tags.

5. In the Paragraph Tags section, click the footnote paragraph tag. The footnote paragraph is displayed in the Paragraphs section.

6. In the Paragraphs section, click the footnote paragraph.

7. In the Format drop-down list, click the footnote cross-reference format you created.

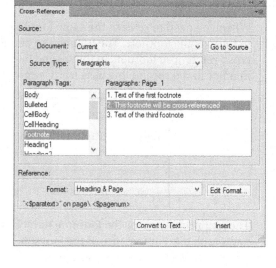

8. Click the Insert button. The cross-reference number is displayed in the text.

For more information on creating cross-reference formats, see "Setting Up Cross-Reference Formats" on page 179. See "Creating character tags" on page 143 for details on setting up a character tag.

Customizing footnote properties

FrameMaker lets you customize footnote formatting and numbering properties. The formatting properties, which you configure in individual files, consist of the footnote prefix or suffix, position of the number, paragraph tag applied to the footnote, and maximum height. You can import the formatting properties into other documents by importing the document properties. The numbering properties determine how footnotes are numbered in a book (or document) and the type of alphanumeric footnote character. If the document is in a book file, you set the number format in the book file setup. If not, you can change the number format in the document's numbering properties.

Modifying Footnote Formatting

To customize footnote formatting, follow these steps:

1. Open the file or book file you want to modify, then select **Format > Document > Footnote Properties** to display the Footnote Properties dialog. (see graphic on page 63 for example)

2. Click the Footnote or Table Footnote tab, depending on which type of footnote you need to modify. You can also click Footnote or Table Footnote in the drop-down list at the top.

3. *(standard footnote only)* To limit the height of the Footnote paragraph, type a number in the Maximum Height Per Column field.

Working with Footnotes

Note In FrameMaker, you can specify the unit of measure used throughout the document—in paragraph tags, text frame properties, table tags, and so on. To change the unit of measure, select **View > Options** and click a different item in the Display Unit drop-down list.

4. To change the paragraph tag applied to new footnotes, type the tag name in the Paragraph Format field. This setting does not change existing footnotes.

Caution If you misspell the paragraph tag name, FrameMaker applies the paragraph tag to new footnotes and considers the paragraph tag an override. In the status bar, the asterisk next to the misspelled paragraph tag indicates the override. You can either correct the paragraph tag in the footnote properties or create a new paragraph tag.

5. To change the format of the footnote character in the main text, do any of the following in the In Main Text section (or the In Cell section of table footnotes):
   - Click a position in the Position drop-down list. (Baseline aligns the character with the main text.)
   - To precede the footnote character with a specific character, type the character in the Prefix field. You can type special characters such as an em space (\sm) or en space (\sn). For more information about typing special characters in dialog boxes, see "Entering Special Characters in dialog boxes" on page 100.
   - To display a specific character after the footnote character, type the character in the Suffix field.

6. To change the format of the footnote character in the actual footnote, change settings in the In Footnote section. See the previous step for descriptions of each option.

7. Click the Set button. Your changes are applied to existing footnotes, except for the maximum height and paragraph tag settings, which are only applied to future footnotes.

Modifying the Footnote Numbering Properties in the Book File

To change the footnote numbering properties in a book file, follow these steps:

1. Open the book file, then select the files whose footnote properties you want to change. Press CTRL+A to select all files in the book; CTRL+click to select specific files.
2. Select **Format > Document > Numbering** to display the Numbering Properties dialog. (see graphic on page 63 for example)
3. Click the Footnote or Table Footnote tab, or click a choice in the drop-down list, depending on the type of footnote you need to modify.
4. Do any of the following:
   - Specify the number of the first footnote in each file by typing in the First Footnote # field.
   - To change the format of the first footnote, click an item in the Format drop-down list that is displayed next to the First Footnote # field. If you click Custom, the Custom Numbering dialog box is displayed. You change the numbering pattern for the footnotes here.
   - According to the following example, FrameMaker assigns the asterisk to the first footnote, the dagger to the second footnote, and the double dagger to the third footnote in the document or book. These symbols are repeated if you have

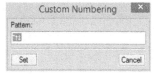

more than three footnotes. For more information on typing special characters in dialog boxes, see "Entering Special Characters in dialog boxes" on page 100.
- To restart footnote numbering on each page, click the Start Over on Each Page radio button, then click the numbering style from the Format drop-down list.
- To continue numbering from the previous footnote, click the Continue Numbering from Previous Footnote in Book radio button.
- To use the footnote numbering properties set up in the file, click the Read from File radio button.

5. Click the Set button. Your changes are applied to the selected files in the book.

Caution   If the file- and book-level footnote properties differ, FrameMaker displays an Inconsistent Numbering Properties warning when you update the book. The warning explains that file-level numbering properties are ignored during the update.

Modifying the Footnote Numbering Properties in Standalone Documents

To modify footnote numbering in individual documents, follow these steps:

1. Select **Format > Document > Numbering** to display the Numbering Properties dialog. (see graphic on page 63 for example)
2. Click the Footnote or Table Footnote tab, or click a choice in the drop-down list, depending on the type of footnote you need to modify.
3. Change the properties as described in "Modifying the Footnote Numbering Properties in the Book File" on page 67, then click the Set button. Your changes are applied to the file.

Importing Text from Other Applications

FrameMaker provides filters that let you convert text from many applications into FrameMaker formats. (Importing graphic files is discussed separately; see Chapter 13, "Importing graphic content.") Depending on the source format, you may have several different options available for pulling text into FrameMaker

MS Word (copy into document)

Beginning with FrameMaker 2015, when converting MS Word documents using the Copy into Document option, you have a robust style mapping wizard that is exponentially faster than other conversion methods. If you are using MS Word with the Import by Reference option, see "Copying, pasting, and converting text" on page 71.

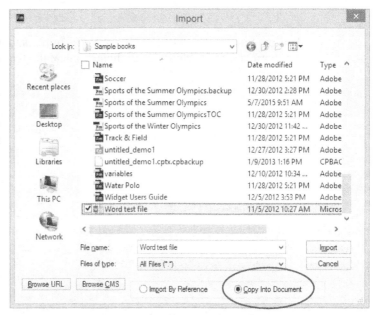

This feature can save a tremendous amount of time, so be prepared to do some tweaking of the settings, especially if you are bringing in a significant amount of content.

To use the mapping functionality, do the following:

1. Create a new document (from your template, if you have one) and save the file to disk.
2. Select **File > Import > File** to display the Import dialog.
3. **Important:** Select the Copy into Document option at the bottom of the dialog. This must be done prior to confirming file choice by double-clicking on the file or choosing the Import button.
4. Select your MS Word file, click the Import button.
5. Choose MS Word in the Unknown File Type dialog and click Convert.
6. If you are importing a DOCX file, consider the warning message provided. I have not needed to downsave my files to

DOC, but you should evaluate your files for proper conversion in this (or any other) conversion process.

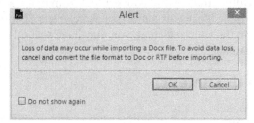

7. Now for the cool part: Use the provided options to map your MS Word paragraph, character, and table styles to their FrameMaker equivalents.

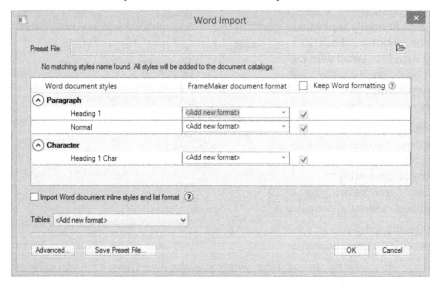

8. Use the Import Word document inline styles and list format option to attempt to rein in ad hoc formatting in your source documents.

Note the Keep Word formatting option. By clearing this check box, you will remove the Word formatting entirely and use only the FrameMaker style definition.

Copying, pasting, and converting text

When you are not using MS Word and Copy into Document, other options include the following:

- **Copying and pasting.** For some applications, you can open the source document, copy the information you want, and then switch to FrameMaker and paste it in. If your selection pastes in as a graphic (or is otherwise unacceptable) select **Edit > Paste Special** for additional options.

Tip   If you pasted from Microsoft Word as Unicode text but there are still gray boxes or question marks appearing instead of the characters you expect, the font you are using in FrameMaker doesn't include the correct Unicode characters. Switch fonts or install the correct font to fix the problem.

To have FrameMaker automatically paste as Unicode text instead of plain text, change the paste order in maker.ini. See "Setting clipboard pasting order" on page 548.

- **Opening the source document.** In FrameMaker, try selecting **File > Open** and then opening the source document (such as a Word file). FrameMaker will filter the document and open it as a FrameMaker document. If FrameMaker does not automatically identify the file type, you are prompted to identify it.

  By default, all filters are installed with FrameMaker, but it's possible to exclude filters from the installation if you do a custom installation. If filters seem to be missing, check to see whether you can install more.

- **Importing the file into a template.** While opening the file directly into FrameMaker will effectively process the content, it will also retain standard FrameMaker components that may not be defined in your template. To keep the number of resources stored in your document to a minimum, use **File > Import > File...** instead of **File > Open**.

- **Creating an intermediate file.** For some applications, such as InDesign, direct import is not available. In most cases, though, you can export the file to an intermediate format (for example, InDesign offers RTF export), and import the intermediate file into FrameMaker. The resulting FrameMaker file usually requires some clean-up.

When importing content from other applications, you might encounter issues with content such as table formatting, graphics, cross-references, and page layouts. You might need to perform additional processing or formatting to move all the content into FrameMaker format.

The amount of additional processing varies depending on consistency and use of styles in the source document, as well as the program in which it was created. Try using one of the provided import filters.

For a detailed list of all available filters provided with FrameMaker, check the *Using Filters* online manual.

Note   At time of writing, this FrameMaker 7 document was available at: help.adobe.com/en_US/FrameMaker/8.0/filters_help.pdf

FrameMaker can import Microsoft Word DOCX files as well as DOC and RTF files. Importing content from Microsoft Word is a relatively smooth process. Text usually converts reasonably well, and paragraph styles applied in Word show up in FrameMaker. Index entries are converted into FrameMaker index markers. The results of the conversion aren't perfect. Here are some things that can improve your import experience:

- If you use FrameScript (a paid scripting environment, replaced somewhat by the ExtendScript package included with FrameMaker), consider purchasing Rick Quatro's Find Change Formats Batch script. I have used and recommended this script many times and it has *always* been worth the cost of acquisition when converting content or managing/cleaning up templates.
- Be sure to accept any revisions in Word *before* importing the document; otherwise, revisions may be displayed as FrameMaker conditional text.
- Graphics are converted, but results are often unusable. Consider removing graphics before conversion, and import the graphics directly into FrameMaker instead of attempting to convert them through the filter. If source files are unavailable, try saving the Word file as HTML and using the resulting PNG, GIF, or JPEG files, which are usually higher quality than the graphics that are produced during Word-to-FrameMaker conversion. You could also save the Word file as a PDF file and export the graphics from it using Adobe Acrobat Professional.
- Tables are converted, but Word formatting is applied as overrides. Clean-up is required. Reapply a table format in FrameMaker and remove custom formatting, as described in Chapter 8, "Understanding table design." The TableCleaner plug-in from Carmen Publishing (www.frameexpert.com) is invaluable for reformatting tables easily.
- Apply your own master pages to fix the headers and footers.
- Master documents, which are Microsoft Word files that embed other Word files, are not converted. Convert the individual files and add them to a FrameMaker book file.

Tip   After you import the Word file into FrameMaker, save it as MIF to remove hidden characters that can cause problems in FrameMaker. This process is often referred to as a MIF wash. See Appendix F, "Maker Interchange Format."

Chapter 4: Establishing a workflow in FrameMaker

Few things can help FrameMaker authors more than an understanding of common FrameMaker workflow and template-based design principles. This chapter describes individual and workgroup workflows, while template concepts are covering in Chapter 5, "Templates."

Here's a look at what's in this chapter:

- Workflow considerations 74
- Planning 75
- Creating and editing content 76
- File organization 76
    - Location of files 77
    - Folder structure 77
    - Book structure 78
- Illustration 78
- Editing/reviewing 78
- Production editing 79
- Indexing 79
- Printing 79
- Single sourcing and digital publishing 80
    - Evaluating methods 80
- Digital publishing workflows 85
- Language support 87
    - Unicode support 88
    - Dictionary support 88

Workflow considerations

An ideal publishing workflow will organize the publishing process into distinct tasks, many of which can be performed concurrently. The tasks involved are fairly standard:

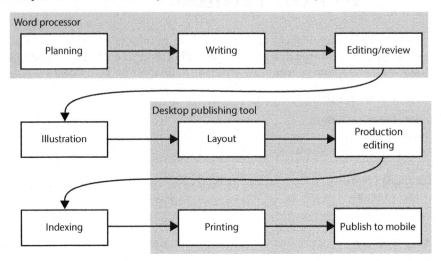

While these tasks were perhaps once performed by different people using different tools, you might required to perform any (or all!) of them. FrameMaker eliminates the need for some of these tools using robust internal options. Most formatting is handled by applying paragraph, character, and object styles. These styles are easily applied using menu items, pods, and keyboard equivalents. Creating content directly in FrameMaker eliminates the layout step from your workflow because layout is done automatically as the author creates information. An editor might review the files prior to publishing, but the vast majority of the layout work is done by the author in the production stage.

As authors create information, they can add illustrations or other graphic objects. These might be created by a graphic artist, but there's no need to wait until after a manuscript review to insert them. This makes it possible for editors and reviewers to see text and graphics in context, on the same page, just as the reader will see them in the final printed version.

Note If your workflow relies upon subject matter experts using something like MS Word, consider creating a word processor template with a style sheet matching your FrameMaker naming conventions. For more information, see Chapter 29, "Content reuse with text insets."

Planning

Given a tight deadline and limited resources, planning is often the first thing to go. However, a minimal amount of time invested at the beginning of the project will make the whole project much more efficient. For short, simple documents, the only thing required in planning might be to create a document template (or identify an existing template that will be used). There are three items that should be created during planning:

- Document templates (if they do not exist already)
- Document outline
- Documentation plan, which includes resources required, a preliminary schedule, a high-level description of the document to be created, and related project management information

More complex projects make planning more critical—failing to address issues before development begins invariably results in more work (and rework) later. When the following factors are present in a project, even more up-front planning is required:

- **Cross-media publishing/single-sourcing.** If your content will be delivered in several different media (print, online, mobile, and so on), detailed planning is required to analyze the documents and map out how information is presented in each medium.
- **Complex document libraries.** If your document library will contain thousands (as opposed to hundreds) of pages of content, planning helps you to avoid redundancy and ensures that all needed topics are covered.
- **Multiple authors.** If multiple authors are working on different books (or portions of the same book), you need to coordinate to make sure that the finished product looks like a single book and not a collection of disjointed sections with different voices.
- **Modular documentation.** Setting up documents with "chunks" of content lets you reuse information instead of writing multiple versions of the same information. It does, however, require careful planning of the document organization, storage and retrieval of the modules, and authoring to ensure that modules are consistent.
- **Distributed or "virtual" workgroup.** A group of authors and editors in a single physical location can communicate informally (by shouting over cubicle walls). When content creators are in different locations all over the world, more formal communication is necessary.
- **Localization and Right to Left (RTL) language support.** Starting with FrameMaker 2015, you can create bidirectional content. This means that you can not only create content in right-to-left (RTL) languages - Arabic, Hebrew, Thai, and Farsi - but a mix of both LTR and RTL languages. Earlier versions of FrameMaker will require more sophisticated workflows to address bidirectional and RTL editing and output.

Creating and editing content

In a FrameMaker workflow, authors write in FrameMaker, not in a separate word processor. Authors should be given a template and some instructions on how to use it. They should also be able to use common features in a FrameMaker template, such as cross-references, variables, and markers.

To ensure that writers can succeed in a new FrameMaker environment, it's important to provide some information about the new environment. With its emphasis on organization, templates, and style sheets, FrameMaker is quite different from other word processor or desktop publishing tools. Authors must understand why style sheets are so critical.

Some writers believe, "I'm a writer, not a publisher." This might translate into disinterest or outright hostility toward templates and strict formatting requirements. However, in a FrameMaker workgroup, adhering to standards laid out in the template is a requirement. Writers who cannot or will not follow templates create documents that will probably print adequately (because they look correct on screen), but the document cannot be reused or processed to produce electronic output. If your group includes such a writer, your options are as follows:

- **Education.** Explain to the writer why following the template is necessary. Show what happens when the template is not followed and how their behavior impacts the automated workflow and the ability to produce other formats (such as DITA, XML, or mobile output) from the FrameMaker files. Explain how the effort of both the author and others in the workgroup can be wasted when styles are not properly used.
- **Persuasion/coercion.** Ensure that there are consequences for not following the template. This approach could include requiring the writer to fix any errors that occur in the final output. Keep in mind, though, that a software-averse writer might not have the skills to correct problems, so additional training might be needed.
- **Cleaning up.** As a last resort, evaluate the writer's value to the organization. Is this person a particularly talented writer with specific knowledge that no one else shares? In that case, it might be worthwhile to assign a person to clean up the writer's files so that their content ultimately conforms to the template.

Caution You may find that some writers moving from a word-processing environment to a single-sourcing workflow will be unwilling or unable to follow strict template rules.

File organization

While there are as many ways to organize a project as there are FrameMaker users, some level of structure will help you more efficiently manage your project.

Location of files

If you have a large project, and especially if you have multiple authors or network backup requirements, you may be tempted to work off of a central server. However, depending on server speeds, storing your files and related assets remotely can result in slow opening and processing speeds.

Content management systems are designed to monitor and improve access to the information, often maintaining up-to-date local copies of content as well as providing check-in and check-out functionality to prevent simultaneous access across authors.

If your organization absolutely won't consider a CMS solution, you can still maintain local up-to-date copies of your content via Dropbox or other cloud file sharing solutions. See "Dropbox and Cloud Collaboration" on page 525 for more information.

Folder structure

Whether you store your FrameMaker files locally, in a CMS, or on a server, take time to organize your projects to take advantage of content and file reuse.

Here's the local folder structure I recommend to clients. I suggest you use this model to create your own optimized structure. Note all directory and filenames are short, and contain only alphanumeric characters. Spaces and punctuation can cause problems in digital publishing output.

C:\FrameMaker-Documents - I place the folder at the root level to ensure short file paths.

> **Templates** - Individual template files simplify style catelogs, master pages, and other assets.
>
>> **TOC**
>>
>> **IX**
>>
>> **Front matter**
>
> **Common-graphics** - Keep logos and branding items in a common location to ensure consistent usage in projects.
> **Project-title-1** -
> **Project-title-2**
> **Project-title-3**
>
>> **Project-title-3.book**
>>
>> **Project-title-3TOC.fm**
>>
>> **Project-title-3IX.fm**
>
> **TopicA.fm** - Resist the temptation to include chapter numbers...you're likely to change the order over time, or add content that throws off your scheme.
>
>> **TopicB.fm**
>
> **Graphics** - Use this directory for graphics used only in this project.

Tip — If your files are in disarray, consider using the Package function to collect them to a specific location. See "Package" on page 412 for more information.
You can also use the Dropbox function to create a folder structure within Dropbox that shows your dependencies. (useful for troubleshooting) See "Dropbox and Cloud Collaboration" on page 525 for more information.

Advantages of this files structure:

- Files in close proximity give you shorter relative paths within your book files.
- Eliminating spaces and special characters (other than hyphens and underscores) help ensure working links for digital publishing.

Book structure

Use book file options to keep your work organized and speed up access to external files and resources. Here is a partial view of the .book file used for this project. Note that the final four files are excluded from processing and output (accomplished by right-clicking on the file within the book window) and the template file is in a directory up one level from the book directory itself. See "Adding special book structures" on page 325 for information on book file folders, groups, and the Exclude command.

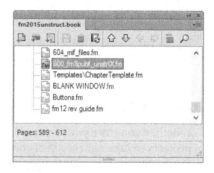

Illustration

Some authors create their own illustrations and other graphic objects, but many work with a graphic artist or technical illustrator. FrameMaker provides a basic set of graphic tools, which may be adequate for simple line art or small flowcharts. To create more complex graphics, use dedicated graphic editing software (such as Illustrator or Visio) and then import the graphic into FrameMaker. FrameMaker supports a wide variety of graphic formats, including native Illustrator (AI), native Photoshop (PSD), PDF, EPS, TIFF, PNG, and GIF. When importing graphics, you can either create a link to the graphic (import by reference) or embed the graphic in the FrameMaker file (copying into the document). Although most organizations use the import-by-reference feature, each has advantages and disadvantages. These are discussed in detail in "Importing a graphic" on page 259.

FrameMaker also allows import of many rich media formats, including native Captivate (CPTX), SWF, MP4 and U3D. While these files will not translate to printed output, your electronic output will contain working video and interactive content, while your print output will still allow for a static "poster" image. See "Set movie poster" on page 285 for more information.

Editing/reviewing

Instead of distributing FrameMaker files, which would require the reviewers to have access to FrameMaker, consider generating PDF files from the FrameMaker source files and using the PDF files for review. Reviewers can mark up changes on a paper copy or use Acrobat's online annotation tools. For an online review, one person initiates the review from Adobe Acrobat Professional, and the reviewers can mark up the file using the free Adobe Reader, even performing reviews on mobile devices. When working with Acrobat 9 (or later) compatible files, the most popular of these comment types can be transferred to FrameMaker automatically. See "Setting up a FrameMaker PDF review" on page 391 for more information.

Regardless of your method of review, be sure to maintain some sort of version control. If you don't have a content management system, you'll want to manage access to the files to ensure accuracy and accountability during the document review. By working on PDF, rather than the source documents, you can more easily see who made the edit, and when it was edited.

Technical editing and copy editing can be done directly in the FrameMaker files, using track text edits, change bars turned on, or other methods, but paper copies or online Acrobat reviews provide much better audit trails for tracking changes.

Production editing

Using many other applications, copy-fitting is a major part of the production edit; that is, pages are checked to ensure they are filled with consistent amounts of text and graphics. Production editors also look for awkward line and page breaks, and they verify that headers and footers, cross-references, tables of contents and indexes, and front and back matter are set up correctly. Because FrameMaker enables automation of many of these tasks, the main task of the production editor is to verify that the template was applied correctly—overrides to template formatting often cause problems elsewhere. For example, applying the wrong paragraph tag to a heading causes that heading to disappear from the table of contents. Starting with FrameMaker 11 you can search for overrides to specific styles, dramatically increasing both speed and accuracy of this process.

Like MS Word, unstructured (standard) FrameMaker still allows you to place things like chapter titles anywhere you like in a document. If you are interested in working with an enforceable content model like XML or DITA, consider using structured FrameMaker.

Indexing

When indexing a book, there are two workflow options:

- **Embedded indexing.** The index is generated from markers that are inserted (embedded) into the document files.
- **Stand-alone indexing.** The index is generated in dedicated indexing software, and the resulting text is inserted into the document.

If you plan to make frequent, incremental updates to the content, creating an embedded index is probably the right choice because you can reuse markers and reduce the amount of work required to update the index. If the book is a one-time project, you might want to create a stand-alone index, especially if you plan to use a freelance indexer. In technical writing groups, embedded indexes are the standard; in publishing environments, stand-alone indexes are more common.

Printing

You can print documents from FrameMaker directly to an office printer, but I recommend first creating a PDF file, using the **File > Save as PDF**. Make sure you use the Adobe PDF print driver (installed with FrameMaker) by selecting Adobe PDF in the **File > Print Setup** dialog. You can then use the PDF to print, post online, or send to a print vendor. For information on the free SetPrint utility, which sets your FrameMaker printer on startup, see page 397.

Printing and other outputs are described in detail in Part V, "Creating Output."

Single sourcing and digital publishing

Adobe defines multichannel publishing as "publishing from a single source to print, PDF, HTML, XML, and more." Many portions of their web site d

nd mobile apps) or multiple documents (for example, a user's guide and an administrator's guide) from one set of files.

Note Single-sourcing can also refer to a document management system (DMS) or content reuse, where content is maintained in one location, and reused wherever needed, much like a graphic file.

Single-sourcing implementation can be simple. For example, a FrameMaker file can both be printed and converted to PDF. A more complex setup might involve a content repository from which you extract relevant information by keyword searches and then assemble and format the information for print or online help.

The basic premise of single-sourcing is that you "write once, publish many." Because you write content once and then use features and tools to process the information for different outputs and media, you save time and money.

For a single-sourcing project, the planning stage is absolutely essential. Each piece of content can be presented differently in each medium, and understanding these relationships before beginning the writing process is critical. To ensure that single-sourcing succeeds, content must conform to templates—overrides and "custom" formatting are usually lost when converting from one format to another. Fortunately, as mentioned earlier, FrameMaker helps in locating and correcting these overrides.

Aside from the increased emphasis on planning and standards, workflow is modified because additional output is generated in the last stage of the project. FrameMaker directly generates output that meets the requirements most users. If your needs are more specific, Adobe RoboHelp will extend those capabilities. For more information, see "Digital Publishing" on page 413.

Evaluating methods

Before starting a single-sourcing project, it's helpful to review the pros and cons of different workflow options. If you need to deliver the same information in different media (for example, procedures in both print/PDF and online help), single-sourcing will probably be the most efficient approach, but there are other options.

Parallel Development (A reasonable starting point)

In a parallel development process, all deliverables are created simultaneously. For example, one writer might create a user's guide in FrameMaker while another writer creates online help in a help authoring tool and yet another creates HTML in a web development tool.

Parallel development has the advantage of allowing writers to specialize in print or online content and optimize the flow of information for each deliverable.

Parallel development works if the information in the deliverables is different. If, however, there is content overlap among the deliverables, there are a number of problems with this workflow:

- Duplicate content creates a labor-intensive process.
- Maintenance is problematic, and differences in terminology or even outright contradictions are common. This confuses the readers.
- Transferring information from one tool to another is time consuming. Formatting is often lost, increasing time spent on conversion.

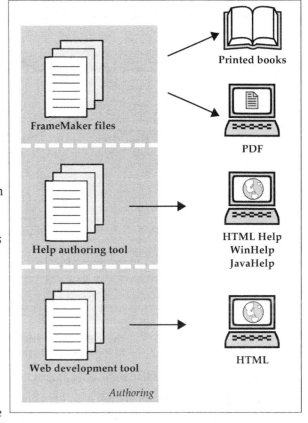

The bottom line: Parallel development makes sense when the information in the print and the online help does not overlap, but when there is overlap, parallel development is time consuming and inefficient.

Serial Development (Better)

Serial development could be considered a type of single-sourcing. You write the information once and then convert it from one format to another. However, serial development typically involves copying the information from one format (for example, the print development tool) to another format (the

Chapter 4: Establishing a workflow in FrameMaker

help authoring tool) and then reworking the information to add formatting. Importing FrameMaker files into RoboHelp (as opposed to dynamically linking the files) is an example of serial development.

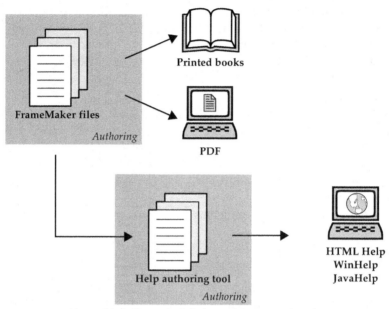

You could also create the help first and then move the information to FrameMaker to create the print and PDF versions.

Serial development has some advantages:

- Because the information is written once, information is consistent across all deliverables.
- The second deliverable is not created until the first deliverable is completed, so the content is finalized.
- Maintenance is simplified because you only convert once per release and do not have to maintain two sets of documentation.
- One writer can create the print output and then the help, so it's less expensive than having two writers working in parallel.

But there are some serious disadvantages:

- Serial development means that one deliverable will lag significantly behind the other. For example, the printed version might be ready three or four weeks before the online help. This leads to scheduling problems before a release because you have to build in several weeks for the conversion and reformatting process.
- Inline formatting can be harder to transfer and control, so some formatting may be lost when you transfer information from one tool to the other. The clean-up that's done in the second version must be repeated for each release (unless you keep the files and only put in the changes, in which case you've reverted to a parallel development process).

Serial development can be viable if you need to transfer only a small amount of content from one deliverable to another, but large-scale manual conversion is tedious, and better options are available.

Modular Development (Best)

Single-sourcing often focuses on creating multiple deliverables in different media, such as print and online help, from a single set of files. Modular development has a slightly different emphasis—creating multiple deliverables in the same medium and reusing overlapping content. For example, if two products share several features, it would be helpful to write about those features only once and reuse the information in both books. Modular development lets you create information that both products share and then reuse that information as needed. At the time of printing, Adobe provides two products for modular development: FrameMaker Publishing Server, and the Adobe Technical Communication Suite (containing FrameMaker and RoboHelp, Acrobat, Captivate, and Presenter).

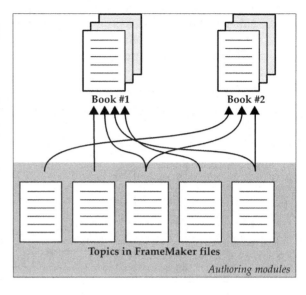

Authoring modules

Here are the FrameMaker features most commonly associated with modular development and single-sourcing:

- Topic-based authoring (structured FrameMaker only, using a standard such as DITA, S1000D, ATA iSpec 2200, or other custom structure model)
- Conditional text
- Text insets
- Variables
- Cross-references
- Content management systems
- Publishing to a variety of electronic formats

These concepts (other than structure) are discussed at length in the appropriate sections of this book.

By reusing modules, you create multiple deliverables from a single set of source files. But modular development does not necessarily include delivery of information in multiple media. Many (but not all) technical publishers use the term single-sourcing to describe producing different media from a single set of source files.

Database Publishing (Different)

Database publishing is used most often for directories or other highly structured content. The information that needs to be published is already stored as records in a database, so you extract information from the database, tag it as appropriate to format it, and create a FrameMaker file. Many databases are now XML-based, and you can extract XML-tagged information to use within FrameMaker. Also, a number of third-party applications, such as Miramo and PatternStream, can create FrameMaker or MIF files from database content.

Database publishing is best suited for working with highly structured information that's already in a database. Information that's output in a particular sequence (for example, alphabetical order for a list of doctors) works well in this environment. Documents that have a well-defined sequence (for example, a programmer's reference with an alphabetical list of commands) can also work well. Less regimented documents, such as user's guides, tutorials, and the like, do not work well in database publishing.

Traditionally, the database serves as the content repository, so you make changes to the information only in the database. The generated FrameMaker files might need formatting tweaks, but any content and formatting changes you make in them are lost the next time you export from the database.

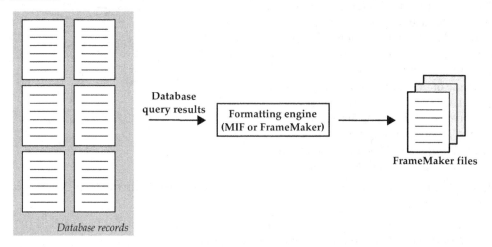

Miramo Personal Edition

If you are interested in checking out database publishing options, you'll find a free personal edition of Miramo available at miramo.com. (Available at time of printing) Functionality is largely intact, though licensing and usage limitations may apply.

Digital publishing workflows

Using the FrameMaker Publish pod

FrameMaker publishes directly to the most widely used electronic formats needed for technical communication projects. When delivering the same content in multiple media, you should first consider output created directly from FrameMaker. See "Digital Publishing" on page 413 for more information.

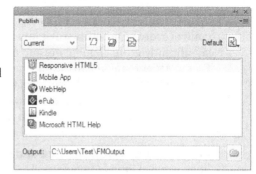

Note If you want to produce XML, use FrameMaker in structured mode.

Creating Online and Mobile Output with Conversion Tools

Some legacy tools are available to create HTML, help and a variety of mobile formats (HTML5, EPUB, Kindle, WinHelp) from FrameMaker files. These tools vary considerably in terms of their intent (stand-alone help authoring versus large-scale conversion), interface, pricing, and licensing. For information, see "Third-Party Tools and Plug-Ins" on page 534.

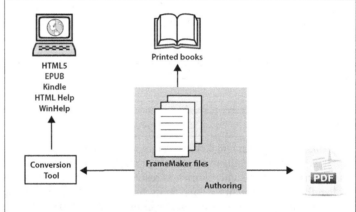

Caution FrameMaker also provides a built-in HTML converter (**File > Save As… > HTML**), but it is quite outdated, and I've never found it appropriate for client output.

Structured Authoring and Single-Sourcing

Structured authoring, which requires setting FrameMaker to run in structured mode, provides a different approach to single-sourcing. You can author XML content in FrameMaker. From the FrameMaker files, you can produce printed books and PDF files as well as using Extensible Stylesheet Language Transform (XSLT) files to process the XML and produce HTML and online help formats.

You could also use external XML editors to modify the XML files and pull information into FrameMaker for publishing. Because of the expense of developing customized XML output tools, some companies use FrameMaker as a publishing engine even if they are using a different editor for XML authoring.

For an overview of structured authoring, see the following sections.

Structured FrameMaker. FrameMaker gives you the choice of working in a structured or unstructured authoring environment. (This book is for people using the unstructured FrameMaker interface.) Unstructured authoring means that you use paragraph and character tags to control formatting. With structured authoring, you define an allowable structure for your document, and FrameMaker automatically handles formatting based on the structure. In the structured authoring environment, you can create both structured and unstructured documents. In unstructured FrameMaker, though, you can create only unstructured documents.

With its enforceable structure rules, structured FrameMaker gives a template designer more control over the author's workflow. However, setting up a structured FrameMaker template is significantly more challenging and more time consuming than creating an unstructured FrameMaker template. In the unstructured version, the template designer creates paragraph tags and other components that control the appearance of the document. For the structured version, the template designer must create a document that describes the structure of the document and then attach formatting information to each element in the structure.

Choosing between structured and unstructured workflows. All of the features available in unstructured FrameMaker are also present in structured FrameMaker, but structured FrameMaker provides additional capabilities. There are some advantages and disadvantages to each type of document. Table 4-1 summarizes the differences between structured and unstructured FrameMaker.

Table 4-1. Comparison of Unstructured and Structured FrameMaker

	Unstructured	**Structured**
Authoring		
	Unstructured FrameMaker is in some ways more flexible than working in structured FrameMaker. Unstructured documents are less restrictive for authors than structured documents, but the content model is unenforceable. For example, rules like "a heading must be followed by a lead-in paragraph" are the author's responsibility. Authors apply style tags which can be similar and require care, such as Note, TableNote, and NoteContinued. Authors must learn which paragraph tag to use in which context.	The content model enforces structuring rules, guiding authors through available options. Authors insert elements and content and the template controls what formatting is applied. Often, formatting of an element is based on the context (or position within the structure) of the element. Structured documents provide a framework which is important in larger workgroups. Structured is more common in large organizations, especially in aerospace, telecommunications, and in government.
Template design		
	Creating templates is easier in unstructured FrameMaker because you do not have to set up a content model.	Creating structured document templates requires XML knowledge, and provides context formatting that is not available in the unstructured model.
Single-source publishing		
	You can export unstructured documents to other formats directly from FrameMaker or through a number of tools.	You can publish to other formats through FrameMaker tools, or process the XML content using a number of industry-standard tools.

Table 4-1. Comparison of Unstructured and Structured FrameMaker (Continued)

	Unstructured	Structured
Learning curve		
	Moderate for document authors; high for template designers.	Low to moderate for document authors; very high for template designers.
Workflow		
	Works well in small groups; more difficult to manage in large groups.	Requires additional up-front planning, but works well for large groups.
XML conversion		
	Export only. No element hierarchy.	FrameMaker works directly with XML files. XSLT allows on-the-fly or programmatic conversion to other content models.

Language support

Starting with FrameMaker 2015, FrameMaker supports Right-to-Left languages like Hebrew and Arabic and allows bidirectional content as well. You can control direction for:

- a document
- a flow
- a paragraph, or paragraph style
- a table, or table style
- a graphic

Bidirectional support extends to electronic output as well. PDF, formats in the Publish pod, and content linked to RoboHelp will all retain their proper direction.

The user interface for FrameMaker is available in several languages—English, French, German, or Japanese—but you can author content in many more languages. FrameMaker supports the Unicode text encoding standard, which is a character set that includes characters for all living languages. With Unicode support, elaborate workarounds are no longer necessary for authoring in Cyrillic languages, such as Russian. You can use Unicode content in document text and in all features including find/change, catalog tags, markers, and so on. You can import Unicode content from other applications and include Unicode content in PDF files and other FrameMaker outputs.

Unicode support

There are several things you must consider if you are going to author in multiple languages in FrameMaker:

- **Fonts.** When working with content in multiple languages, you need to install fonts for those specific languages or a Unicode font that includes all the characters you need. If you try to work with a language when you don't have a font installed to display that text or don't have the correct font applied to the text, the characters are replaced with placeholders such as question marks or gray boxes.

Tip   Question marks, gray boxes, or incorrect glyphs may also be displayed, depending on the capabilities of the printer you have selected. To avoid this problem, convert to Adobe PDF using the Adobe PDF printer driver prior to printing.

- **Text entry method.** To type Unicode content, you can copy and paste from FrameMaker's Character Palette utility (**File > Utilities > Character Palette…**) or from any Unicode application, including Microsoft Word or FrameMaker. However, to enter any significant amount of text, configure your system to use either a language-specific keyboard, and then use that keyboard to enter characters, or set up an Input Method Editor (IME), which is software that lets you insert characters phonetically that aren't on the keyboard. (IMEs are primarily used for authoring in Korean, Japanese, and Chinese.)

Note   For more about the Unicode text encoding standard, go to www.unicode.org.

If characters look wrong after you paste text from Microsoft Word, a browser, or other application, try the paste again with **Edit > Paste Special** and select Unicode Text. The text is pasted into FrameMaker and maintains all the Unicode characters from the original source.

Tip   To have FrameMaker automatically paste as Unicode text instead of plain text, change the paste order in maker.ini. See "Setting clipboard pasting order" on page 548.

Dictionary support

FrameMaker installs a number of dictionaries for spell-checking and hyphenating. FrameMaker continues support for Proximity dictionaries, and now has support for the Open Source HunSpell dictionaries. To author in languages without a provided dictionary (including programming code) set the language to None in your paragraph and character styles. This ensures that FrameMaker won't attempt to spell-check or hyphenate the text.

Chapter 5: Templates

Effective FrameMaker workflows rely on style sheets. By storing formats in an external template file, multiple authors can create documents with consistent formatting and pagination. FrameMaker files do not refer to the external template file; rather, once the template is set up, you import formatting information from a template file into document files. Importing in this way updates specific formatting information from the template file across the document files. Any document can theoretically be used as a template, but it's a good idea to create a single, recognized template file for each of the major parts of your book (chapter, table of contents and index at a minimum) and keep it in a safe place. All formatting changes or additions should be made to those files. Without templates, documents will eventually start to diverge as you or others add tags to individual files (a phenomenon some call "tag creep").

Here's a look at what's in this chapter:

User roles	89
Definition of a FrameMaker template	90
Why you should care about templates	90
Getting started with templates	91
Importing settings from a template file	91
What's imported?	92
What makes a "good" template?	92
Understanding template interactions	93
Tips and tricks	98
Paragraph tags	98
Character tags	99
Table tags	100
Reference pages	100
Entering Special Characters in dialog boxes	100
Naming conventions	104
Capitalization	105
Special characters	105
Taking keyboard shortcuts into account	105
Separating out "housekeeping" tags	106
Documenting your template	107
Using single-purpose templates	107

User roles

Setting up FrameMaker templates requires more detailed knowledge about FrameMaker than does using templates after they are developed. Many organizations identify internal specialists or hire an outside consultant to set up their templates (I'm one of those specialists).

Generally, the responsibilities in a writing group are divided into two different roles:

- **Template designer:** The template designer(s) needs advanced FrameMaker skills. The designer creates FrameMaker templates. Since template design is on an *as needed* basis, the template designer(s) also perform authoring duties.

- **Content creator (author):** The author needs basic-to-intermediate FrameMaker skills. Authors use established FrameMaker templates to create and edit information. Authors need to know how to apply formats and insert special content, such as cross-references.

In recent years, even large publications groups have combined most technical communication functions into one role. Fortunately, FrameMaker gives you tools to reuse content from subject matter experts (like MS Word content mapping), importing of video and interactive content, automatic TOC and index maintenance, and publishing to a wide array of electronic formats, including mobile apps for iOS and Android. Using thoughtful templates and settings files, FrameMaker gives you an end-to-end publishing system that can't be beat!

Often in small organizations, one person performs all of these tasks. Even in a *single author* scenario, you will do well to follow this template workflow. For example, without a template file, if you modify a paragraph format in your content, you will be less likely update that style across other files in a book, resulting in inconsistent formatting across chapters or entire documents.

This chapter describes some of the issues you must consider as a template designer.

Definition of a FrameMaker template

A FrameMaker template is *any* FrameMaker document used to store formatting information. You transfer formatting information from the template file to a source document by using the **File > Import > Formats** menu option.

Why you should care about templates

A well-designed template increases productivity because it reduces time spent formatting, and increases consistency across chapters and books. Table tags, for example, let you define ruling and shading properties in a table so that when an author inserts a table, the table automatically uses the correct settings. Without a table tag, authors or production editors would have to spend time formatting each table manually. Running headers and footers save time by accurately picking up information from the main body text, such as the current first-level heading, and automatically displaying it at the top or bottom of the page.

Reference page definitions control how generated tables of contents and indexes are formatted. Predefined cross-reference formats anticipate references that the author might need to insert and provide various options for the cross-reference text.

It's *possible* to create consistent documents without templates, but it's not efficient. In an environment where you are producing hundreds or thousands of pages of documents every year, working without an established template is a lot like walking a tightrope without a safety net—and wearing clown shoes.

Getting started with templates

The easiest way to begin creating templates is to start with existing content and styles, modifying it to suit your requirements. When you install FrameMaker, several templates are put in your installation directory in the templates folder. These templates are also available when you select **File > New > Document** by selecting the Explore Standard Templates button. In fact, even the "default" FrameMaker document (the Blank Paper: Portrait option in the new document dialog) has formatting definitions, and thus is a workable template file.

Adobe also has additional templates online. They are a bit dated, but try searching the web for "framemaker template pack" if you are curious. If you don't have the time to create your own templates, many FrameMaker consultants offer template design services to rapidly match your corporate branding standards.

Importing settings from a template file

When you import formats from a template file to a document, you can specify which catalogs you want to update. In each catalog, FrameMaker performs an additive merge—that is, tags with matching names are updated, and any tags that exist in the template but not in the document are added to the document. Importing formats *does not* remove unused formats from the current document's catalogs.

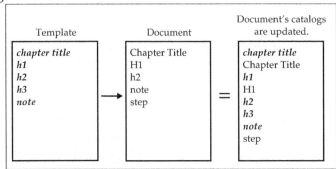

To apply a template, follow these steps:

1. Open the template file.
2. Open your working file (the file you want to apply the template to).

   Tip   To import formats into several files in a book file, open the book, then select the files in the book that you want to update.

3. Select **File > Import > Formats** to display the Import Formats dialog. (see graphic on page 90 for example)
4. In the Import from Document drop-down list, select your template file. Only open files are shown in this list.

   Tip   One quick way to remove overrides from a document is to import formats from the Current file and check all the available checkboxes.

5. In the Import and Update section, check all of the items you want to update from the template. Generally, you will want to update everything, but you can also perform selective updates. Click the Deselect All button to uncheck every option; click the Select All button (which toggles with the Deselect All button) to check every option.
6. *(optional)* To remove page breaks that were set with overrides, check the Manual Page Breaks checkbox.
7. *(optional)* To remove overrides other than page breaks, check the Other Format/Layout Overrides checkbox.
8. Click the Import button to import the specified items from the template.

What's imported?

While at first glance this section seems redundant, some of the options are not labeled as directly as you might think. Specifically, the Page Layouts option controls Master Pages, and the Document Properties is a wide array of settings, including page and chapter numbering.

Each check box in the Import and Update section of the Import Formats dialog box (see page 91) controls a specific set of formatting catalogs. They are as follows:

- **Paragraph Formats:** Imports the contents of the Paragraph Catalog.
- **Character Formats:** Imports the contents of the Character Catalog.
- **Page Layouts:** Imports the master pages.
- **Table Formats:** Imports the table catalog; the list is in the Table Designer.
- **Color Definitions:** Imports the color catalog.
- **Document Properties:** Imports several miscellaneous settings. They include custom marker definitions, change bar settings, numbering properties, text options, and PDF settings.
- **Reference Pages:** Imports the reference pages.
- **Variable Definitions:** Imports the variables and their definitions.
- **Cross-Reference Formats:** Imports cross-reference formats.
- **Conditional Text Settings:** Imports conditional text tags, conditional expressions, and show/hide settings.
- **Math Definitions:** Imports custom math element definitions (used in equations).
- **Combined Fonts:** Imports fonts created with the **Format > Document > Combined Fonts...** command.
- **Object Styles:** Imports the content of the Object Styles Catalog (first available in FrameMaker 11).

What makes a "good" template?

Because templates involve making design decisions, beauty may well be in the eye of the beholder. There are, however, features that can make some templates more functional than others. The purpose of a template is to enable authors to create documents that have a consistent look and feel. A good template will assist in this process and help automate some tasks.

Consider, for example, the use of paragraph tags. It's possible to create two pages that look the same without using paragraph tags—the author could use the Format menu's Font selections to match formatting of one page based upon the formatting of another. This would be tedious and time-consuming. It is more efficient to create paragraph tags to control types of paragraphs across your project. Of course, paragraph tags are useful only if the authors choose to use them!

Designing a good template requires more than just implementing paragraph tags. It also requires you, the template designer, to balance complexity and completeness. A *complete* template is a file that contains all of the tags needed to format content consistently. But if that template is too *complex*, authors will face a steep learning curve in figuring out all the available tags and how to use them. Authors generally respond to an overly complex template by ignoring it.

Perhaps the best definition of a good template is any template that authors are willing and able to follow consistently, and can be used to produce consistent results.

Understanding template interactions

There is no single correct way to build a template. Some template designers start with master pages and page settings, and then add paragraph tags to the overall page design. Others start with character tags because they are used by other features (such as paragraph numbering, variables, and cross-references). Another approach (my favorite) is to define naming conventions for all items listed in the Import Formats dialog, and create items as needed—if a paragraph tag needs a character tag for autonumber formatting, the designer jumps over to the Character Designer, creates the character tag, and returns to the Paragraph Designer to complete the paragraph definition.

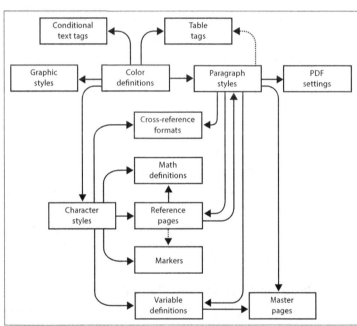

Make a list of all of the needed formats and check them off as you build them so that you can keep track of what's needed and what's been created.

As you create your template, keep in mind that most of the template features are interdependent. Variables use character tags for formatting. Paragraph tags use the color catalog for available colors. Cross-references can use content in specific paragraphs. The configuration of master pages affects how paragraph tags appear on the page. Master pages use system variables for headers and footers. All of these interactions can make designing a template rather like a crossword puzzle. Try to work in too linear a process, and you're sure to get frustrated.

When importing formats, it's important to note that most template items contain dependencies on other template items. This means, for example, that importing a new or modified paragraph format might also require importing of a character format used in autonumbering, as well as a color definition used in either the autonumber character format or in the Basic properties of the Paragraph Designer.

The following sections explain the dependencies within template components.

Paragraph Formats

Settings in the Paragraph Designer have three dependencies: color, reference pages, and character tags. For color, the situation is identical to the Character Designer; you can only assign colors that are defined in the color catalog. Character tags are available to format autonumbering. On the Advanced properties sheet, the reference pages are involved because the options available in the Frame Above and Frame Below drop-down lists correspond to the named graphic frames defined in the reference pages.

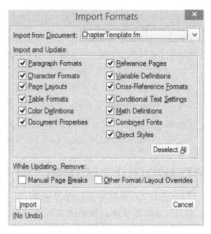

Paragraph tags are used by several other template components. Master page maps on reference pages use paragraph tags to determine which master page should be assigned to which page, variables call paragraph tags and use their text for running headers and footers, and so on. See Chapter 6, "Formatting text with paragraph tags," for details.

Character Formats

Character tags have only one dependency,; they depend on the colors available in the Color drop-down list (the color catalog). To assign a custom color to a character tag, you must first define the color in the color definitions.

Character tags are widely used in other template components, including paragraph tags, variables, cross-references, and reference pages. They are also used in markers to assign formatting.

Page Layouts (Master Pages)

Master pages almost have a dependency on variables. The Current Page # system variable, for example, is on nearly every master page definition. Running header and footer variables are also common.

You can assign master pages based on paragraph tags that occur on the body pages, making the master pages dependent on paragraph tags and the reference pages.

Note The mapping table that establishes these links is stored on the reference pages. In that case, master pages are dependent on the reference pages as well

Table Formats

Table tags use the color catalog and, under the covers, paragraph tags. The color definitions are available in several locations.

Understanding template interactions

When you define shading colors, the Table Designer uses the color catalog. To get clean colors in your table shading define lighter, solid colors that you can use at 100%. See "Managing color definitions" on page 432 for details.

Custom ruling and shading (**Table > Format > Custom Ruling & Shading**), also relies on the color catalog. Finally, if you edit a ruling style (click the Edit Ruling Style button on the Custom Ruling and Shading dialog box), you again have access to the color catalog.

The relationship between table tags and paragraph tags is more subtle. The table definition stores a list of default paragraph tags used when you create new tables. For details, see "Applying paragraph tags" on page 162.

Color Definitions

The color catalog is rarely the first item created in a template, but perhaps it should be. The color definitions are accessible to a number of other items, but color definitions do not have any dependencies on other template items.

Tip If you discover invisible text in your document, check the **View > Color > Views** settings. You may find that your color has inadvertently been set to Cutout or Invisible.

Document Properties

While not containing dependencies, per se, the Document Properties option controls a number of significant items. Among the properties controlled are:

- volume, chapter, page, and paragraph numbering options
- text options
- change bars
- footnote and table footnote properties and numbering options
- marker types
- PDF setup options
- view options

When working on lengthy documents, I choose either to avoid importing document properties across my chapters, or I use Rick Quatro's Import Formats Special ES extendscript for all the regular options and to break out these options into eleven separate choices.

95

Marker Types (Stored in Document Properties)

When you create marker text, such as index entries, you can use character tags to control the formatting of the file generated with the marker information. The appearance of the marker text is further governed by the information set on the reference pages in the generated file.

PDF Settings (Stored in Document Properties)

The PDF Setup dialog box (**Format > Document > PDF Setup**) can create bookmarks and tags based on available paragraph tags. As a result, PDF Setup might be one of the last items you'll configure in your template.

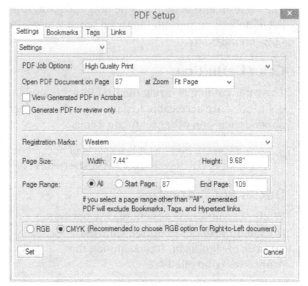

The Bookmarks sheet lets you specify which FrameMaker paragraph tags should be displayed as bookmarks in the PDF file; you also specify their hierarchy.

The Tags sheet lets you specify which FrameMaker paragraphs to include to create tagged PDF files, necessary for electronic PDF reviews.

See Chapter 22, "Print, PDF output, and package" for tips on PDF settings.

Reference Pages

Reference pages use information from the character tags and paragraph tags, and in turn are used by paragraph tags. There is also a relationship between reference pages and markers. When you set up flows for generated files, you work on the reference pages. These flows use character tags for formatting and refer to paragraph tags and sometimes markers.

The Paragraph Designer uses reference page information for Frame Above and Frame Below settings.

Graphic Objects (stored on Reference Pages)

When you create various graphic objects using FrameMaker's drawing tools, the colors you can assign are determined by the color catalog.

Text lines can be formatted with character tags. Text in text frames is formatted with paragraph tags along with possible character tag formatting. See Chapter 16, "FrameMaker's graphics tools" for details.

Variable Definitions

Variable definitions, like cross-reference formats, use character tags for formatting and paragraph tags to determine what information to display. When you edit a system or user variable definition, you can assign character tags to all or part of the text.

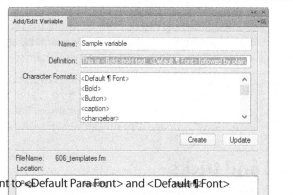

Tip The </> building block is equivalent to <Default Para Font> and <Default ¶ Font> building blocks, but takes up less space.

Some variables, such as running headers and footers, let you reference specific paragraph tags with syntax such as:

<$paratext[ChapterTitle]>

If you import variables from another FrameMaker document, both the system variables and user variables are imported together. System variables should remain constant across all your files. For example, the system variable Running H/F 1 should have the same value across all the files in your book. Otherwise, when you import variable formats, the headers or footers in some of the files may be wrong. On the other hand, user variables (product names, etc.) can be handled separately through the use of a MIF snippet or a third-party plug-in. (see "Using MIF fragments to update catalog settings" on page 568)

Cross-Reference Formats

Cross-references use character tags for formatting and paragraph tags to get information. When you set up a cross-reference definition, character tags are available in the list of building blocks. The items available are taken from the document's character catalog.

Several of the building blocks, such as <$paratext>, can be set up to refer to a specific paragraph tag.

Conditional Text Tags

The colors available for conditional text tags are determined by the color catalog.

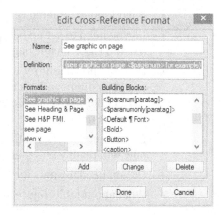

You might want to set up conditional text in the template, perhaps to include certain items in online output that don't show in print, or to insert comments that you can show in drafts and hide in the final output. Plan which conditions you need and how you want them to interact. You can choose whether you want conditions to Show as per Condition or Show as per Expression, which is as confusing as it sounds. See Chapter 25, "Setting up conditional text," for details about conditional text.

Math Definitions

Math definitions are custom components for equations. For the most part, equations stand on their own separately from other document objects. Equations do, however, use character tags for formatting.

Equations can also use information from reference pages. You can define a custom equation object on a FrameMath reference page. This object is then available in the Equations Palette. See "Writing equations" on page 499 for details on using equations.

Tips and tricks

This section describes some design tricks that automate certain features and make templates easier to use.

Paragraph tags

When creating paragraph tags, pay close attention to naming conventions. Users spend more time using paragraph tags than any other item in the template.

Here are some issues to consider as you create paragraph tags:

- A prefix will visually organize your catalog, and speed tag assignment via the Smart Insert function
- How many heading levels does the document need?
- How many types of body text does the document need?
- How many levels of indent?
- What kind of numbered lists?
- What kind of bulleted (unordered) lists?
- Do you need to set up header and footer styles? Are they different on left and right master pages?

- Does the document require an overall numbering scheme? Will you use <$volnum> and <$chapnum>? How will these integrate into the autonumbering you need to define? Will your book include folders ( ) or groups ( )? If so, be familiar with the <$sectionnum> and <$subsectionnum> variables as well.
- Do you need notes, cautions, warnings, tips, and other asides? If producing online formats, how will readily will these convert?
- Do you need formats for sideheads?
- What about table, figure, or example captions?

The following items will make your tags more useful:

- **Use Keep with Next/Previous (Pagination properties) to avoid awkward page breaks.** Headings usually belong with the paragraph beneath them, so set heading formats to keep with the next paragraph. This eliminates unsightly headings hanging at the bottom of a page. Consider whether other paragraph tags have similar requirements. Second level index entries and sublists may benefit from a Keep with Previous setting. List introduction paragraphs are another format that benefits from a Keep with Next.

- **Control spell-checking with the Language setting (Default Font properties).** A language of None causes a paragraph to be skipped by the spell-checker, which can be useful for code. If you have multiple languages in a single document, create paragraph tags for each language and set the Language attribute for each one. FrameMaker will use the dictionary and hyphenation settings of the specified language as it spell-checks each paragraph.
- **Set Widow/Orphan Lines (Pagination properties) to reduce awkward page breaks.** Use a widow/orphan setting of at least 2 throughout the paragraph tags to prevent solitary lines at the beginning or end of a page. If you want to keep an entire paragraph on the same page, use a very high number (for example, 99 or 999).
- **Use autonumbering for bullets, steps, notes, cautions, and any other repeated text.** Autonumbering is wonderful for numbered headings, figure captions, and tables, but don't forget "text-based" autonumber options, which can also save significant amounts of time.
- **Reset numbered lists with headings and intro paragraphs.** If your content model specifies an introductory paragraph to lists, the intro paragraph can reset the list, eliminating the need for a special list start tag to do that.
- **Use the Start setting (Pagination properties) to set up paragraph tags that should always start at the top of a page or column.** This might be your chapter title (Start Top of Right Page) and your first-level headings (Start Top of Page), but also may apply to a rotated table page, or a full-page graphic.
- **Use the Next Pgf Tag setting (Basic properties) when there is a logical sequence of paragraphs.** Heading paragraphs may be followed most often by a body or firstbody paragraph. Most step 1 paragraphs are followed by step 2. By setting the default, you'll save your users the step of applying the tag manually.

Character tags

In addition to providing character-level formatting in regular text, character tags are referenced by several other features. Refer to this preliminary checklist as you build your list of required character tags:

- Character-level formatting
- Variables
- Formatting for autonumbers
- Cross-references
- Generated files
- Markers, especially index markers
- Math definitions
- Reference pages
- Text lines

As you create your character tags, keep in mind the following issues:

- In general, it's best to use the As Is setting throughout the character tag and assign specific settings only to the items you want to change.

- The Language attribute in the character tag overrides the Language attribute assigned by the paragraph tag. By setting language to None, you can use a character tag to turn off spellchecking for certain things.
- Consider creating special one-letter character tags that make inserting formatting in markers faster. For example, while you could italicize a word in an index entry with this syntax:

    cross-stitching needle: <Emphasis>see<Default Para Font> tapestry needle<$nopage>

    But if you define an "i" character tag, you could use this syntax for the same result:

    cross-stitching needle: <i>see</>tapestry needle<$nopage>

    These shorter tags will also be very helpful on the reference pages. (Notice that these tags do break the rule against naming tags by their formatting.)

Table tags

In addition to the ruling and shading attributes, keep in mind that you can set several properties as defaults (although not in the Table Designer). The defaults are saved based on the settings in the selected table when you click the Update All button in the Table Designer. The settings saved as defaults are the number of rows and columns, the column widths, and the default paragraph tags for the heading rows, body rows, and footing rows.

Reference pages

When you create a graphic frame on the reference pages, you must assign a name to that frame. New graphic frames are not, however, labeled by default, so to figure out which frame has which name, you click the frame so see its name. To avoid this, add a text label next to each graphic frame as you create it.

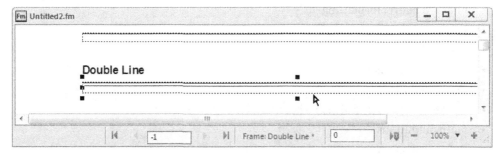

Entering Special Characters in dialog boxes

Template designers often need to enter special characters in dialog boxes. If you need to enter a special character that isn't on your keyboard into a dialog box or pod, you cannot use the standard Control key keyboard sequences or the Symbols list available from the toolbar. The easiest way to enter special characters into dialog boxes is to copy and paste them from a document, but there are other methods you can use:

- Windows Character Map utility
- Unicode code points or hexadecimal codes
- Special escape sequences that FrameMaker provides
- ANSI codes
- FrameMaker Character Palette
- FrameMaker Hex Input Pod

Pasting into Dialog Boxes from the Windows Character Map Utility

You can cut and paste from the Windows Character Map utility (see page 37) into a dialog.

For example, if you are adding a variable in which you want to use one or more characters from the Wingdings2 font, you can define a character tag named WD2 for the font. The variable definition includes the character tag, the special character, and the building block for the default paragraph font. The definition looks like this in the dialog box for editing the variable:

```
<WD2>u</>
```

The formatting from the character tag is not applied until you insert the variable in your document. In this example, the variable looks like this in the text:

```
u
```

Remember Characters pasted into dialog boxes and pods such as the Find/Change pod retain their formatting. Characters in variables, character tag autonumbering, and markers display in the default system font until you insert them in your document.

Using Unicode Code Points or Hexadecimal Codes

You can type Unicode code points or a hexadecimal codes into dialog boxes like those for cross-reference formats and autonumbering properties. You can find the codes in the Windows Character Map utility. Supported hexadecimal codes are also listed in the *Adobe FrameMaker 8 Character Sets* online manual, available on the web.

To insert a character using a Unicode code point or hexadecimal code, follow these steps:

1. Open the Windows Character Map utility (see "Pasting from the Windows Character Map Utility" on page 37) and locate the code for the special character you want to insert:
   a. In the Font drop-down list, select the font.
   b. Check the Advanced View checkbox.
   c. If the Character Set drop-down list is available, select Unicode. (The Character Set is selectable when the font has Unicode code points associated with it. Selecting Unicode ensures that the correct characters are displayed in the dialog. If the field isn't selectable, you can ignore this setting because the field has no effect on the characters shown in the dialog.)

d. Click the special character you want to use. Its code is displayed at the bottom of the window. The code is a Unicode code point if it begins with a U (such as U+1EBB). The code is a hexadecimal code if it doesn't have a U at the beginning (such as 0x65).

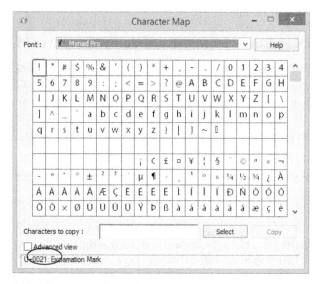

In the FrameMaker dialog box, type the code:

- **Unicode code point.** Type the Unicode code in the format \uXXXX. where XXXX is the code point. For example, to enter the Arial character ẻ (U+1EBB), type **\u1EBB**.
- **Hexadecimal code.** Type the hexadecimal code preceded by \x. (The *Adobe FrameMaker 8 Character Sets* online manual includes the \x with the codes.) If you found the code in the Windows Character Map utility, exclude the leading zero. For

example, to enter the Webdings ○ symbol (the Webdings representation of the q character, hex code 0x71), type **\x71** in the appropriate FrameMaker dialog.

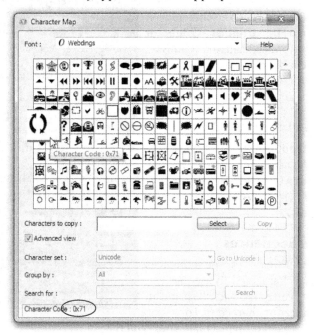

In areas such as the Find/Change pod, the pod displays the character itself instead of the code. In dialog boxes and pods that allow formatting with a character tag—such as cross-reference definitions, variables, and paragraph autonumbers—the character is displayed with the system's default font, so the character may not look the way you expect until you insert it in the document with font family applied via a character tag.

Note If the text is not using a Unicode font, you may see a question mark (?) displayed in the text instead of the character. Apply a Unicode version of the font to fix the problem. You may also see an unexpected character if the Adobe PDF printer is not selected in print setup.

Using the FrameMaker Character Palette

To use the character palette utility, position your cursor where you need the character in your file, and go to **File > Utilities > Character Palette**. Select the appropriate font family, navigate to the character needed, and click once on the character to insert it into your document. The Character Palette is resizable

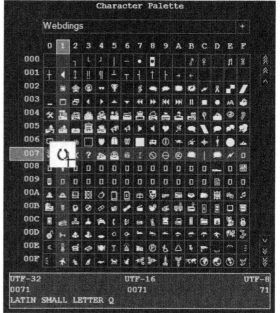

Using the Hex Input Palette

To use the Hex Input palette, position your cursor where you need the character in your file, and go to **File > Utilities > Hex Input**. See "Insert Characters from the Hex Input Palette" on page 37.

Using Escape Sequences

FrameMaker also provides escape sequences for typing special characters. To use the escape sequence, you type a backslash and an additional character or characters. For example, to type an em dash in a dialog box, you type:

\=

The escape sequences for typing special characters in dialog boxes are documented in the FrameMaker online help.

For most special characters, the actual character (and not the backslash and other characters you typed) is displayed when you reopen the dialog. For example, if you type \=, the dialog box contains the actual em dash character (—). However, for characters that would be invisible or confusing if the actual character were displayed in the dialog box, the escape sequences (such as \sm for an em space, \+ for a nonbreaking hyphen, or \t for a tab) continue to be displayed when you reopen the dialog.

Using ANSI Codes in Dialog Boxes

You can use ANSI codes to insert special characters in dialog boxes just as you would within your FrameMaker document. See "Inserting special characters" on page 35. After you have added the character using either the ANSI code, you will no longer see the ANSI code in the dialog. The dialog box will display the character itself.

Naming conventions

The names you assign to tags can make your template significantly more usable. A carefully chosen set of tag names will be easy to understand and lend themselves to using keyboard shortcuts.

Capitalization

For the most part, the capitalization you choose for tags can be left to your personal preference. There are, however, a few things to consider:

- Be consistent! As a template designer, you are likely to type the names of your tags into may places, including the TOC reference page, and the definitions of variables and cross-references. Using consistent naming conventions will help if/when you update the template, as well as help authors in applying your formats while writing.
- FrameMaker's default tags have an initial capital letter (Heading1, Body, and so on). If you plan to use the default tags, you may want to match other tags to use similar capitalization.
- If you do not plan to use FrameMaker's default tags, consider creating only lowercase tag names in your template so that you can easily distinguish your tags from any "renegade" tags.
- Default Word template styles use initial capital letters. When you convert Word files to FrameMaker, the style names are converted to paragraph tag names, and similar considerations apply as for the default FrameMaker templates.

Special characters

FrameMaker *will* generally allow you to use special characters, such as asterisks (*), question marks (?), and pound signs (#), as part of your tag names. These characters can cause major problems, though, when you attempt to reference the tags. We strongly recommend that you use only alphanumeric characters (a–z, A–Z, and 0–9), underscores, and spaces. If you ultimately plan to migrate to structured authoring, you may want to avoid spaces to help smooth the conversion from unstructured to structured content.

Taking keyboard shortcuts into account

Authors can apply many tags using keyboard shortcuts. In FrameMaker, we have the Smart Insert feature, which brings up a pod at your insertion point, allowing navigation via the keyboard or mouse to quickly apply the tag (In previous versions, pressing a key sequence displayed a similar tag list in the status bar). To assist authors using these shortcuts, it's helpful to create tag names whose first few letters are unique. A good example of how *not* to do this are the default heading names in FrameMaker: Heading1, Heading2, and HeadingRunIn. Instead, you might use 1Head, 2Head, and 3Head. Some template designers start each tag name with a short, unique prefix, as in this example.

Most tag names are based on the format or function of the tag. For example, an Italics character tag is a format-based name; a PrintOnline conditional text tag is a function-based name.

Template designers generally recommend that you use names that describe function. There are several reasons for this:

- **Tags named by function are easier for an author to learn.** For example, if a book title occurs in a document, the author has an easier time remembering to apply the BookTitle character tag than remembering that book titles will be formatted with the Italics character tag.

- **You can change formatting associated with functional tags without making the tag name obsolete.** Consider a character tag that indicates menu selection. In the past, your style guide indicated that the menu choice must be boldface; the new standard is to make the menu choice italic. You can change a MenuItem character tag from boldface to italics; if your tag was named Bold, changing it to actually apply italics would cause great confusion.
- **Functional names are better suited to output/conversion to structured documents.** A paragraph tag labeled Helv10pt would need to be mapped to an HTML tag of <p> or an XML element named <para> or something similar. Using functional names in both FrameMaker and XML makes it easier to understand the relationship between the two formats.

There are a few occasions where a combination of functional and formatting names might be useful. For example, you might create a Body paragraph tag (functional) and a second BodyBold paragraph tag (functional and formatting). BodyBold tells the author very quickly what the difference is between the two paragraph tags. A more functional name, such as BodyImportant, might not deliver the message as quickly.

When implementing functional names, keep in mind that some authors may still tag content by format and not by function. You might, for example, have several different functional items that need to be italicized:

- BookTitle
- ForeignWord
- Emphasis
- FileName
- URL

Some authors will learn quickly that BookTitle creates italics, and will use BookTitle any time they need italics, even though other tags are more appropriate. Be sure to reinforce the importance of tagging by content.

Separating out "housekeeping" tags

Every template will have a number of tags that are intended for use "under the covers" and should not be applied by authors. These might include paragraph tags for the header and footer paragraphs and other information that occurs on the master pages. Consider using a special prefix for those tags. For example, naming them with a z_ prefix ensures that these tags are sorted to the bottom of the Paragraph Catalog, which makes it less likely that an author will accidentally use one of those tags. Many tags may only be applied or modified by the template designer on the master or reference pages. To further simplify, you might delete the tags from your paragraph catalog (deleting from the catalog does not remove the tag from your content), and use the Paragraph Designer to Update All instances of the tag without actually storing the paragraph tag definition in the catalog itself. Another option is to use the Paragraph Catalog options to create a customized list of styles to display. See "Using the Paragraph Catalog options" on page 111 for more information.

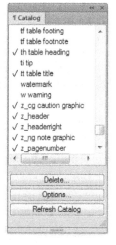

Documenting your template

Please document your templates. Documentation doesn't have to be elaborate, but a basic list of tags and how they are used will be extremely helpful to authors. It also helps to enforce consistency.

The best place to provide documentation for the template is in the template file itself. You can list paragraph tags in one table, character tags in another, conditional text settings in a third, and so on. Providing an example of each tag to show what it looks like is helpful. If you store documentation on a custom reference page, importing formats into documents will update that documentation each time the reference pages are updated.

Template documentation is a good place to list autonumbering streams. This helps you keep track of which series labels have been used and avoid collisions. For example, you might have something like the following table:

Tag	Autonumbering Definition
ChapterTitle	H:<$chapnum>
Heading1	H:<$chapnum>.<n+>
Heading2	H:<$chapnum>.<n>.<n+>
Figure	F:Figure <$chapnum>-<n+>
Table	T:Table <$chapnum>-<n+>
Step1	S:Step <n=1>
Step2	S:Step <n+>
SubStepA	U:<a=1>.
SubStepB	U:<a+>.

We recommend providing a list of tags for each type of formatting item. You may also want to provide additional documentation, such as standard settings for graphics (what resolution should be used when importing bitmap images?), instructions for creating books, tables of contents, and indexes, and perhaps even some basic style guidelines.

Remember When you add to or modify the tags in your template, don't forget to update the template documentation.

Using single-purpose templates

Some publishing applications have special documents that manage settings called *control files*. When you make a change in the control file, it is automatically applied to all of the files in the book.

FrameMaker does not have actual control files. You can, however, mimic this feature by using a single-purpose template, a document that contains a limited number of settings. You might, for example, create a template file that does not contain user variables. If your template did contain user variables, reimporting the template might reset the unique variables in a project back to their initial values.

Instead of defining the variables, you can create a MIF fragment document that contains only specific custom definitions for your user variables. Now, by opening the MIF fragment and importing variables into your project, you add the custom variables. Further, if you've created a MIF fragment containing the proper values for the variables, you now have a "variables template" which allows you to reset your variables whenever needed by reimporting the MIF fragment.

When you are ready to set the variables to the values needed for a book, or for one or more chapters within a book, follow these steps:

1. Open the variable control file. (usually a MIF fragment)
2. Open the book file.
3. Select all of the files in the book.
4. Select **File > Import > Formats** to display the Import Formats dialog.
5. In the Import from Document drop-down list, click your variable control file.
6. In the Import and Update section, click the Deselect All button to uncheck all the boxes. (The button label toggles between Deselect All and Select All.)
7. Check only the Variable Definitions checkbox.
8. Click the Import button to import variable definitions from the selected file into all of the files in the book.
9. Close the variable control file without saving changes.

You could accomplish the same thing by modifying the variables in a chapter file and importing the formats from that file to the others in the book. Maintaining a separate variable-only file makes it easier, though, to segregate the changes you want to make (variable definitions) from any other changes that might have crept into the other chapter files.

Tip   To avoid importing settings you don't want, you can create a MIF fragment that contains only the variable definitions instead of using a full-blown FrameMaker file. For details, see Appendix F, "Maker Interchange Format."

Chapter 6: Formatting text with paragraph tags

FrameMaker provides several different ways to format text. *Paragraph styles* (commonly referred to as paragraph tags in FrameMaker) let you save a group of formatting choices and apply them in one step. In a single, named paragraph tag (for example, "Body" or "ChapterTitle"), you can store information including font name and size, tabs, indents, lines above or below, and hyphenation settings.

Here's a look at what's in this chapter:

The purpose of paragraph tags	109
Applying paragraph tags	110
Using the paragraph catalog to apply a tag	110
Using the formatting bar	111
Selecting a menu choice	112
Using Smart Insert for paragraph formats	112
Avoiding formatting overrides	112
Modifying paragraph tags	113
Paragraph Designer settings	114
Creating paragraph tags	126
Renaming paragraph tags	127
Setting properties across the entire paragraph catalog	128
Updating selected paragraphs globally	128
Deleting paragraph tags	129
Autonumbering details	130
Basic autonumbering	131
Numbered steps (1, 2, 3 and a, b, c)	132
Incorporating chapter and volume numbers	134
Creating numbered headings	136
Marking the end of a story	137

The purpose of paragraph tags

When you apply the Body paragraph tag to a paragraph, all of the stored settings are applied, which is much more efficient than assigning each item individually.

Efficiency, however, is only half the story. Using paragraph tags ensures consistency across an entire document. Every time you apply the Body tag, FrameMaker assigns the same formatting. The result is that all of your body paragraphs use identical settings. In a two-page document, this is nice. In a 200-page document, it's critical. When you update the Body tag's settings, FrameMaker immediately updates every body paragraph in the document.

When you assign a paragraph tag, it's applied to the paragraph where the cursor is located or to the paragraph(s) selected; you do not have to select the entire paragraph. Unlike other applications, FrameMaker *will not* apply a paragraph tag to a portion of a paragraph (a few words or characters), paragraph tags are always applied to the entire paragraph. If you need to assign special formatting to a few words or characters within the paragraph, use character tags. These are discussed in Chapter 7, "Formatting text with character tags."

Applying paragraph tags

You can apply paragraph tags in a number of ways, including:

- Using the Paragraph Catalog
- Using a Smart Insert shortcut
- Using the Formatting Bar
- Selecting a menu choice
- Using the right-click pop-up menu

Note   You can also use the Paragraph Designer to apply paragraph tags, but because you could inadvertently change the paragraph tag definition, I do not recommend this approach.

Using the paragraph catalog to apply a tag

Paragraph tags are listed in the Paragraph Catalog. If not currently displayed, select **Window > Pods > ¶ Catalog**.

To apply a paragraph tag using the Paragraph Catalog, follow these steps:

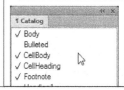

1. In the text, click in a paragraph or select a portion of a paragraph.
2. In the Paragraph Catalog, click a paragraph tag. The selected tag is applied to the current paragraph.

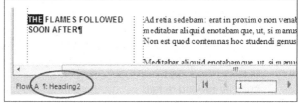

The lower-left corner of the status bar shows the paragraph tag name.

Deleting paragraph styles from the Paragraph Catalog

Paragraph style settings are stored both in the catalog, and in each instance of text containing the style. If you have unneeded styles in your paragraph catalog, you can either hide them from view or remove the definitions from the catalog entirely. To delete a style from the catalog, follow these steps:

1. Click the Delete… button in the Paragraph Catalog. The Delete Formats from Catalog dialog box is displayed

Caution   Do not click first on the paragraph style to be deleted. Clicking first on the style to delete will apply the paragraph style to the currently selected paragraph, and will not have any effect on the deletion process.

Applying paragraph tags

2. Select the paragraph style you want to delete. Multiple selections in this dialog are not allowed—you must delete each tag one at a time.
3. Click the Delete button.
4. Repeat for any additional styles to delete.
5. When finished, click the Done button.

Using the Paragraph Catalog options

A few organizational tools are available within the Paragraph Catalog. To explore the most useful of these options, follow these steps:

1. Click the Options button in the Paragraph Catalog to display the Set Available Formats dialog.
2. Select one of the options below:

   - Click the Show used before unused checkbox to force unused paragraph styles to the display after those currently used in the document.
   - Click the Delete all unused formats to quickly remove legacy tags from your catalog.
   - Click the Edit button in the Customized List option to create a specific list of paragraph tags to display. This might be useful if you wanted to hide template designer tags from authors, or if you have authors with only specific tasks to perform on a document. Move any undesired tags into the Don't Show column and click the Set Button.
3. Click the OK button.
4. Repeat steps 1 through 3 as needed for other tasks.

Using the formatting bar

You can use the drop-down list on the Formatting toolbar to apply paragraph tags. The list of tags is identical to the list in the Paragraph Catalog.

To apply paragraph tags using the Formatting Bar, follow these steps:

1. Display the Formatting Bar by selecting **View > Toolbars > Paragraph Formatting**.
2. In the text, click in a paragraph.
3. In the Formatting Bar, select a tag in the drop-down list.

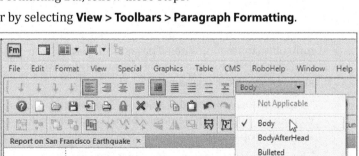

111

Selecting a menu choice

To apply paragraph tags using a menu choice, follow these steps:

1. In the text, click in a paragraph.
2. Select **Format > Paragraphs**, then select a paragraph tag from the list that is displayed in the submenu.

Using Smart Insert for paragraph formats

You can use keyboard shortcuts to assign paragraph tags. To do so, follow these steps:

1. In the text, click in a paragraph.
2. Press CTRL+9 or F9 to display the Smart Insert dialog.
3. Navigate to the tag you want by doing any of the following:
   - Use the arrow keys to scroll through the list of available tags.
   - Type the name of the tag until the tag is selectable.
   - Type the first few letters of the tag name.
4. Press ENTER.

All versions of FrameMaker allow for some level of insert using the CTRL+9 or F9 keys. Play with the options to see how your version may differ from FrameMaker 2015.

Tip   You can use the ESC J J command to apply the last paragraph style used.

Note   Some of the keys on your computer may be specially programmed. If so, one or more keyboard shortcuts may not work as described. Consult your system documentation for help.

Avoiding formatting overrides

Paragraph tags give you complete control over the appearance of the text in your document. When you update paragraph tag settings, FrameMaker automatically applies those changes to every instance of that paragraph in the entire document.

Formatting overrides occur when the tag settings are changed for a single paragraph. Overrides cause inconsistency in the document and make it more difficult to maintain. When a paragraph has an override, the status bar usually indicates this with a star in front of the tag name.

To quickly locate overrides, use the Find/Change command and select Paragraph Format Override from the Find drop-down.

The fastest way to remove overrides is to reapply your template and tell FrameMaker to remove all formatting overrides (see "Importing settings from a template file" on page 91).

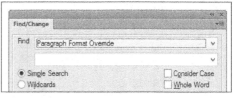

A better approach is not to introduce overrides at all. When you apply paragraph tags from the Smart Insert dialog or the Paragraph Catalog, you are less likely to introduce overrides. Problems occur when you go into the Paragraph Designer and make changes "just this once" to customize the paragraph for a special formatting requirement. Applying font, font size, or font style attributes from the Format menu will also create an override; avoid these if at all possible. Similarly, assigning tabs by dropping them on the ruler creates overrides. You can make tabs part of the tag definition by setting the tab in the ruler and then clicking the Update All button in the Paragraph Designer. The Bold, Italics, and Underline buttons on the Formatting Bar are also overrides; use character tagging instead.

Modifying paragraph tags

Applying paragraph tags is useful, but it probably won't be long before you need to modify the tags. You can make changes in the Paragraph Designer.

Caution In many companies, templates are set up and maintained by a standards group or template designer. If you have an official template, you're probably not supposed to make changes to it—unless, of course, *you* are the template designer.

To make changes to an existing tag, follow these steps:

1. Display the Paragraph Designer by selecting **Format > Paragraph > Designer** or by pressing CTRL+M or ESC O P D.
   You can also CTRL+CLICK a paragraph tag in the Paragraph Catalog to display the Paragraph Designer for that tag.

2. For each of the eight property sheets, change the properties as needed, then click the Update All button. You can move from sheet to sheet using the buttons at the top of the window but if you've made changes, you'll need to click Update All before changing sheets.

   The Update All button applies your change to all of the existing paragraphs that use the current tag. It also saves the changes into the Paragraph Catalog, so that any paragraphs you create in the future will use the same formatting.

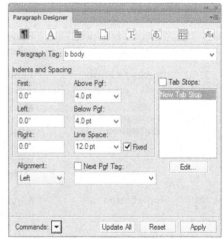

Note Clicking the Apply button instead of the Update All button creates an override, which is not recommended.

The following sections describe each setting in the Paragraph Designer.

Paragraph Designer settings

Tag settings are stored in the Paragraph Designer's eight property sheets. To navigate from one sheet to the next, click the tab at the top of the pod. The sheets are as follows:

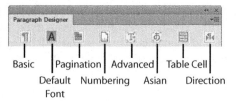

Options available on every sheet

The items on the bottom of the Paragraph Designer are displayed on every sheet. Table 6-1 lists these properties.

Table 6-1. Paragraph Designer Common Properties

Item	Description
Paragraph Tag drop-down list	Lists the paragraph tags defined in the catalog.
Commands pop-up menu	Provides advanced commands, such as global updates (see "Updating selected paragraphs globally" on page 128).
Update All button	Applies the changes you've made to all paragraphs that use the current tag and updates the Paragraph Catalog. Use the Update All button whenever you want to make a global change.
Reset button	If you change properties and then decide not to update the format, you can reset the properties by clicking this button. You can also choose Reset Window From Selection from the Commands menu in the designer. Selecting the Apply button will disable the Reset button.
Apply button	Applies the changes you've made to the current selection. Use the Apply button only when you want to create an override.

Basic sheet

The options on the Basic sheet (See "Modifying paragraph tags" on page 113) let you control the paragraph's positioning, such as indents and space above and below the paragraph. You'll find additional positioning choices on the Pagination sheet.

Table 6-2. Basic Sheet

Item	Description
Indents	All indents are measured from the edge of the text frame, not from the edge of the paper. (The text frame is represented by a dotted line surrounding the "live" area where you can insert text.) See "Setting up indentation" on page 117 for details. • **First:** Sets a different left margin for the first line of the paragraph. • **Left:** Sets a left margin for the second and subsequent lines of the paragraph. This indent does not affect the first line. • **Right:** Sets the right margin for each line of the paragraph.
Alignment	Sets the alignment—left, center, right, or justified—for the paragraph.
Space Above Pgf	Sets the amount of white space between the current paragraph and the preceding paragraph. To determine spacing between paragraphs, FrameMaker uses the larger of the space above the current paragraph or the space below the previous paragraph. They are *not* added together. The Space Above setting is ignored for a paragraph at the top of a page. If you need to move a paragraph down from the top of the page, either move the text frame down or use a Frame Above to provide spacing (see "Advanced sheet" on page 124 for details). Negative values are valid, which allows some interesting effects with the Frame Above Pgf and Frame Below Pgf.
Space Below Pgf	Sets the space below the paragraph; see previous entry for how total space between paragraphs is calculated.
Line Spacing	Sets the *leading*—the vertical space between lines inside the paragraph. When you change the font size (on the Default Font properties), the leading automatically changes (approximately 15 to 20 percent more than the font size value), so set the font size before setting the line spacing.

Fixed	When checked, this option sets line spacing from baseline to baseline. (The *baseline* is the bottom of a line of text, not including characters that drop below this invisible line. The letters g, p, q, and y are all *descenders;* they extend below the baseline.) If not checked, line spacing is adjusted to accommodate superscripts and subscripts. Compare the following two paragraphs:
	• This bullet uses fixed line spacing.
	  Superscripts don't have much[1] of an effect.
	· In this bullet, line spacing is not fixed.
	  The superscript does make a[1] difference here.

Table 6-2. Basic Sheet (Continued)

Item	Description
Tab Stops	Lets you set tab stops for the paragraph. In FrameMaker, paragraphs do not have tab stops by default. Until you add tab stops, pressing the TAB key in a paragraph has no visual effect. See the following section, "Setting tab stops," for details on how to control the tab settings.
Next Pgf Tag	Sets the default tag for the next paragraph tag. When you press ENTER to start a new paragraph, this choice determines what paragraph tag is set by default. For example, a Heading tag could default to Body as the next tag. You can change the Body tag to something else later. If you do not specify a setting for the next paragraph tag, FrameMaker uses the same tag as the current paragraph.

Note In FrameMaker, you can specify the unit of measure used throughout the document—in paragraph tags, text frame properties, table tags, and so on. To change the unit of measure, select **View>Options** and click a different item in the Display Unit drop-down list.

Setting tab stops. FrameMaker's tab stops are measured from the edge of the text frame, not the edge of the page. By default, paragraphs do not have any tab stops defined; until you create them, pressing the TAB key doesn't have any visual effect. FrameMaker's tab stops act a little differently than other applications. For example, if a paragraph has tab stops defined at 1 inch and 5 inches, two tab characters will still be needed if the tab occurs beyond 1 inch from the edge of the text frame.

To set up tab stops, follow these steps:

1. In the Paragraph Designer's Basic sheet, make sure that New Tab Stop is selected in the Tab Stops section. (If a specific tab location is selected, the following steps will change the selected tab, rather than create a new tab stop.)
2. Click the Edit button in the Tab Stops section to display the Edit Tab Stop dialog box.

3. In the New Position field, type the position you want to use for the new tab stop.
4. In the Alignment section, specify one of the following:
   - **Left:** Text after the tab is left-aligned starting at the tab's position.
   - **Center:** Text after the tab is centered at the tab's position.
   - **Right:** Text after the tab is right-justified from the tab's position.
   - **Decimal:** Text after the tab is aligned so that the decimal point (a period by default in the English version) is aligned on the tab. To align based on a different character, type that character in the Align On field.
5. In the Leader section, specify whether you want any characters preceding the tab. Most often, you use "dot leaders" in tables of contents. To create a different leader, type one or more characters, spaces or symbols in the Custom field.

Modifying paragraph tags

6. If you want to create several tabs at once, specify at what increments you want to repeat the tab in the Repeat Every field.
7. Click the Continue button to return to the Paragraph Designer. Note the Delete and Delete All buttons in the Edit Tab Stop dialog box in case you made mistakes.
8. Click the Update All button to save your changes in the paragraph definition.

Note If you prefer to use the ruler to create tabs, set up the tab in the ruler at the top of the window (click and drag to position a tab on the ruler), display the Paragraph Designer, and then click the Update All button to save the tab in the paragraph definition.

Setting up indentation. Getting your indents positioned correctly can be confusing because FrameMaker calculates indentation differently from most other applications. All indents are calculated from the edge of the text frame (not the edge of the page). The Left indent setting is applied only to the second and subsequent lines of a paragraph, never to the first paragraph—the first line uses the First indent setting. The following illustration shows a few examples of the interaction between the First and Left indent settings.

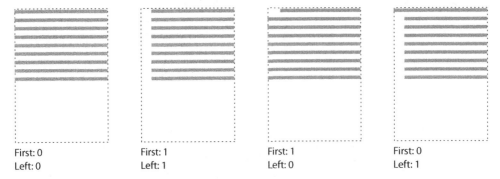

First: 0 First: 1 First: 1 First: 0
Left: 0   Left: 1   Left: 0   Left: 1

The Right indent always applies to all lines in the paragraph and is measured from the right side of the text frame.

Default Font sheet

The Default Font sheet lets you specify font, font size, and other typographical controls.

The following table lists the Default Font sheet items.

117

Table 6-3. Default Font Sheet

Item	Description
Family	Sets the base typeface, such as Garamond, Palatino, Helvetica, or Times.
Size	Sets the size of the type. The smallest allowed value is 2 points.
Angle	Sets the angle, such as Italic or Oblique.
Weight	Sets the weight, such as Bold, Black, Medium, or Light.
Variation	Some fonts have variations available. These might be choices such as Small Caps or Expert Numbers.
Color	Sets the color of the text. The choices come from the color catalog. See Chapter 24, "Color output," to learn more about the color catalog.
Spread	Adjusts the space between letters. The default is 0%. Increasing this value increases the space between letters; a negative value reduces the space between letters. Spread is similar to tracking in other applications.
Stretch	Adjusts the width of the letters to the specified percentage. The default is 100% (normal shape). A stretch of 200% results in letters that are twice their normal width. A condensed or expanded font is usually a better choice than using the Stretch attribute.
Language	Determines which dictionary is used when the paragraph is hyphenated or spell-checked. To prevent spell-checking, set the Language to None. A setting of None is useful for paragraphs that contain code samples.
Underline drop-down list	• **Underline:** Applies a line under the text. • **Double Underline:** Applies a double line under the text. • **Numeric Underline:** Similar to a regular underline, but spaced a little farther from the text.
Overline	Applies a line over the text.
Strikethrough	Applies a strikethrough line (a line through the middle of the text).
Change Bar	Applies a vertical line in the margin of the document. You can create automatic change bars; for details, see "Using change bars" on page 58.
Superscript	Sets text above the baseline.
Subscript	Sets text below the baseline.

Table 6-3. Default Font Sheet (Continued)

Item	Description
Capitalization drop-down list	All of the capitalization choices change the appearance of the text but preserve the capitalization of the underlying content. For example, you can display the text as ALL CAPS, but a cross-reference or TOC would display whatever content had been entered originally. • **Small Caps:** This option does not use any small cap glyphs found in a font. Rather, it scales standard glyphs according to specs in the Text options. For more information, see "Formatting superscripts, subscripts, and small caps" on page 54. • **Lowercase:** Sets the text in lowercase letters. • **Uppercase:** Sets the text in uppercase letters.
Pair Kern	Adjusts the spacing between each pair of letters to improve readability. In general, Pair Kern should be turned on for most paragraphs.
Tsume	Excerpted from the FrameMaker 9 help system: To move a Japanese character closer to the characters next to it, select Tsume. The amount of space a variable-width character, such as a parenthesis, can move is determined by the metrics for that character.
Background Color	Creates a block of color behind the text. Spaces show the background color, tabs do not.

Pagination sheet

The Pagination sheet sets the paragraph's position on the page. For example, you can set a particular paragraph style to start at the top of a page.

The following table lists the Pagination sheet items.

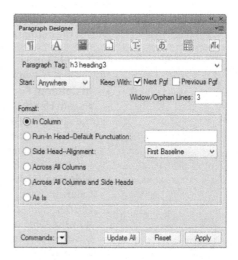

Table 6-4. Pagination Sheet

Item	Description
Start	Lets you set where on the page the paragraph begins. • **Anywhere:** Starts in the next available space. This would be more accurately labeled "Anywhere it fits". • **Top of Column:** Moves the start of the paragraph to the top of the next column. (In a single-column document, this is the same as Top of Page.) • **Top of Page:** Starts at the top of the next page. • **Top of Left Page:** Starts at the top of the next left page. • **Top of Right Page:** Starts at the top of the next right page. (This setting is very common for chapter titles, which are often required to be on a right page.)
Keep With	Controls how the paragraph interacts with the preceding and following paragraphs. Use with discretion, as too many "Keep With" settings can dramatically impact pagination. • **Next Pgf:** If checked, the current paragraph is attached to the next paragraph. Useful for preventing "widowed" headings, or headings that separate from their content. The Keep With setting forces at least the number of lines specified in the Widow/Orphan Lines setting to be placed on the same page as the next paragraph. • **Previous Pgf:** If checked, the current paragraph is attached to the previous paragraph. Good for second or third level lists.
Widow/ Orphan Lines	Sets the minimum number of lines that can appear before or after a page break. For example, if you type **2**, a paragraph must always have at least two lines at the bottom of the page or at the top of the page.
Format	Determines the paragraph's general positioning in relation to the text column, side head, and following paragraph. See Figure 6-1 on page 121 for examples. • **In Column:** Positions the paragraph in the main text column. • **Run-In Head:** Positions the paragraph so that the next paragraph starts on the same line. Useful for glossary entries and for lower-level headings. In the Default Punctuation field, you can specify any punctuation you want after the heading, such as a period. In this book, the fourth-level headings (such as "Setting tab stops" on page 116) use a run-in heading. • **Side Head:** Positions the paragraph so that it is on the same line as the following paragraph but in a separate column. Requires a side head area (discussed in Chapter 12, "Text flows"). The drop-down list lets you choose how the side head is aligned in relation to the next paragraph: First Baseline, Top Edge, or Last Baseline. • **Across All Columns:** Spans all text columns (but not the side head area). • **Across All Columns and Side Heads:** Spans all columns and the side head area. • **As Is:** Equivalent to "Ignore", this setting is useful only when you are performing global updates ("Setting properties across the entire paragraph catalog" on page 128).

The following figure illustrates the available pagination format options.

Figure 6-1. Understanding pagination format options

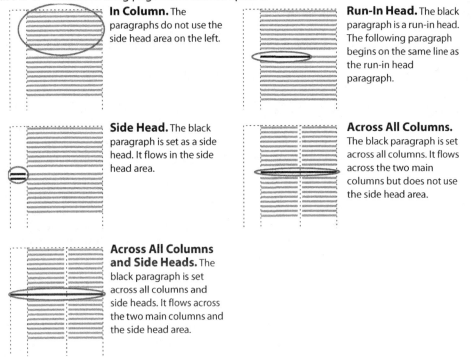

In Column. The paragraphs do not use the side head area on the left.

Run-In Head. The black paragraph is a run-in head. The following paragraph begins on the same line as the run-in head paragraph.

Side Head. The black paragraph is set as a side head. It flows in the side head area.

Across All Columns. The black paragraph is set across all columns. It flows across the two main columns but does not use the side head area.

Across All Columns and Side Heads. The black paragraph is set across all columns and side heads. It flows across the two main columns and the side head area.

Numbering sheet

The Numbering sheet lets you set up automatic numbering for your paragraph. This includes step numbers, chapter numbers, bullets, and repeated words (such as "Note:") at the beginning (or end) of a paragraph.

Table 6-5 describes the Numbering sheet options.

Table 6-6 on page 122 lists the building blocks you can use in the Autonumber Format field.

Table 6-5. Numbering Sheet

Item	Description
Autonumber Format	Sets the definition of the numbering for the paragraph. The Autonumber Format checkbox determines whether the autonumbering definition (typed in the field that follows the checkbox) is used. See page 130 for a full discussion.
Building Blocks	Lists the most commonly used numbering blocks used in the Autonumber Format field. See Table 6-6 for a description of each building block. You can also type the building blocks instead of selecting them from the list. Notably missing from the list is the < =0> reset code, which is helpful in complex numbering schemes.
Character Format	Displays the character tag used to format the automatically generated number. (the number portion, not the entire paragraph) If the style indicated isn't available, you must create the character tag, or the Character Format setting will be ignored. See Chapter 7, "Formatting text with character tags," for details.
Position	Determines whether the autonumber occurs at the beginning or the end of the paragraph. Steps, bullets, and the like are normally at the beginning of the paragraph. You might use the end of paragraph position for a graphic "bug" that marks the end of a section. The autonumber is always right-aligned when positioned at the end of a paragraph.

Using counters, you can create several different types of numbering—numeric, alphabetic, and Roman. The counters come in groups of three, where the three members of each group have these functions:

- Display current value (for example, <n> or <a>)
- Increment counter and display new value (for example, <n+> or <a+>)
- Set or reset counter (for example, <n=1> or < =0>)

Table 6-6. Building Blocks for Autonumbering

Name/Function	Counter	Description
Bullet	• or \b	Inserts a bullet.
Tab	\t	Inserts a tab. (You cannot use the TAB key inside a dialog.) In the Basic sheet of the Paragraph Designer, specify tab position, type, leader, and the hanging indent. (see "Setting tab stops" on page 116)
Non-breaking space	\	Type backslash-space to insert a nonbreaking space.

Table 6-6. Building Blocks for Autonumbering (Continued)

Name/Function	Counter	Description
Displays the current value	<n>	Numeric value (1, 2, 3)
	<a>	Lowercase alphabetic value (a, b, c)
	<A>	Uppercase alphabetic value (A, B, C)
	<r>	Lowercase Roman value (i, ii, iii)
	<R>	Uppercase Roman value (I, II, III)
Increments by 1 and displays the new value	<n+>	Numeric value (1, 2, 3)
	<a+>	Lowercase alphabetic value (a, b, c)
	<A+>	Uppercase alphabetic value (A, B, C)
	<r+>	Lowercase Roman value (i, ii, iii)
	<R+>	Uppercase Roman value (I, II, III)
Sets to the specified value and displays that value	<n=1>	Numeric value (1, 2, 3)
	<a=1>	Lowercase alphabetic value (a, b, c)
	<A=1>	Uppercase alphabetic value (A, B, C)
	<r=1>	Lowercase Roman value (i, ii, iii)
	<R=1>	Uppercase Roman value (I, II, III)
Volume number	<$volnum>	Displays the current value of the volume number.
Chapter number	<$chapnum>	Displays the current value of the chapter number.
Section number	<$sectionnum>	Displays the current value of the Section, as defined in the Numbering Properties.
Subsection number	<$subsectionnum>	Displays the current value of the Subsection, as defined in the Numbering Properties.
Placeholder counters	< > (includes a space between brackets)	(angle brackets with a space) Suppresses display of a counter but does not change its value. This building block has a space between the angle brackets. This counter was common in numbering series developed in FrameMaker version 5.5 or earlier.
Reset	< =0> (a space preceeds equal symbol)	(space=0) Suppresses display of a counter and resets counter to zero. This counter isn't listed as a building block; you have to type it in the Autonumber Format field.
Asian language building blocks	(varied)	Includes zenkaku, kanji, daiji, hira, kata, full-width and Chinese characters.
Right to Left	(varied)	Includes Indic, Farsi, Hebrew, Abjad, and Alif Ba Ta

For more information on typing special characters in the Paragraph Designer, see "Entering Special Characters in dialog boxes" on page 100. See "Autonumbering details" on page 130 for several examples of autonumbering.

Advanced sheet

The Advanced sheet lets you set hyphenation controls, justification, and lines above or below the paragraph.

Table 6-7 on page 124 lists the Advanced sheet options.

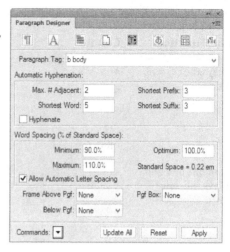

Table 6-7. Advanced Sheet

Item	Description
Automatic Hyphenation	• **Max. # Adjacent:** Sets the number of hyphens that can occur in consecutive lines in the paragraph. For example, 2 means that if you have two lines with hyphens at the end, the next line cannot be hyphenated. • **Shortest Prefix:** Sets the minimum number of letters for a hyphenated beginning fragment. For example, 3 means that the prefix must have at least three letters—"pre-fix" could be hyphenated, but "re-create" could not. • **Shortest Word:** Sets the minimum length of a hyphenated word. For example, 8 means that FrameMaker cannot hyphenate "pre-view." • **Shortest Suffix:** Sets the minimum number of letters for a hyphenated end fragment. For example, 5 means that FrameMaker cannot hyphenate "pre-tend."
Hyphenate	If checked, the paragraph allows hyphenation. That is, if a word doesn't quite fit at the end of a line, FrameMaker may break it with a hyphen (for example, "acro-batic"). Processing is based on the language that's selected on the Default Font sheet for this paragraph. If no language is selected, the paragraph is not hyphenated.
Word Spacing	Word Spacing controls the changes FrameMaker can make when justifying text in a paragraph. • **Minimum:** Sets the minimum amount of word spacing that's allowed when justifying text and avoiding hyphenation. For example, 90% means that each space between words must be no smaller than 90 percent of a normal space. • **Optimum:** Sets the target amount of word spacing. The default is normally 100%. • **Maximum:** Sets the maximum amount of word spacing allowed.

Table 6-7. Advanced Sheet (Continued)

Item	Description
Allow Automatic Letter Spacing	If checked, FrameMaker adds space between characters (not just words) when justifying text.
Frame Above Pgf	Lets you specify a graphic frame that's inserted above this paragraph. The graphic frames are stored on the reference pages. See "Placing graphics above and below the paragraph" on page 275 for details.
Below Pgf	Lets you specify a graphic frame that's inserted below this paragraph.

Asian Fonts sheet

The Asian Fonts sheet extends options for Asian font spacing and punctuation.

Table 6-8 lists the Asian Font sheet options.

Table 6-8. Asian Font Sheet

Item	Description
Western/Asian Spacing	Allows setting of minimum, maximum and optimum values as a percentage of font size.
Asian Character Spacing	Allows setting of minimum, maximum and optimum values as a percentage of font size.
Asian Punctuation	Squeeze as necessary, never squeeze, always squeeze.

Table Cell sheet

The Table Cell sheet sets properties for paragraphs inside tables.

The cell margin properties interact with the default cell margins set in the Table Designer. You can either add to (or subtract from) the default margins in the Table Designer, or you can override the Table Designer's margins entirely with the settings in these paragraphs.

These properties affect only paragraphs in a table. Table 6-9 on page 126 lists the Table Cell sheet options.

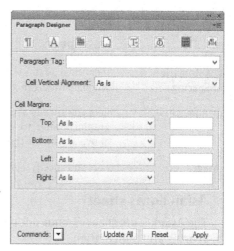

Table 6-9. Table Cell Sheet

Item	Description
Cell Vertical Alignment	Sets the vertical position of the paragraph in the table cell. The Top setting applies only to the first paragraph in a cell. The Bottom setting applies only to the last paragraph in a cell. The Left and Right settings are used for every paragraph.
Cell Margins (Top, Bottom, Left, Right)	Adjusts the cell margins, when tag appears within a table cell. You can add to the default margins (From Table Format, Plus), or you can completely ignore the default margins (Custom). Negative values allow positioning text directly at the edge of the table cell, but not beyond it.

Creating paragraph tags

The easiest way to create a new paragraph tag is to start with a tag that's close to what you need. If, for example, you want to create an indented paragraph, you could start with the Body tag—Body and Indent are probably identical except for the first and left indent settings.

To create a new paragraph tag, follow these steps:

1. Click in the paragraph needing the new paragraph tag.
2. Apply the paragraph tag most similar to what you need.
3. Display the Paragraph Designer.
4. Click the Commands button, then select New Format from the pop-up menu to display the New Format dialog.

Caution By default, the Store in Catalog and Apply to Selection checkboxes are checked. The latter is irrelevant unless you selected text before creating the tag. Make sure that Store in Catalog is checked, or the tag will not be included in the Paragraph Catalog.

5. In the Tag field, type a name for the new paragraph tag (such as Indented). Verify that the Store in Catalog checkbox is checked. If you are using the current paragraph as a

template for your new tag (and thus you don't want to change the current paragraph), then uncheck the Apply to Selection checkbox.

6. Click the Create button. FrameMaker creates a new paragraph tag. At this point, the tag's definition is identical to the tag you started with.

Tip Create a new tag faster by changing the label in the Paragraph Tag box and clicking Apply. Rename a tag by changing the label in the Paragraph Tag box and clicking Update All.

7. In each sheet of the Paragraph Designer, make the needed changes. Click the Update All button before moving from one sheet to the next. If you forget, the Apply Changes dialog box will appear.

   If you see this dialog box, do one of the following:

   - Click the Apply Changes button to apply your changes to the current selection only.
     This creates a formatting override.
   - Click the Update All button to update the tag definition and all paragraphs that use it.
   - Click the Don't Apply button to throw away your changes and go to the next sheet.
   - Click the Cancel button to discard your changes and go back to the current sheet.

8. When you have finished, close the Paragraph Designer. The new paragraph tag is now available in the Paragraph Catalog.

Renaming paragraph tags

You can rename any paragraph tag. When you rename a paragraph tag, FrameMaker does the following:

- The original tag name (A) is changed to the new tag name (B).
- Tag A is removed from the Paragraph Catalog. Tag B is displayed in the Paragraph Catalog.
- All paragraphs that use tag A now use tag B.

Note You can also make a copy of an existing paragraph tag. In this case, the old tag is retained in the Paragraph Catalog. For details, see "Creating paragraph tags" on page 126.

To rename a paragraph tag, follow these steps:

1. Display the Paragraph Designer.
2. In the Paragraph Tag drop-down list, select the paragraph tag you want to rename.
3. In the Paragraph Tag field, type the new name of the paragraph tag.
4. Click the Update All button and confirm that you want to rename the tag.

All the paragraph tags are renamed for you. Notice that the old tag name is no longer available in the Paragraph Catalog.

Setting properties across the entire paragraph catalog

Changing properties for a single paragraph tag is fairly quick, but if you need to make the same change for all of your paragraph tags, changing them one by one is tedious.

FrameMaker gives you a way of making global changes within your paragraph tags. For example, assume that your original files use GimzelFlopper as the official font, so all of your paragraph tags use GimzelFlopper. But now, the new corporate guidelines dictate that GimzelFlopper is no longer allowed. Instead, you need to switch to the new, official font MegaCorpo.

You could make this change by displaying each paragraph tag in turn, changing the font, and selecting the Update All button. But you can also do this with a global change, even though your paragraph tags use all sorts of different settings.

Caution   This is not for the faint of heart. Back up your files before trying this!

To change a paragraph property for all the tags in your document, follow these steps:

1. Display the Paragraph Designer. It doesn't matter which paragraph tag is selected.
2. Display the Default Font sheet.
3. Click the Commands button, then select Set Window to As Is from the pop-up menu (or click in the page margin outside your text frame). This sets all of the properties in the Default Font sheet to "ignore this setting." This is useful when you're trying to preserve conflicting settings in different paragraphs.
4. In the Family field, select a different font.
5. Click the Update All button (or click the Commands button, then select Global Update Options from the pop-up menu) to display the Global Update options dialog.
6. In the Use Properties in the Paragraph Designer section, click the Default Font Properties Only radio button (or the button corresponding to the properties you want to change).
7. In the Update Paragraph Formats section, click the All Paragraphs and Catalog Entries radio button.
8. Click the Update button. Every paragraph tag in your document now uses the new font.

Updating selected paragraphs globally

Along with modifying a property of all of the tags in the Paragraph Catalog, you can change specific paragraphs. To update several tags at once, use the All Matching Tags in Selection radio button in the Global Update Options dialog. Before you begin, set up several paragraphs in a row, and apply a paragraph tag you want to change to each paragraph (for example, Heading 1, Heading 2 and Heading 3). Then, follow these steps to modify the tags globally:

1. Select the paragraphs you created. Every paragraph tag in this selection will be changed, so you select only tags you want to change.

2. Display the Paragraph Designer. Notice that where the various paragraph tag settings conflict, the As Is setting is used.
3. For each sheet of the Paragraph Designer, follow these steps to make changes:

    a. Modify the sheet as needed. If you want the paragraphs to retain different settings for an item, leave the As Is setting intact.

    b. Click the Commands button, then select Global Update Options from the pop-up list. This displays the Global Update Options dialog.

    c. In the Use Properties in the Paragraph Designer section, select the current sheet only (in our example, Default Font Properties Only).

    d. In the Update Paragraph Formats section, click the All Matching Tags in Selection radio button.

    e. Click the Update button.

Deleting paragraph tags

You can remove paragraph tags from the catalog. After deleting, these tags are removed from the Paragraph Catalog, but not from paragraphs that tag still have the tag applied to it. Now, however, the paragraph appears to have an override. After you delete the tag, you should apply a different tag to these "orphaned" paragraphs. Alternatively, you can reassign a new tag to the paragraphs that use the tag you plan to delete. It's usually more efficient to apply the new tag before you delete a paragraph tag.

To delete paragraph tags, follow these steps:

1. Display the Paragraph Catalog.
2. Click the Delete button at the bottom of the Paragraph Catalog. to display the Delete Formats from Catalog dialog.

3. For each tag you want to delete, select the style, then click the Delete button. To delete all of the tags, press and hold ALT+E until all of the tags are gone.
4. When you have finished, click the Done button.

Caution   If you make a mistake, you must click the Cancel button and start over.

After removing the paragraph tag, you'll probably want to assign new names to those paragraphs (see "Creating New Formats" on page 130).

To locate paragraph and character format overrides, use the Find/Change pod and note the new options provided with FrameMaker 11 and later.

New options introduced in FrameMaker 11:

- Paragraph Format Override
- Character Format Override
- Table Format Override
- Object Style Tag
- Object Style Format Override

Tip Users of FrameScript will find Quatro's FindChangeFormatsBatch script useful and worth many times the investment.

Creating New Formats

If you inherit FrameMaker files riddled with overrides, paragraph and character overrides may be the rule and not the exception. In this case, you may want to identify all the overrides. FrameMaker provides a way to do this—select **File > Utilities > Create & Apply Formats**.

The Create & Apply Formats feature does the following for paragraph and character tags:

- If a tag is used by a paragraph or character but doesn't appear in the appropriate catalog, Create & Apply Formats adds it to the catalog.
- If a paragraph or character uses a tag that's in the catalog, but the tag is used with an override, Create & Apply Formats creates a new tag for that item and adds it to the catalog. The tag name is based on the original name, so if the tag was Code, then any new tags would be Code1, Code2, and so on.
- IMPORTANT (AND VERY COOL): If an override exists, but there is an appropriate tag in the catalog, FrameMaker will apply the tag to the text.

If new tags are created, you can rename them or apply the correct tags globally to remove the overrides.

Autonumbering details

FrameMaker's autonumbering lets you set up automated numbering, such as step numbers and chapter numbers. But you can also use autonumbering for repeated text (Note:) or symbols (•) at the beginning or end of the paragraph. It's much more efficient to define a paragraph tag that automatically inserts these items because you don't have to type in the text over and over again or keep track of the last number used.

Autonumbering details

Note Because autonumbering is formatting, applied via a paragraph tag, you can neither select it nor delete it. To remove autonumbering, either redefine the paragraph tag or apply a different (non-numbered) paragraph tag.

This section describes how to set up autonumbers with varying degrees of complexity. In some cases, the autonumbers interact with other tags, so you have to set up several paragraph tags to get the result you want.

Some of the examples build on concepts introduced in the early examples, so I recommend that you read this entire section through the first time.

Basic autonumbering

As mentioned in "Numbering sheet" on page 121, you use the Autonumber Format for bullets and text-based formats, such as notes, cautions, and warnings. To set up these autonumbers, you insert the text you need in the Autonumbering Format field. When you apply the paragraph tag, the autonumber codes and text are inserted automatically.

Bullets

FrameMaker includes a bullet in the list of building blocks. You can also use other characters, such as hyphens, em dashes, or wingdings, to create bullets.

To create a basic bullet, click the bullet building block, then click the tab (\t) building block to insert the following in the Autonumbering Format field:

•\t

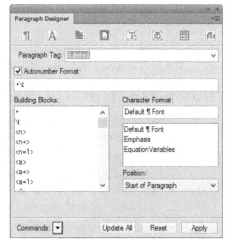

For more information on typing special characters in the Paragraph Designer, see "Entering Special Characters in dialog boxes" on page 100.

Next, set up the Basic properties to allow for the tab and a left indent. The definition and the result are shown in the following illustration.

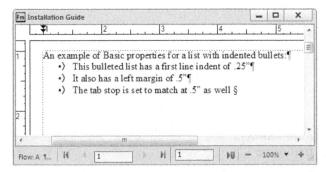

Another way to set up bullets involves taking advantage of the character formatting that's available. You can, for example, create a character tag that applies the Zapf Dingbats or Wingdings font, and then use one of the dingbats or wingdings to make a special bullet (such as check marks, diamonds,

131

or pointing hands). Because symbols are mapped to ASCII characters, when you enter the character in the Autonumber Format field, no formatting is displayed. However, when you apply the paragraph tag to text, the character is displayed using the font from the character tag you selected in the Character Format field. For an example of placing "Smiley Face" bullets, see Figure 6-2.

Figure 6-2. Creating a Special Bullet

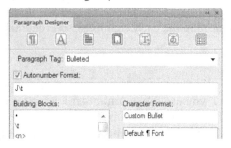 

Notes, cautions, and warnings

You can set up notes and similar formats with autonumbering.

> **Note:** This is a note. It uses autonumbering so that you don't have to type in the word "Note:" and apply formatting for each occurrence.

The previous example uses a character tag to apply formatting only to the autonumber portion of the paragraph, and Basic settings for the hanging indent.

Numbered steps (1, 2, 3 and a, b, c)

When you create step formats, you must tell FrameMaker when to increment (or count) steps and when to reset them. There are two ways of doing this:

- You can use paired step formats: In FrameMaker's default paragraph catalog (in a document created by selecting **File > New > Document > Use Blank Paper: Portrait**) these tags would be called Numbered1 and Numbered. Numbered1 resets your steps to start at 1 and is always used for the first step in a series. Numbered is used for all subsequent steps.
- You can create a single step format, Step. In this approach, you must use another format, such as Body, Body Intro, or Heading, to reset your step series.

Working with paired step formats

The advantage of using a two-step approach is that you explicitly control how the steps are reset. The disadvantage is that you have to remember to reset the step list. Rearranging the steps may require different paragraph tag assignments to ensure that step (or list) paragraph tag is applied correctly. Here is another example of two-step autonumbering:

Paragraph Tag	Autonumber Definition
StepFirst	S:<n=1>
StepNext	S:<n+>

The S: is called a *series label*. It tells FrameMaker that the counters in the two paragraph tags are related, and isolates the autonumbers from other series in your document. You can use any character for a series label, but all tags related to the series must have the same series label. You might use S: for steps and F: for figures, but neither could call upon autonumbers from a separate series.

Tip An autonumber format without a series label is considered part of the unlabeled series. Unlabeled series are treated as any other series, and can conflict with each other.

Fortunately, the <$chapnum> and other variables are not defined in a paragraph format autonumber. If the series labels match, then the <n=1> and <n+> counters will affect each other and the result looks like this:

Paragraph Tag	Autonumber Definition	Result
StepFirst	S:<n=1>	1
StepNext	S:<n+>	2
StepNext	S:<n+>	3
StepNext	S:<n+>	4
StepFirst	S:<n=1>	1
StepNext	S:<n+>	2

Working with paired formats, if you need second-level steps (a, b, c) in addition to the first-level numbered steps, you can set up two more tags, StepLetterA and StepLetterNext. Since we're using the two-step method, we'll use a different series label for these substeps as well.

Paragraph Tag	Definition
StepLetterA	T:<a=1>
StepLetterNext	T:<a+>

To set up similar numbering with Roman numerals, use the <r> and <R> building blocks for lowercase and uppercase Roman numerals.

As you can see, manually setting and restarting your numbering can be tedious, and introduces potential for operator error. See the next section for my recommended single step format numbering scheme.

Working with less step formats

If you want to minimize the number of paragraph tags in your document, you can use a single step format. In this case, you need to use other paragraph tags to reset your steps. This approach requires use of a separator paragraph (Body, in the example) to reset numbering between your lists. As with the paired step approach, if you don't use the reset format between lists, the numbering won't work properly.

Note Although the single-step method is slightly more complicated to set up, it incorporates principles of content modeling and *"pseudo-structure"* which generally pay big dividends in the long run.

Set up the paragraphs as shown here:

Paragraph Tag	Autonumber Format
Step	S:<n+>
Body	S:< =0>

Here, the Step tag is set to <n+>, increasing each time you use it, and the S: series is reset to 0 each time a Body paragraph is used. Don't forget the space!

Tip Here is a little-known autonumbering feature:
When no counters are listed in the definition, they are reset to zero!

You could use a similar approach for your substeps, by assigning the StepA reset function to the Step paragraph tag.

Paragraph Tag	Autonumber Format
Step	S:<n+>< =0>
Body	S:< =0>
StepA	S:< ><a+>

To ensure that the Step tag resets the StepA tag, you must use the same series label for those tags. But at the same time, the StepA tag should not impact the Step numbering, so you have to insert a blank counter < > (with a space between the brackets) at the first position. I call this < > building block a *placeholder*. This tells FrameMaker not to display a counter at that position but to preserve the value of that counter.

Incorporating chapter and volume numbers

In FrameMaker version 5.5, Adobe introduced two new system variables: <$chapnum> and <$volnum>. Starting with FrameMaker 11 we have <$section> and <$subsection> variables. These are a departure from paragraph-based autonumbering. In effect, these variables are *global* building blocks—they are available in autonumbers, cross-references, variables, and several other places. Within an autonumber, <$chapnum> returns the current value of the chapter number and <$volnum> returns the value of the volume or part number. These are normally set in your book file (see "Managing numbering" on page 332 for details).

Autonumbering details

Unlike other building blocks, you can only change the value of these items at the document or book level. Furthermore, you can use them in several different autonumbering definitions that are *not* tied together with a series label, along with using them on master pages and reference page items like TOC and IX generation.

To reference the chapter number (for example, in your chapter title paragraph tag), you can use the <$chapnum> building block in your paragraph tag autonumber.

The value and display of these variables is defined in the Numbering Properties dialog box, which you can access by selecting **Format > Document > Numbering**.

You can the value for <$chapnum> inside the file, but a book setting will override the value set in the file. The rules for <$volnum> are the same; it uses the Volume sheet on the Numbering dialog.

The example that follows shows how you can take advantage of <$chapnum> to set up independent figure, table, and example numbering that uses the chapter number.

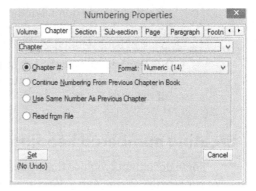

Paragraph Tag	Definition	Result
ChapterTitle	<$chapnum>:	1: , 2: , 3:
FigureTitle	F:Figure\ <$chapnum>\+<n+>	Figure 1-1, Figure 1-2, Figure 1-3
TableTitle	T:Table\ <$chapnum>\+<n+>	Table 1-1, Table 1-2, Table 1-3
ExampleTitle	E:Example\ <$chapnum>\+<n+>	Example 1-1, Example 1-2, Example 1-3

The backslash-space sequence produces a nonbreaking space; backslash-plus produces a nonbreaking hyphen. Using these special characters prevents unattractive line breaks. The nonbreaking spaces and hyphens are also picked up by any cross-references to these items.

> ### Formatting Ideas for Paragraph Tags
>
> FrameMaker's Paragraph Designer offers some standard formatting and interesting customization options. Here are some ideas:
>
> - **Inserting lines above or below a paragraph.** On the Advanced sheet, use the Frame Above and Frame Below settings to set lines or other graphics above or below your headings. If you don't like the provided lines, go to the reference pages and create your own graphic frames (see "Placing graphics above and below the paragraph" on page 275 for details).
> - **Forcing headings to the top of a page.** On the Pagination sheet, use the Start drop-down list to specify that a particular heading must always start at the top of the page.

Creating numbered headings

Some documents require that you number your headings and perhaps even the body paragraphs. FrameMaker autonumbering can handle this situation for you. You use the higher-level headings to restart the lower-level headings. The following example shows a typical setup.

Paragraph Tag	Definition	Result
ChapterTitle	H:<$chapnum>	1, 2, 3
Heading1	H:<$chapnum>.<n+>	1.1, 1.2, 1.3, 1.4
Heading2	H:<$chapnum>.<n>.<n+>	1.1.1, 1.1.2, 1.1.3, 1.1.4
Body	H:<$chapnum>.<n>.<n>.<a+>	1.1.1.a, 1.1.1.b, 1.1.1.c, 1.1.1.d, 1.1.1.e

In this example, the ChapterTitle tag carries a series label because it resets the second counter in the Heading1 format to zero.

You might also want to set up numbered headings for an outline, as shown in the following example.

Paragraph Tag	Definition	Result
Heading1	H:<R+>	I, II, III, IV
Heading2	H:<R>.<A+>	I.A, I.B, I.C, I.D
Heading3	H:<R>.<A>.<r+>	I.A.i, I.A.ii, I.A.iii, I.A.iv
Heading4	H:<R>.<A>.<r>.<n+>	I.A.i.1, I.A.i.2, I.A.i.3, I.A.i.4

Marking the end of a story

In many magazines, the end of a story is indicated by a small graphic "bug," which gives the reader a visual indication that the article is done. This is very helpful when the story jumps over pages of advertisements and the like. You can automate the insertion of this graphic by setting up a special paragraph tag for it. You might, for instance, create a paragraph tag called BodyEnd. In it, you could set up the autonumbering format to insert the graphic. You may need to apply a character tag to use a font such as Zapf Dingbats.

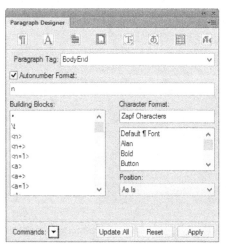

More Formatting Ideas for Paragraph Tags

- **Formatting your glossary with run-in headings.** Use run-in headings (set on the Pagination sheet) for glossary terms; use body text for the definitions. Set the glossary term in bold. (This also makes it easier to cross-reference the glossary later because the term is in its own paragraph. If you use a single paragraph tag and bold the glossary term, you cannot pick up just the term in a cross-reference.)
- **Eliminating spell-checking.** For text that should not be spell-checked (such as code listings), set the Language to None on the Default Font sheet.
- **Stacking lines with negative line spacing.** On the Basic sheet, set the space below paragraph one to a negative number and the space above paragraph two to a negative number, you can force FrameMaker to write two paragraphs in the same location.
- **Change bars.** For typing comments in a file, create a comment paragraph tag with the Change Bar property selected on the Default Font sheet.

Chapter 6: Formatting text with paragraph tags

Chapter 7: Formatting text with character tags

FrameMaker distinguishes between paragraph and character range formatting. *Paragraph tags* always format an entire paragraph. *Character tags* provide a way to format one or more characters within a paragraph. For example, you can create formats for italicized, underlined, or bold text, and create formats that apply and control any combination of these properties at once.

Here's a look at what's in this chapter:

      Maintaining consistency with character tags 139
      Applying Character Tags 140
            Selecting character format with the Character Catalog 140
            Selecting character format with a menu choice 140
            Using Smart Insert for character formats 140
      Modifying character tags 141
            As Is character properties 142
            Avoiding character tag overrides 143
      Creating character tags 143
            Renaming character tags 145
            Updating character properties globally 145
      Removing character tag formatting 146
      Deleting character tags 147
      Additional character tags tips 148

Maintaining consistency with character tags

Character tags allow for rapid text formatting, but more importantly, they make it easy to update formatting within your document.

Suppose your corporate branding changes and no longer permits display of menu items in boldface text. If your menu items use a Menu character tag, you could change the tag settings to format menu items accordingly. Update the tag's definition globally in your docs, and you're done.

If instead you had used the Format menu (or B button, or CTRL-B) to emphasize menu items, you would have to scan your book for each bold menu item and manually remove the bold formatting.

Character tags also act as building blocks for other FrameMaker features. Using character tags, you can assign color or other formatting to a portion of a cross-reference or a variable.

Applying Character Tags

The *character catalog* lists the character tags in a document. If needed, display the catalog by selecting **Window > Pods > f Catalog**.

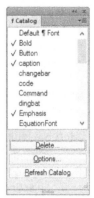

Note We'll apologize on behalf of Adobe for this obscure pod title...it dates back to a *florin* (or *f*), used in the character catalog title bar long before Adobe purchased FrameMaker.

Although you can apply multiple character tags to achieve formatting effects, FrameMaker recognizes just the last tag. If, for example, you apply an italicized character tag and then apply a boldface tag, depending on the use of As Is, the text may display as boldface, italicized text. FrameMaker, however, recognizes only the last character tag you applied. A better approach is to create a single character tag that combines both properties.

Selecting character format with the Character Catalog

To apply a character tag through the Character Catalog, follow these steps:

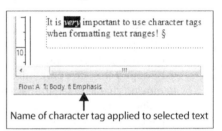

Name of character tag applied to selected text

1. Select the text you want to format.
2. In the Character Catalog, click a character tag. The character tag is applied to the selected text. In the status bar, the character tag name indicates that you applied the tag.

FrameMaker provides other ways to apply a character tag, including:

- Selecting a menu choice
- Using the right-click pop-up menu
- Using keyboard shortcuts

Note You can also apply character tags by using the Character Designer, but this isn't recommended because you could accidentally modify the character tag definition.

Selecting character format with a menu choice

To apply a character tag through a menu choice, follow these steps:

1. Select the text you want to format.
2. Select **Format > Characters**, then select a character tag from the submenu.

Using Smart Insert for character formats

To apply a character tag using a keyboard shortcut, follow these steps:

1. Select the text you want to format.

2. Press F8 or CTRL+8 to display the Smart Insert menu.
3. Display the character tag you want by doing any of the following:
   - Use the arrow keys to scroll through the list of available tags.
   - Type the first few letters of the character tag name.
4. Press ENTER.

Note   Some of the keys on your computer may be specially programmed. If so, one or more keyboard shortcuts may not work as described. Consult your system documentation for help.

Modifying character tags

In the Character Designer, you modify character tag properties, such as font size, color, and angle. When you save your changes, you Update All character tags in the document at once. Content formatted by the updated character tag is automatically reformatted.

The Character Designer consists of the same choices as the Paragraph Designer's Default Font sheet. For a description of each item, see "Default Font sheet" on page 117. The commands on the left side of the Character Designer also work very much like the commands in the Paragraph Designer.

To modify a character tag, follow these steps:

1. Deselect your text by clicking outside your text frame.
2. by Select **Format > Characters > Designer** (or press CTRL+D or ESC O C D) to display the Character Designer
3. Select the character tag you want to modify in the Character Tag drop-down list.
4. Properties set to As Is will be controlled by the *paragraph* tag. If you allow things like font size to be controlled by the paragraph tag, you will need fewer character tags. See the following section, "As Is character properties," for details.
5. Use the drop-down lists and checkboxes to specify the properties you need.

Note   If the color you need isn't listed, modify your document's color definitions (**View > Color > Definitions**), and the new color will be displayed in the Character Designer. For details, see Chapter 24, "Color output."

6. Click the Update All button to modify all instances of the character tag in your document.

As Is character properties

Optimally, character tags don't use all of the properties available in the Character Designer. For example, an Emphasis character tag might have the Angle set to Italic (or Oblique, depending on the font). The remaining properties—font family, font size, weight, variation, color, word spread, underline, and so on—are set to As Is, and thus come from the paragraph tag applied to the paragraph. Using the As Is setting for the other properties saves time and helps you build character tags that format content properly.

Figure 7-1 illustrates the benefit of using As Is properties. Most properties have been modified in the first example, which improperly formats the text. In the second example, properties except weight are set to As Is, so the text formats correctly. All of the checkboxes are filled, not checked, and the Size, Spread and Stretch options are blank. These indicate that those properties have been set to As Is. When you click As Is in the Size drop-down list, the field is cleared. See the sidebar "Understanding Checkboxes in FrameMaker" on page 144 for more information.

Figure 7-1. Using As Is properties

Note If you have trouble seeing the As Is settings in your own work, try clicking outside the text frame before selecting your character tag.

Avoiding character tag overrides

When you modify a tag in the Character Designer and click the Apply button instead of clicking the Update All button, you create an *override*. An asterisk displays next to the character tag name in the status bar. Applying a character tag to an entire paragraph will also create an override. If you need to change the style of only one entire paragraph, you should create a new paragraph tag instead of applying a character tag.

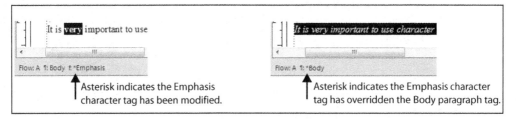

Asterisk indicates the Emphasis character tag has been modified.

Asterisk indicates the Emphasis character tag has overridden the Body paragraph tag.

You can remove character tag overrides by importing the template (see "Importing settings from a template file" on page 91) with the Remove Format Overrides option checked, by reapplying the character tag, or by selecting Default ¶ Font from the Character Catalog. To remove a paragraph tag override, apply the Default ¶ Font character tag or reapply the paragraph tag.

Overrides defeat the goal of maintaining consistently formatted documents and can cause problems with electronic output like WebHelp, HTML5, and EPUB. Although FrameMaker provides ways to remove overrides, it's better to avoid creating them in the first place.

Creating character tags

In addition to modifying tags in the Character Designer, you can also create them. When you create a character tag, it's best to follow a specific naming convention. Typically, you should name character tags based on their function rather than their style properties. For example, your template may include a character tag that formats computer commands with the Courier font. Instead of naming the tag Courier, you name it Command. If you decide later to format commands in bold, you can update the Command character tag without changing the tag name. If you had named the tag Courier, the character tag name would not describe the current properties.

To create a character tag, follow these steps:

1. Click outside your text frame.
2. Display the Character Designer. If you skipped step 1, the settings of the currently selected text (or of the text at the insertion point) are displayed.

Caution When your cursor is in text formatted with a character tag, the Character Designer displays that character tag's name. However, the settings shown in the Character Designer are not necessarily accurate. Many character tags have items set to As Is, so you may be seeing the underlying paragraph tag properties. To see the actual settings, click the tag name again in the Character Tag drop-down list.

3. If needed, click the Commands button and select Set Window to As Is from the pop-up menu. This resets the character properties. The As Is concept is discussed in more detail in "As Is character properties" on page 142.

4. Click the Commands button, then select New Format from the pop-up menu to display the New Format dialog.

5. Type the name of your new tag in the Tag field, then click the Create button. The new tag is displayed in the Character Catalog.

Caution By default, the Store in Catalog and Apply to Selection checkboxes are checked. Make sure that Store in Catalog is checked, or the tag will not be available in the Character Catalog.

6. Change the properties as needed, then click the Update All button to save your changes. Notice that all of the As Is settings disappear again. To verify that the tag is set correctly, click its name in the Character Tag drop-down list again.

Understanding Checkboxes in FrameMaker

Most programs have only two settings for a checkbox—checked or unchecked. In FrameMaker, there are three values: checked, unchecked, and As Is, as shown in the illustration.

To clear the As Is setting, select the box once. To change the value from As Is to checked, select the box twice.

Want more control over superscript characters?

Not everything you can do in FrameMaker is available through the format designers. For some things you'll need to go "under the hood" and noodle with MIF code. See "Creating a character tag for a vertical baseline shift" on page 567 for more information.

Renaming character tags

The Character Designer makes it easy to change the name of a character tag you've already created. For example, you may need to rename the Emphasis character tag to Strong. After you rename Emphasis, the character tag name will change to Strong where you've applied it, and the previous tag name will no longer be displayed in the Character Catalog.

Renaming character tags also lets you globally assign a new character tag to text formatted by a deleted tag. When you select the formatted text, the deleted character tag name is still displayed in the status bar. You can "rename" the deleted tag in the Character Designer. This applies the new character tag where the deleted tag is applied in your document.

Caution When you rename a character tag, FrameMaker updates the catalog definition and the locations where you applied the tag; however, references to the old tag name (for example, in variables and cross-references) are not updated. You'll need to search for the old tag name in these components and change the name yourself.

To rename a character tag, follow these steps:

1. Display the Character Designer.
2. Click outside of your text frame (or otherwise set the window to As Is).
3. Click the tag you plan to rename from the Character Tag drop-down list. If you're renaming a deleted tag, select the text that was previously formatted by the tag. (The deleted tag name is marked in the status bar with an asterisk.)
4. In the Character Tag field, type the new name over the existing name.
5. Click the Update All button to display a confirmation dialog.
6. Click the OK button. The renamed character tag is applied in your document and listed in the Character Catalog in place of the initial tag.

Note The Global Update Options dialog box also lets you rename a tag, but the method involves more steps. See the next section for details.

Updating character properties globally

FrameMaker's global update feature provides several ways to modify more than one character tag at once. For example, you can replace all instances of magenta text with black text, apply a different character tag in place of the old one throughout the document, and more.

There are three global update options:

- **All Characters and Catalog Entries:** You can update specific properties in all character tags, paragraph tags, and text lines. Use this option if, for instance, you need to change the default font size in all tag definitions and text lines.
- **All Matching Tags in Selection:** You can modify character tags that have been applied to different words in a paragraph. For example, suppose the Menu Item character tag is applied to one word and the Command character tag is applied to a different word in the same paragraph. You can select the paragraph and remove bold formatting from both character tags.
- **All Tagged:** Applies a new character tag in place of the selected tag. For example, you can use this option to apply the Hyperlink character tag instead of the Underline character tag in an entire document.

Changing specific properties

You can modify specific character properties in a document, whether the properties are in character tags, paragraph tags, or text lines. This is handy for removing color from all characters or only from selected character tags and changing the default font.

To modify selected properties, follow these steps:

1. Display the Character Designer.
2. Click the Commands button, then select Set Window to As Is from the pop-up menu. This resets all character properties.
3. Set the properties you want to update.
4. Click the Commands button, then select Global Update Options from the pop-up menu to display the Global Update Options dialog.
5. Do *one* of the following:
    - To update specific properties in *all* character tags, paragraph tags, and text lines, click the All Characters and Catalog Entries radio button. If you selected a tag name before performing this step, FrameMaker gives you the option to apply the tag to all characters in the document. Typically, you want to avoid this, so click the No button to update only the properties you changed.
    - To update specific properties in character tags applied to *selected text*, click the All Matching Tags in Selection radio button.
    - See the following section for use of the All Tagged: radio button
6. Click the Update button. FrameMaker globally makes the appropriate changes.

Reassigning a character tag globally

FrameMaker provides a way to assign a new character tag in place of the selected tag. For example, you can swap each instance of the Button character tag with the Key character tag.

To reassign a character tag globally, follow these steps:

1. Display the Character Designer.
2. In the Character Tag drop-down list, click the tag name you want to apply.
3. Click the Commands button, then select Global Update Options from the pop-up menu to display the Global Update Options dialog.
4. Click the All Tagged: radio button, then click the tag you want to replace from the drop-down list.
5. Click the Update button to display a confirmation dialog.
6. Click the OK button to update the tag.

Removing character tag formatting

After you apply a character tag, you can remove the tag by selecting text and applying the Default ¶ Font option in the character catalog. To reformat a text range with the current paragraph's tag properties follow these steps:

1. Display the Character Catalog.
2. Select the text you want to reformat.
3. In the Character Catalog, click Default ¶ Font. The text reverts to the original paragraph style. In the left status bar, the character tag name is no longer displayed. If a star displays next to the paragraph tag name, the paragraph still contains a formatting override. See "Avoiding character tag overrides" on page 143 for more information.

If a character tag has been applied to an entire paragraph, you can remove the character tag formatting by reapplying the paragraph tag.

Tip   Instead of using the Character Catalog, you can press F8 or CTRL+8 to invoke the Smart Insert pod. Select Default ¶ Font to remove the character formatting.

Deleting character tags

All items can be deleted from the Character Catalog except the Default ¶ Font. When you delete a tag, the catalog is updated, but variables, cross-references, and other components may still refer to the deleted tag name. Deleting a tag also does not remove the character tag from text to which it was applied. The name of the deleted character tag is still displayed in the status bar, along with an asterisk to indicate an override. To prevent the override, globally replace each instance of the character tag you plan to delete with another tag. See "Reassigning a character tag globally" on page 146 for details.

Tip   If you delete the tag without globally applying another one, you can select the text formatted by the deleted tag and assign a new one in its place. For more information, see "Renaming character tags" on page 145.

To delete a character tag from the Character Catalog, follow these steps:

1. Display the Character Catalog.
2. Click the Delete button to display the Delete Formats from Catalog dialog. You can also use the Commands button in the Character Designer to access this dialog.
3. Click the tag you want to delete, then click the Delete button. To delete all character tags, press and hold ALT+E. This method quickly deletes the tags. You can delete tags from the Paragraph Catalog the same way.
4. Click the Done button. The character tag you deleted is no longer displayed in the Character Catalog. If you make a mistake, click the Cancel button and start over. There is no Apply button.

Additional character tags tips

This chapter has covered applying character tags to text directly, but you can also use character tags to format text in variables, cross-references, autonumbered paragraphs, markers, and other FrameMaker components. Table 7-1 describes more advanced uses for character tags. You'll read more about each component in the corresponding chapter (for example, cross-references formats are discussed in the cross-reference chapter).

Table 7-1. Character Tag Tips

Item	Description
Variable	Create a book title variable and include a character tag to italicize the title. For details see Chapter 10, "Storing content in variables."
Autonumbered paragraph	Create a bullet character tag that uses the Wingdings font and use the character tag in the autonumber format of a bulleted paragraph tag. See "Autonumbering details" on page 130.
Cross-reference format	Add color to a cross-references in a PDF file by inserting a character tag in the cross-reference format. When printing, set the character tag to Black. When creating PDF, set the color to Blue or another color. See "Formatting cross-references" on page 180.
Marker	Italicize *See Also* in an index entry by including a character tag in the index marker. See Chapter 19, "Creating indexes."
Text line	Draw a text line using the Text Line feature in the Tools panel and format the text with an italicized character tag. See Chapter 16, "FrameMaker's graphics tools."
Spell checker	Skip spell-checking for specific text by setting the language to "None" in a character tag. See Chapter 3, "Word-Processing Features."
Change bar	Mark text you modify with a change bar character tag instead of or in addition to relying on automatic change bars. See Chapter 3, "Word-Processing Features."
Reference page	Create highlighted page numbers for TOC entries in a PDF file by applying a blue character tag to <$pagenum> building blocks on the TOC reference page. See Chapter 18, "Creating tables of contents."
Hyperlink	Create a character tag with all properties set to **As Is** and apply to hyperlinks. The format marks the beginning and end of the hyperlink without modifying the character properties. See Chapter 27, "Creating Interactive Content with Hypertext."
Table of Contents	Standardize the tab leaders and page references by applying a character tag to the tabs and <$pagenum> building blocks, or typing the tag name in brackets (e.g. <Emphasis>) before the tab character.

Chapter 8: Understanding table design

FrameMaker lets you quickly use sophisticated preformatted table formats using *table tags*. The table tags store formatting such as alignment, spacing, cell margins, pagination, ruling, shading, and the default number of columns and rows. When you need a table, you select a table tag and insert the table in your document. Table tags also store paragraph formatting, so when you insert a new table, the paragraph tags for text in the table are already applied to the table title and cells.

Table 7: Results of Annual Tri-City Dog Show

Ribbon	Dog Name	Breed
Red	Charlie Chan's Revival	Toy Poodle
Blue	Trixie the Trotter	Pomeranian
Purple	Island Girl Blanc	Maltese
Yellow	Jake's Nine Lives	Pointer

Judge R.C. Smith presided.

Here's a look at what's in this chapter:

```
Table tag advantages . . . . . . . . . . . . . . . . . . . . . . . . . . . . . . . . . . . . . . .149
Inserting tables . . . . . . . . . . . . . . . . . . . . . . . . . . . . . . . . . . . . . . . . . .150
    Selecting cells . . . . . . . . . . . . . . . . . . . . . . . . . . . . . . . . . . . . . . . .151
    Moving and deleting tables . . . . . . . . . . . . . . . . . . . . . . . . . . . . . .151
Changing the assigned table tag . . . . . . . . . . . . . . . . . . . . . . . . . . . . .152
Modifying a table tag . . . . . . . . . . . . . . . . . . . . . . . . . . . . . . . . . . . . .152
    Changing table designer settings . . . . . . . . . . . . . . . . . . . . . . . . . .154
    Globally updating table tags . . . . . . . . . . . . . . . . . . . . . . . . . . . . .158
Creating a table tag . . . . . . . . . . . . . . . . . . . . . . . . . . . . . . . . . . . . . . .158
Customizing tables outside the Table Designer . . . . . . . . . . . . . . . . . .158
    Adding and deleting rows and columns . . . . . . . . . . . . . . . . . . . . .159
    Resizing columns . . . . . . . . . . . . . . . . . . . . . . . . . . . . . . . . . . . . .160
    Applying paragraph tags . . . . . . . . . . . . . . . . . . . . . . . . . . . . . . .162
Customizing cell ruling and shading . . . . . . . . . . . . . . . . . . . . . . . . . .163
Merging table cells . . . . . . . . . . . . . . . . . . . . . . . . . . . . . . . . . . . . . . .165
Rotating table cells . . . . . . . . . . . . . . . . . . . . . . . . . . . . . . . . . . . . . . .167
Sorting table data . . . . . . . . . . . . . . . . . . . . . . . . . . . . . . . . . . . . . . .168
Deleting table tags . . . . . . . . . . . . . . . . . . . . . . . . . . . . . . . . . . . . . .169
Creating tables from text . . . . . . . . . . . . . . . . . . . . . . . . . . . . . . . . . .170
Converting a table to text . . . . . . . . . . . . . . . . . . . . . . . . . . . . . . . . .172
```

Table tag advantages

The biggest advantage of table tags is that they help maintain consistency. If the corporate style guide changes or your department decides to redesign tables, you update the FrameMaker template and import the new table tag into your document. Tables tagged with the modified table tag are updated instantly. Some properties, however, are only applied when creating new tables, not when modifying table formats or changing the table tag applied to an existing table. For example, if you update the paragraph tag associated with a heading or body cell, existing content is unaffected. The changes apply only to new tables.

You can also modify the specifics of a table without changing the applied table format. Adding rows and columns, inserting a table title, shading specific cells, merging cells, modifying borders, and other reformatting can be done on a table-by-table basis. Some of these changes, such as shading of rows and columns, can be saved in the table tag; others apply only to the selected table. For example, you can't save merged table cells or irregular ruling patterns in a table tag.

Inserting tables

Tables consist of four basic components—the table title, heading rows, body rows, and footing rows. When you insert a table into your document, you specify the number and types of rows, the number of columns, and the table tag. Table title visibility and placement are determined by the table tag.

To insert a table, follow these steps:

1. Position your cursor where you want to put the table.
2. Select **Table > Insert Table** to display the Insert Table dialog.
3. In the fields on the right, type the number of rows and columns you want for the table.

You can add three kinds of rows to a table:

- **Heading row.** Useful for naming each column and repeats if table breaks beyond the current page.
- **Body row.** Displays main content.
- **Footing row.** Similar to heading row, and useful for repetitive content.

4. Select a table tag in the Table Format list.
5. *(optional, starting with FrameMaker 2015)* Select Table Continuation and Table Sheet variable options as needed. Note that the variables are inserted into the table title, so choosing a table format that includes a table title is necessary for this feature to function as expected.
6. Click the Insert button to display the table in your document.

With the FrameMaker text symbols displayed, you'll see an upside-down **T** (or table anchor) above the table. The anchor locks the table to the paragraph. As the paragraph fills the page, the table moves down. (Floating tables are a bit different. See "Basic sheet" on page 154.)

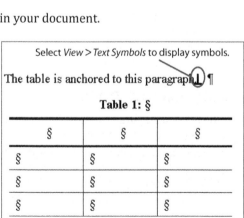

After you insert a table, you type text in the table cells and press the TAB key (or your cursor keys, starting with FrameMaker 2015) to move the cursor from cell to cell. The cursor moves horizontally to the end of the row and then to the first cell in the next row until reaching the end of the table.

Starting with FrameMaker 2015, pressing the TAB key at the end of the table will add a row. In all versions of FrameMaker, CTRL-ENTER adds a table row below the selected cell.

Inserting a tab in a table cell is also different. To insert a tab, position your cursor in the cell and press ESC TAB. The paragraph tag applied to the table cell must include a tab stop, or text will not appear to be after a tab; however, a tab is still inserted and you can see the tab symbol if you view text symbols.

Selecting cells

Clicks in a table cell select the contents of the cell, not the actual table cell. The quickest way to select a table cell is by CTRL+clicking the cell you want to select. A small node, or *handle*, is displayed on the right border of the selected cell. You select and drag a handle to resize the selected column or columns manually.

You can also select a cell by using the Object pointer ( ) in the Tools panel or by pressing the mouse button and dragging the cursor over the table cells.

To select a row or column, you CTRL+CLICK twice in a specific location. The border closer to your cursor before you select the cells determines whether the row or column is selected. If the cursor is closer to the left or right cell border, you select the row. If the cursor is closer to the top or bottom cell border, you select the column.

CTRL+click near the left or right border to select the cell.	CTRL+click near the left or right border twice to select the row.	CTRL+click near a top or bottom border twice to select the column.

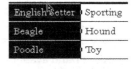

In addition, **View > Toolbars** offers a table formatting toolbar, from which you can select a column, a row, column body cells, or the whole table.

Moving and deleting tables

You can move and delete tables by selecting the table anchor or selecting the table itself. The document symbols must be displayed for you to see anchors (select **View > Text Symbols**), and anchors can be difficult to select. If there are two adjacent tables, the anchors may be stacked on top of one another on the same line. In this case, selecting the table is easier than selecting the anchor, but both methods are described here.

Note   If you don't select the entire table before cutting and pasting or selecting the anchor, a warning dialog box is displayed with the choice to leave the cells empty or to remove the cells. Click the Cancel button and start over.

To select a table without highlighting an anchor, follow these steps:

1. Position your cursor in the table you want to move.
2. CTRL+CLICK three times to select the table.

To select a table anchor, follow these steps:

1. Position your cursor next to the anchor. You might have to view text symbols and/or change the page magnification rate to 150 or 200 percent to see the anchor.
2. Press SHIFT-LEFT ARROW or SHIFT-RIGHT ARROW key to select the anchor until the table is highlighted. If more than one table is selected, the two anchors may be stacked. CTRL+TRIPLE CLICK in the desired table to select the table.

You can now delete, cut, or copy the table as needed. Starting with FrameMaker 2015, you can also drag the table where needed. However, at time of writing, this leaves a blank copy of the table which must be deleted.

To delete a table, follow these steps:

1. Select the table.
2. Press the DELETE key, or select **Edit > Clear**.
3. The table is deleted from the document.

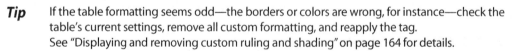

Changing the assigned table tag

Starting with FrameMaker 11, you can apply table tags to existing tables using the Table Catalog. This lets you quickly reformat the selected table. In all versions of FrameMaker, the table catalog is also displayed in the Table Designer and available when you insert a table.

To apply a table tag, follow these steps:

1. Select **Table > Format > Catalog** or otherwise display the Table Catalog.
2. Place your cursor anywhere inside a table.
3. Click the appropriate tag in the Table Catalog.

Tip If the table formatting seems odd—the borders or colors are wrong, for instance—check the table's current settings, remove all custom formatting, and reapply the tag.
See "Displaying and removing custom ruling and shading" on page 164 for details.

Modifying a table tag

Most table tag properties are configured in the Table Designer. When you make changes in the Table Designer and click the Update All button, two things happen—all tables of that type in the document are updated where you've inserted them, and the table catalog is updated so that new tables of that type will format properly. When you modify table formatting outside the Table Designer, the changes aren't applied throughout the document; they apply only to the selected table.

In the Table Designer, you can change the following properties:

- Indents
- Space above and below the table
- Cell margins
- Vertical and horizontal table alignment
- Cell sequence for autonumbers
- Title position and gap
- Number of orphan rows
- Borders
- Shading
- (starting with FrameMaker 2015) Direction of text (Left-to-right, right-to-left)
- Initial paragraph tag pattern (as defined in first row of body, heading, and footing rows)

Note   If you make changes to first row paragraph tags applied to table content without selecting the Update All button in the Table Designer, those changes are not stored in the table tag, applying only to the selected table, not to tables subsequently added to the document.

Outside the Table Designer, you can modify the following properties:

- Column width
- Number of columns
- Number of rows (for heading, body, footing rows)
- Paragraph tags applied to table content

To modify a table tag in the Table Designer, follow these steps:

1. Place your cursor anywhere in a table and display the Table Designer. The current table format will be displayed in the Table Designer.
2. Change any desired settings, then click the Update All button. Repeat for the other sheets. Your changes are saved in the table tag and applied to all existing tables using that format.

Note   Clicking the Apply button instead of the Update All button creates a formatting override. Minimize format overrides of all types.

The following sections describe each setting in the Table Designer. For details on changing table tags outside the Table Designer, see "Customizing tables outside the Table Designer" on page 158.

Changing table designer settings

The Table Designer has three property sheets. To view a sheet, click its tab or click the sheet in the Properties drop-down list. This section describes the sheets.

Basic sheet

On the Basic sheet, you change indents, spacing, pagination, numbering, orphan rows, and other properties.

Table 8-1 describes the options.

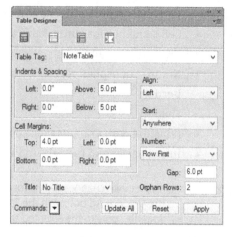

Table 8-1. Table Designer, Basic Sheet

Item	Description
Indents & Spacing	Adds space above, below, and to the left and right of the table and the rest of the document. If used, the table title is part of the table unit, so space above/below is added above/below the table title. As with paragraph format spacing above/below, the actual space above/below will be the greater of (not the sum of) the spacing specified by the table tag and the paragraphs above or below the table.
Cell Margins	Sets the minimum padding between table cell border and content. Interacts with settings in the Paragraph Designer Table Cell sheet, where you can customize table cell margins and alignment for a paragraph tag. (See "Table Cell sheet" on page 126 for details.)
Alignment	Sets the horizontal position of the table relative to the text frame. This does not set the alignment of text within table cells, which is set within the paragraph tags applied to table content. Alignment respects the pagination settings of the paragraph tag applied to the table anchor. For example, if the table is in a paragraph with side head alignment, the table stays in the side head. • **Left:** Aligns the table with the left side of the column. • **Right:** Aligns the table with the right side of the column. • **Center:** Places the table in the center of the text area. • **Side Closer to Binding:** Aligns the table closer to the bound edge of the current page. • **Side Farther from Binding:** Aligns the table farther from the bound edge of the current page.

Table 8-1. Table Designer, Basic Sheet (Continued)

Item	Description
Start	Sets the allowable start position on the page. • **Anywhere:** Places the table "anywhere it fits" on the page, subject to allowable table breaks. If no break is allowed, the table moves to the next column or page. • **Top of Column:** Forces the start of the table to the top of the next column. (Same as Top of Page in one-column documents.) • **Top of Page:** Moves the table to the top of the next page. • **Top of Left Page:** Places the table at the top of the next left page in a double-sided document. • **Top of Right Page:** Places the table at the top of the next right page in a double-sided document. • **Float:** Moves the table to the first available position. If the table doesn't fit on the current page, the table floats to the next available position, and text following the table anchor may fill in the space after the anchor.
Numbering	Organizes autonumbered paragraphs either horizontally or vertically: • **Row First:** Numbers paragraphs horizontally, as in: 	1. Go	2. Yield	3. Stop
---	---	---
4. Merge	5. Slow	6. Exit
7. Caution	8. School	9. Turn | • **Column First:** Numbers paragraphs vertically, as in: | 1. Go	4. Yield	7. Stop
---	---	---
2. Merge	5. Slow	8. Exit
3. Caution	6. School	9. Turn	
Title Position and Gap	Positions the table title. • **No Title:** Indicates the table has no title. • **Above Table:** Displays the title above the table. • **Below Table:** Places the title below the table. The gap is the space between the table and the table title. (Spacing in the table title paragraph itself does not apply.)
Orphan Rows	Specifies the minimum number of table rows that must stay together prior to a table break across a page. To force the entire table to fit on the same page, specify up to 255 rows.

Ruling sheet

The Ruling sheet contains border styles for rows, columns, heading and footing rows, and the outside edges of the table.

You can arrange rulings in a variety of patterns. For example, row headings can have a thick rule, and the other columns can have a thin border or no border at all. Five default ruling styles—double, medium, thick, thin, and very thin—are available. For details on creating, editing, and deleting ruling styles, see "Changing ruling options" on page 164. Table 8-2 describes the Ruling sheet.

Table 8-2. Table Designer, Ruling Sheet

Item	Description
Column Ruling	Specifies a unique border for the *n*th column and a repeating border for the remaining columns. Useful if the table cell headings are vertical instead of horizontal.
Body Row Ruling	Specifies a ruling for every *n*th body row and a border for the remaining body rows. This is independent of Heading/Footing ruling.
Heading and Footing Ruling	Separates the heading and footing rows from the table and provides a separator in the event of multiple heading or footing rows.
Outside Ruling	Specifies borders around the outside of the table, with an option to draw the bottom border only on the last sheet of a table (for tables that break across pages).

Shading sheet

On the Shading sheet, you choose the color and shading amount for heading and footing rows and patterns for body cell shading.

For example, your table can have a 30 percent blue heading row and alternate 10 percent yellow and white body rows. The fill percentages (3, 10, 30, 50, 70, 90, and 100) are defaults and cannot be modified. It's advisable, however to set up additional colors in the document's color definitions, to improve the choices and allow for higher quality tints. See Chapter 24, "Color output," for details. Table 8-3 describes the Shading sheet.

Table 8-3. Table Designer, Shading Sheet

Item	Description
Heading and Footing Shading	Applies one fill percentage and color to *both* the heading and footing rows, if present.
Body Shading	Applies fill percentage and color to *either* the rows or the columns and lets you define shading patterns for rows or columns.

Direction sheet

Right-to-left support for tables is available, starting with FrameMaker 2015. You can set the table tag to use to RTL, LTR, or inherit the direction of the parent.

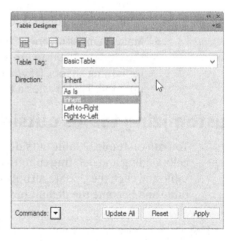

Globally updating table tags

You can update table tag properties across the entire table catalog, just as you do with the paragraph and character tags. For example, you can change the body row ruling in all table definitions to None, and all table catalog entries and corresponding tables in the document will be updated.

The global update options in the Table Designer let you change either all properties, or the properties of the currently selected sheet. So if you click the Ruling sheet before displaying the Global Update Options dialog box, you have the option to change just the ruling properties. With the Shading sheet displayed, you have the option to change properties for the shading sheet only.

For more information about global updating, see "Setting properties across the entire paragraph catalog" on page 128.

Creating a table tag

New table tags are based on existing tags—you must position your cursor in a table before creating a new tag.

To create a table tag, follow these steps:

1. Position your cursor in a table with the tag that looks closest to the table you want to create.
2. Display the Table Designer, then type a new name in the Table Tag field.
3. Choose Commands: New Format or click the Apply button to display the New Format dialog.
4. Verify the name you typed, then click the Create button. Your table format is displayed in the Table Designer.
5. Make changes on one of the Table Designer sheets, then click the Update All button to save your changes. Repeat for remaining sheets. (If you click Apply instead of Update All, the table definition is not updated. This creates a formatting override.)

Customizing tables outside the Table Designer

You can customize a table outside the Table Designer by applying paragraph tags to the text in the cells, adding and deleting columns or rows, and resizing columns. If you make such changes in a table and click the Update All button in the Table Designer, your changes are made to the current table and become the default properties for all future tables of that type.

Note Unlike changes you make in the Table Designer, changes you make to a table outside the Table Designer are applied only to the current table. However, if you click the Update All button in the Table Designer, many of those changes will be stored in the Table Catalog and applied to tables you insert in the future using that tag.

Adding and deleting rows and columns

Although there's only one type of column, there are three types of rows:

- **Heading row.** Repeats column headings.
- **Body row.** Contains main content.
- **Footing row.** Repeats text at the bottom of the table.

When your cursor is in a table and you click the Update All button, the table tag uses the number and location of rows and columns in that table as the default the next time you add a table of the same type. For example, if the current table consists of one heading row, five body rows, and two columns, those defaults are stored in the updated tag. When you insert a new table using the updated tag, the default row and column settings are included unless you change them in the Insert Table dialog. You can, however, add rows and columns to an existing table and save your changes as the new defaults for that tag.

To change the number of rows and columns in a table, follow these steps:

1. Place your cursor in a table you want to modify.
2. Select **Table > Format > Add Rows or Columns** to display the Add Rows or Columns dialog.
3. To add rows, type the number of rows in the Add field, then click the row location in the Row(s): drop-down list. You can also add a row below the current row by pressing CTRL+ENTER.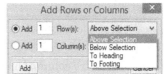
4. To add columns, type the number of columns in the Add field, then click the column location in the Column(s): drop-down list.
5. Click the Add button. You can click the Cancel button if you change your mind.
6. *(optional)* To save the current number of rows and columns as the default for this table tag, keep your cursor in the table, display the Table Designer, and click the Update All button.

To delete rows and columns, do the following:

1. Select the row or column you want to delete. If you select only one cell, you'll delete the contents of the cell, not the cell itself.
2. Press the DELETE key to display the Clear Table Cells dialog.
3. Click the Remove Cells from Table radio button to delete the cells, then click the Clear button. The row or column is removed from the table.

   Note   If you click the Leave Cells Empty radio button, the contents of the row or column are deleted, not the row or column.

4. *(optional)* Click the Update All button in the Table Designer to save your changes in the table tag. When inserting new tables, this will be the new default for number of rows and columns.

Resizing columns

Table tags store default column widths. Each time you insert a particular table style, the columns are the same size from table to table. You can resize table columns either manually or by specifying a width. When you resize one column, the other columns remain the same size.

Original column widths	Equipment	Aisle	Stock Number
	Lantern	3B	730-439H93

Column 2 is resized while maintaining the table width.	Equipment	Aisle	Stock Number
	Lantern	3B	730-439H93

The table widens when column 2 is resized without maintaining the table width.	Equipment	Aisle	Stock Number
	Lantern	3B	730-439H93

Note To maintain the table width when you manually resize a column in FrameMaker, press SHIFT while dragging. The width of the column on the right is automatically adjusted to maintain the original table width.

After resizing columns, you can update the table tag to save the new default column widths. This updates the table tag but not tables already in the document.

Manually resizing columns

You can resize table columns manually and use Snap and the FrameMaker ruler as a guide. As you drag the column border, the ruler marking moves. If you turn on FrameMaker's snap feature, the border moves to the nearest ruler marking. This helps you precisely resize the columns. To drag the column border between markings, you need to turn Snap off. Select **Graphics > Snap** to toggle the snap feature. See "Working with grids" on page 297 for details.

To resize columns manually, do the following:

1. In the column you want to resize, either CTRL+click twice to highlight the column or select a few cells in the column. The sizing arrow and handles are displayed.

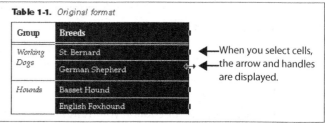

2. Select one of the handles and drag the column to the left or right. When the column is adjusted to the correct width, release the cell handles.

3. *(optional)* To save the current column widths as the default for this table tag, keep your cursor in the table, display the Table Designer, and click the Update All button. All new tables inserted using the modified tag will have the new column widths.

Note   FrameMaker always resizes the entire column. You cannot create a table with an irregular grid, except by straddling cells. For more information about straddling cells, see "Merging table cells" on page 165.

As you resize a table column, the width displayed in the status bar also changes. The dimensions reflect the width of all selected columns.

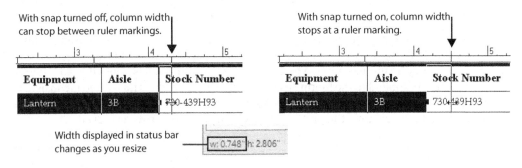

Specifying the column width

You can drag columns to resize them, or you can specify the column width as follows:

1. Select one or more columns to resize.
2. Select **Table > Format > Resize Columns** to display the Resize Selected Columns dialog.
3. Type the appropriate measurement next to the method you want to use:

   - **To Width:** Type the column width. If you select two columns, each column is resized to the specified width.
   - **By Scaling:** Change the width by a specific percentage. For example, you can scale the selected column 200 percent to double the width.
   - **To Width of Column Number:** Match the column width to the specified column.
   - **To Equal Widths Totalling:** Resize two or more columns to equal a specific width.
   - **By Scaling to Widths Totalling:** Resize two or more columns a specific percentage while maintaining the original column proportions.
   - **To Width of Selected Cells' Contents:** Resize the cell to fit its contents.
4. Click the Resize button. The selected table columns are updated.
5. *(optional)* To save the current column widths as the default for this table tag, keep your cursor in the table, display the Table Designer, and click the Update All button. All new tables inserted using the modified tag will have the new column widths.

Note In FrameMaker, you can specify the unit of measure used in paragraph tags, text frame properties, table tags, and so on. To change the unit of measure, select **View > Options** and click a different item in the Display Unit drop-down list.

Applying paragraph tags

Paragraph tags format the text in each table cell and in the table title. The default FrameMaker template includes paragraph tags such as TableTitle, CellBody, CellHeading, and TableFooting, but you can use any paragraph tags you choose. The table tag stores a pattern based on one or more paragraph tags used in the first row (and only the first row) of the heading, body, and footing. New tables include the paragraph tags stored from the table tag's pattern.

If you use different paragraph tag patterns to other rows in a table, those patterns cannot be saved in the table tag. For example, in the following illustration the table on the left is consistently formatted—CellHeading is applied to the heading row, CellBody is applied to all body rows, and TableFooting is applied to the footing row. The table on the right is formatted inconsistently. Even though the CellBullet paragraph tags are applied in the second body row, this pattern can't be saved in the table tag; only the pattern of tags in the *first* body row are saved.

Consistent application of paragraph tags

Inconsistent CellBullet tags aren't saved in the table tag.

Table 6: TableTitle

CellHeading	CellHeading	CellHeading
CellBody	CellBody	CellBody
CellBody	CellBody	CellBody
CellBody	CellBody	CellBody
TableFooting		

Table 6: TableTitle

CellHeading	CellHeading	CellHeading
CellBody	CellBody	CellBody
• CellBullet	• CellBullet	CellBody
CellBody	CellBody	CellBody
TableFooting		

Tip You can insert tables that have complex formatting by selecting the table from your own customized menu created with the third-party tool, Auto-Text (www.siliconprairiesoftware.com). See Appendix A, "Resources," for more information.

To apply paragraph tags to cells in a table, follow these steps:

1. Put your cursor in a cell (or select several cells), then apply a paragraph tag.

Note Formatting overrides (indicated by the star displayed in the status bar by the paragraph tag name) are saved with the table tag. If you modify a paragraph tag in the table, be sure you avoid overrides by updating the paragraph tag definition.

2. *(optional)* To save the paragraph formatting of initial rows in the table tag, display the Table Designer, then click the Update All button.

Customizing cell ruling and shading

FrameMaker's custom ruling and shading feature lets you apply ruling and shading to particular cells in a table, providing more flexibility than the Table Designer properties. You apply the formatting only when you need it (for example, to highlight important content). Starting with FrameMaker 2015, table cell fills display and print as they should. Earlier versions didn't work quite as well, requiring custom colors for best results. For example, a cell with a 10% black fill was difficult to read. As a workaround, you might define a lighter solid color called Black 10 Percent that can be used at a 100% tint in the table tag. See Chapter 24, "Color output," for details.

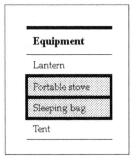

Caution By applying custom ruling and shading to table cells, you create a formatting override. Typically, you can remove overrides by reimporting the template with the Remove Layout/Format Overrides setting checked. To strip custom ruling and shading, however, you must remove the formatting manually (as explained later on page 164). Modifying tables outside the table tag makes it difficult to maintain consistency, so you should customize tables sparingly.

To apply custom ruling and shading, follow these steps:

1. Highlight the cells you want to format.
2. Select **Table > Format > Custom Ruling & Shading** to display the Custom Ruling and Shading dialog.
3. To apply custom ruling, complete the following:
   a. Click a border style in the Apply Ruling Style list.
   b. *(optional)* If you do not want to change the custom cell shading when you change the ruling, uncheck the Custom Cell Shading checkbox.
   c. Click the cell borders you want to reformat in the To Selection section. The border types are displayed in the following table, which assumes that the entire table is selected.

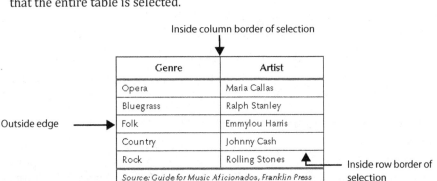

4. To apply custom shading to selected cells, do the following:
   a. Click the type of shading in the Fill drop-down list. You can choose specific percentages or solid white.

    b. *(optional)* If you do not want to change the custom cell ruling when you change the shading, uncheck the Custom Cell Ruling checkbox.

    c. Click a color in the Color drop-down list (unless you set the fill to White, None, or As Is).

    d. Click the Apply button. Your changes are made to the selected table cell or cells.

Changing ruling options

You can edit, create, and delete ruling styles, including custom colors, and the styles will be available in both the Custom Ruling and Shading dialog box and in the Table Designer.

To edit and create ruling styles, follow these steps:

1. Position your cursor within a table.
2. Select **Table > Format > Custom Ruling & Shading** to display the Custom Ruling and Shading dialog.
3. In the Apply Ruling Style list, click the style you want to edit or base the new style on, then click the Edit Ruling Style button to display the Edit Ruling Style dialog.

4. Change the appropriate properties:
   - Click a color in the Color drop-down list.
   - Click a fill percentage (or White) in the Pen Pattern drop-down list.
   - Type a new line width from .015 to 360 points in the Width field.
   - If choosing the Double-Gap radio button, specify the space between the lines (from .015 to 360 points).
5. If you're creating a new style, type a new name in the Name window.
6. Click the Set button. The ruling style is added to your choices for custom cell ruling.

To delete a ruling style, follow these steps:

1. Display the Custom Ruling and Shading dialog.
2. Click the style you want to delete, then click the Edit Ruling Style button.
3. Click the Delete button. If the ruling style is used in the document, a warning is displayed.
4. Click the OK button to confirm the change. The ruling style is removed from the Custom Ruling and Shading dialog box and from table cells to which you applied the border.

Displaying and removing custom ruling and shading

In addition to customizing table cells, you can also display and remove formatting properties for selected table cells. Showing the current settings indicates whether cells are formatted by the table tag or the custom ruling and shading properties. If table tag properties aren't displaying as you expect in a particular table, you can show the current settings through the Custom Ruling and Shading dialog box and reset the properties.

To display current selection settings, follow these steps:

1. Select specific table cells to view their settings, or select the entire table to view all settings.
2. Display the Custom Ruling and Shading dialog box, then click the Show Current Settings button to display the Current Selection's Settings dialog.
3. Click the Done button.

To restore table tag defaults, do the following:

1. Select the appropriate table cells, then display the Custom Ruling and Shading dialog.
2. Check the Custom Cell Shading and Custom Cell Ruling checkboxes.
3. In the Apply Ruling Style list, click From Table, then check all checkboxes in the To Selection section.
4. In the Fill and Color drop-down lists under the Custom Cell Shading checkbox, click From Table.
5. Click the Apply button. The default borders and shading are restored.

Tip You might find this feature useful for tables imported from other applications. Restoring table tag defaults is a quick way to make a clean table with no overrides.

Merging table cells

You merge, or *straddle*, table cells to create one cell across two or more rows or columns. By straddling cells, you can modify the table design to display content more clearly. For instance, the following tables list two dog groups—working dogs and hounds—and two breeds for each group. Unless you know your dog breeds, the two blank cells in the Group column of Table A are confusing. Should the blank cells contain content or remain empty?

Table B (on the right) divides the space more efficiently. In the first column, the first and second body cells—and then the third and fourth body cells—have been straddled. Now it's evident that two breeds belong to each dog group.

Table A

Group	Breeds
Working dogs	St. Bernard
	German Shepherd
Hounds	Basset Hound
	English Foxhound

Table B

Group	Breeds
Working dogs	St. Bernard
	German Shepherd
Hounds	Basset Hound
	English Foxhound

Displaying document borders makes it easier to see straddled cells.

Note Straddled table cells cannot break across pages. This can cause unsightly page breaks. When a straddled cell fills with text, the entire row moves to the next available page or text column. If you have this problem, consider techniques other than straddled cells.

Before straddling table cells, consider the following:

- You can straddle two or more contiguous cells; however, you cannot straddle across different row types—a body row and footing row, for instance.
- You cannot sort across straddles. For example, the breeds in Table B cannot be alphabetized until you unstraddle the rows in the Group column. You can, however, alphabetize the columns, which would place the Breeds column before the Group column. See "Sorting table data" on page 168 for more information.
- Display the document borders before straddling table cells (select **View > Borders**). It's easier to see which cells you've selected.
- You cannot save straddled table cells in a table definition.

To straddle table cells, follow these steps:

1. Select the table cells you want to join.
2. Select **Table > Straddle**. The straddled table cells appear as one cell.

Group	Breeds
Working dogs (two breeds are provided as examples)	St. Bernard
	German Shepherd

To unstraddle table cells, follow these steps:

1. Select the merged table cells.
2. Select **Table > Unstraddle**. The merged cells are restored as separate cells. All text moves to the first unstraddled cell, which expands to accommodate the text.

The first unstraddled cell expands to accommodate the text.

Group	Breeds
Working dogs (two breeds are provided as examples)	St. Bernard
	German Shepherd

Caution If you straddle two cells that have text in each cell, then unstraddle the cells, the text remains in the first cell that was straddled. Be careful when straddling and unstraddling cells.

Rotating table cells

You can rotate table cells to change the direction in which content is displayed. For example, to prevent text in a heading row from wrapping, you could rotate the text vertically. The table cell expands vertically to display the line of text. The width of the rotated table cell, however, does not change. It is still determined by the column width.

FrameMaker rotates table cells in 90° increments. In the following Table A, the heading cells were rotated counterclockwise 90° to prevent text from wrapping.

Table A

Table 6: The Sweet Shack Inventory of Desserts

	Warehouse A	Warehouse B	Warehouse C
Strawberry Pie	54	29	3
Angel Food Cake	32	94	43
Chocolate Mousse	5	23	382

Table B

Table 6: The Sweet Shack Inventory of Desserts

	Warehouse A	Warehouse B	Warehouse C
Strawberry Pie	54	29	3
Angel Food Cake	32	94	43
Chocolate Mousse	5	23	382

To rotate a table cell, follow these steps:

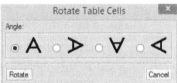

1. Select the cell or cells you want to rotate.
2. Select **Graphics > Rotate** to display the Rotate Table Cells dialog.
3. Click one of the radio buttons to indicate the direction you want to rotate the cell. To rotate text counterclockwise 90°, as shown in the previous table, click the last radio button.
4. Click the Rotate button. The text is displayed in the new orientation. If you make a mistake, you can select **Edit > Undo**.

Note You can rotate only table cells this way; tables themselves are reoriented by applying a rotated master page. For details, see "Creating a landscape master page" on page 221.

If you want to display text at an angle other than the four angles in the Rotate Table Cell dialog box, you can insert an anchored frame in the cell, and then put text inside the anchored frame and rotate that text. You can rotate this text at any angle. For more information, see "Rotating" on page 315. Note that if you save the file to HTML, the rotated text is saved as a graphic and not as text.

Chapter 8: Understanding table design

Sorting table data

FrameMaker can sort table data in a number of ways:

- **By row or column.** Sorted vertically (by row) or horizontally (by column). When sorting by row, you can sort specific rows or all rows in the table.
- **In ascending or descending alphanumeric order.** Sorted from A to Z and 1 to 9 (ascending), or Z to A and 9 to 1 (descending).
- **Considering case sensitivity.** If capitalization is considered, *snake* is sorted before *Snake*.
- **Second- and third-level sorting.** After sorting the table by the initial criterion, FrameMaker performs the second and third sorts. For example, FrameMaker can sort a table of university departments and courses in which departments are sorted first and then each course within the department is sorted.

Caution   You cannot undo a table sort, so save the file before you begin. You can select **File > Revert to Saved** to restore the original file if necessary.

You can sort a table that contains Unicode content, but note that the sorting performed is based on the location setting of your system. In Windows 7, location is specified in the Control Pod in the Clock, Language and Region dialog. Make sure before you sort that your location setting is the best match for your content to ensure correct sorting.

To sort a table, follow these steps:

1. Click in the table you want to sort.
2. To sort only part of the table, select those rows. If you do not select anything in the table, the entire table is sorted.
3. Select **Table > Sort** to display the Sort Table dialog.

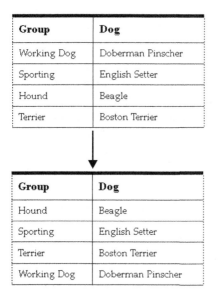
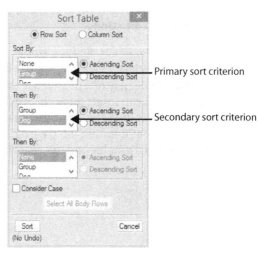

4. Click the Row Sort or Column Sort radio button to indicate the type of sort you want.

5. In the Sort By window, click the appropriate column or row you want to use to sort by.
   - To rearrange data by rows, specify the column to be sorted. In the preceding table, FrameMaker alphabetized the table rows by the Group column.
   - For a column sort, specify a row to sort on. If row one is empty, for example, specify row two.
6. Indicate whether the sort should be in ascending or descending order by clicking the appropriate radio button.
7. *(optional)* For secondary and tertiary sorting, change settings in the Then By sections. In the following table, rows in the first column were sorted first, followed by rows in the second column, then rows in the third column. For the secondary sort, courses within each department are alphabetized. For the tertiary sort, course numbers in the third row were alphabetized within each department.

Caution If you click the None setting for the primary sort, you cannot change the secondary and tertiary sort options.

Table A

Department	Course	Number
Art History	Rothko	GT4901-F
Art History	Rothko	DS2031-S
English	Film Noir 101	AT8032-E
English	French Poetry	Z87322-N

For the secondary sort, courses within each department were sorted.

Table B

Department	Course	Number
Art History	Rothko	DS2031-S
Art History	Rothko	GT4901-F
English	Film Noir 101	AT8032-E
English	French Poetry	Z87322-N

For the tertiary sort, course numbers for the identical Rothko courses were sorted.

8. To consider the capitalization of words during the sort, check the **Consider Case** checkbox.
9. *(optional)* If you selected specific rows to sort and change your mind, click the Select All Body Rows button to highlight all rows in the table. This option is grayed out if you clicked in one cell instead of selecting rows.
10. Click the Sort button. The table data is rearranged to your specifications.

Deleting table tags

You can delete table tags using the Table Catalog. Existing tables, however, still retain the tag assignment. This is considered an override. Before deleting a table tag, you can assign a different tag to the tables globally, just as you can with paragraph and character tags. "Reassigning a character tag globally" on page 146 describes a similar process related to character tags.

To delete a table tag, follow these steps:
1. Display the Table Catalog.
2. Click the Delete button.

3. Click the tag you want to delete, then click the Delete button. To delete all table tags quickly, press and hold ALT-E.

4. Click the Done button. The tags are removed from the table catalog.

If you find a table still formatted by a deleted tag, you have several options. You can globally rename the deleted tag. This applies the new table tag where the deleted tag is applied in your document. "Renaming character tags" on page 145 describes the equivalent process for character tags. Another option is to use the Find/Change pod to replace deleted tags with current ones in your catalog.

Creating tables from text

Instead of inserting an empty table and typing text into the cells, you can convert imported text files or existing text to tables. The pieces of text must be separated by one of the following:

- Tabs
- Two or more spaces
- Commas or another character used only for that purpose in the text
- Paragraphs

The following illustration shows text separated by all four methods and the table created by converting the data. The converted data is displayed identically despite the differences in the original formats.

Tabs	**Spaces**	**Commas**	**Paragraphs**
Fruit Color¶	Fruit Color¶	Fruit,Color¶	Fruit¶
Pear Green¶	Pear Green¶	Pear,Green¶	Color¶
Apple Red¶	Apple Red¶	Apple,Red¶	Pear¶
Pomegranate Purple¶	Pomegranate Purple¶	Pomegranate,Purple¶	Green¶
Banana Yellow¶	Banana Yellow¶	Banana,Yellow¶	Apple¶
			Red¶
			Pomegranate¶
			Purple¶
			Banana¶
			Yellow§

Fruit	Color
Pear	Green
Apple	Red
Pomegranate	Purple
Banana	Yellow

Creating tables from text

After you convert text to a table, you can convert it back to paragraphs. This option lets you undo a table conversion or format text to import into other applications. See "Converting a table to text" on page 172 for details.

Caution Table conversion can be have unpredictable results, so consider saving your file or creating a backup copy before the table conversion. You can probably undo any mess you might create, but it's best not to rely on FrameMaker's undo and history features.

To create a table from text, follow these steps:

1. Set up the text using one of the four methods described above.
2. Select the text, then select **Table > Convert to Table** to display the Convert to Table dialog box
3. In the Table Format list, click a format for the table.
4. Click the radio button that describes how your text is currently formatted.
    - For text separated by tabs, click the Tabs radio button.
    - For text separated by spaces, click the Spaces radio button and type the number of spaces.

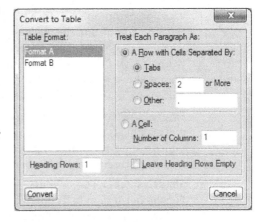

Caution Be careful when using spaces as delimiters. If the text you are converting to a table uses different amounts of space between columns, you can get unexpected results—especially if Smart Spaces is turned on. For more information, see "Smart spaces" on page 53.

  - If another character (such as a comma) separates the text, click the Other radio button and type the character.
  - If the text is in paragraphs, click the A Cell radio button and type the number of columns you want in the table.
5. Define the heading row.
    - To type the column headings after the table is created, check the Leave Heading Rows Empty checkbox and type the number of heading rows you want in the Heading Rows field.
    - To convert existing text to column headings, type the number of heading rows and uncheck the checkbox.
    - To create a table with no heading, type **0** in the Heading Rows field. You can add a heading row later if you change your mind. See "Customizing tables outside the Table Designer" on page 158 for details.
6. Click the Convert button. The text is formatted with the selected table tag.

Converting data may take more than one pass and a bit of luck, so don't worry if your table doesn't convert correctly the first time. Undo your changes, or if you created a backup file, open the file and start over.

Handy Table Tags

Most FrameMaker documents include table tags with basic features such as heading rows and cell ruling. You can, however, use tables for other purposes, such as for lists of interface buttons and labels, a warning paragraph with a graphic, and more.

- **List of icons.** Insert a two-column borderless table with narrow cell margins. You apply paragraph tags and save them in the table tag. After inserting the table in your document, you can import the icons in the first column and type the icon names in the other column (as shown on the right).
- **Warning paragraph.** Create a two-column, one-row table with the graphic inserted in the left column, and the warning paragraph tag applied to the right column. Add the warning graphic on the Reference page in a graphic frame named something like *Warning icon*. Create a 2-point font paragraph tag and specify *Warning icon* as the Frame Above Paragraph drop-down list. See "Placing graphics above and below the paragraph" on page 275 for details.
- **Figure table.** Insert an invisible one-cell table into which you import screen captures and graphics. Apply a figure caption paragraph tag to the table title section. Use this table to keep a floating graphic with the caption (otherwise, the caption floats away from the graphic, which is a bad thing).
- **Shaded sidebar.** For sidebars (such as this one), create a two-column, one-row table. For the borders, set all rulings to None. For the shading, specify a fill for the second column. (this color is 15 percent black)

Converting a table to text

When you convert a table to text, the table content (including the title) is converted to separate paragraphs. The original paragraph tags are still applied to each paragraph. To convert a table to text, you specify conversion by row or column. Tables that read across must be converted by rows to display correctly in paragraph format. Tables that read down must be converted by columns. The following table was converted using both methods.

Dog	Group
English Setter	Sporting
Beagle	Hound
Boston Terrier	Terrier

Converted by rows	Converted by columns
Dog	**Dog**
Group	English Setter
English Setter	Beagle
Sporting	Boston Terrier
Beagle	**Group**
Hound	Sporting
Boston Terrier	Hound
Terrier	Terrier

To convert a table to text, follow these steps:

1. Select the table, then select **Table > Convert to Paragraphs**. to display the Convert to Paragraphs dialog.
2. Choose to convert by Rows or Columns.
3. Click the Convert button. The table is displayed in the selected format. If necessary, you can undo the conversion by selecting **Edit > Undo**.

Reformatting Word Tables

Starting with FrameMaker 2015, converting Microsoft Word documents and tables is much easier. Using the **File > Import > File > Copy into Document** option, you can map Word table styles directly into a FrameMaker table style of your choosing. Once you've imported the tables you may need to do one or more of the following:

- **Assign a different table tag.** If your documents use more than one tag, use the Find/Change and the Table Catalog to quickly assign specific table tags to content. Adding or removing table titles are a common reason for assigning different tags.
- **Adding a heading row.** Heading rows in a Word document convert to body rows. You may need to add a heading row to the table, and cut/paste column headings into the heading row.
- **Fixing the borders and shading.** After you convert a table and apply a table tag, you may need to reset the ruling and shading properties to the tag's defaults. You can apply custom ruling and shading after you apply the default settings (see "Displaying and removing custom ruling and shading" on page 164).
- **Assigning paragraph tags.** You may need to apply paragraph formats after conversion for irregular tag patterns.

A third-party FrameMaker plug-in, TableCleaner, from Rick Quatro, automates these functions; it can add or convert heading rows and automates many other functions, such as applying paragraph tags and removing custom ruling and shading.

In previous versions, FrameMaker choked a bit on documents with table footnotes. An error dialog box displays the message "The filter encountered an error and could not complete the conversion." If you see this, , type the table footnote as regular text below the Word table and hard-code the footnote number in the table cell. You can create a real table footnote in the FrameMaker table after conversion.

Chapter 8: Understanding table design

Chapter 9: Cross-references

Cross-references direct readers to other parts of a document. Often, cross-references start with "see" or "refer to."

> see page 217
> refer to "All About Chocolate" on page 82

You could type these references, but setting them up with cross-references lets FrameMaker update the page numbers and referenced text for you automatically. This is much more efficient (and accurate) than checking all the references manually after final pagination of your document.

Here's a look at what's in this chapter:

Advantages of cross-references	175
Inserting Cross-References	177
Creating paragraph-based cross-references	177
Creating spot cross-references	178
Setting Up Cross-Reference Formats	179
Formatting cross-references	180
Cross-reference building blocks	181
Cross-reference examples	184
Deleting cross-reference formats	184
Updating cross-references	185
Updating Cross-References in a Book	185
Forcing a cross-reference update	185
Automatic updating problems	186
Preventing automatic updates	186
Renaming files without breaking cross-references	187
Correcting unresolved (broken) cross-references	187
Changing the referenced file	188
Replacing broken cross-references	189
Converting cross-references to text	189

Advantages of cross-references

Another major advantage to using cross-references is that they become live hyperlinks when you convert the FrameMaker file to electronic delivery formats. They are also live links while editing within FrameMaker documents.

When you set up a cross-reference, you create a link between the cross-reference and the referenced location (see page 176). FrameMaker calls this referenced location the "source"—the place where the information is coming from.

When you insert a cross-reference, it is formatted as part of the current paragraph, unless your cross-reference format specifies something different. Several default cross-references formats are provided, and you can create additional ones as necessary.

Cross-references look like text, but they act a little differently:

- **You cannot edit cross-references as part of the text.** You can't even click inside the cross-reference text. When you click on text that's part of a cross-reference, FrameMaker selects the entire cross-reference. To change it, you must edit the cross-reference inside the Cross-Reference pod. This behavior is useful for identifying something as a cross reference. Just click on the text and see whether the entire reference highlights. If you can position the cursor in the text, then it's regular text. If the reference is selected as a block, then it's a cross-reference. (though, technically, it could also be a variable)

A real cross-reference is selected as a single block.   "How to consume chocolate truffles" on page 88

Regular text   "How to consume chocolate truffles" on page 88

- **When searching for text, cross-references are ignored.** If, for example, you search for the word "page," none of the occurrences inside cross-references will match. The Find/Change pod does, however, give you the option of searching for cross-references, unresolved cross-references, or cross-references that use a specific format.

- **Cross-references are hyperlinks inside a FrameMaker document.** CTRL+ALT+CLICK on the cross-reference to jump to the source of the cross-reference.

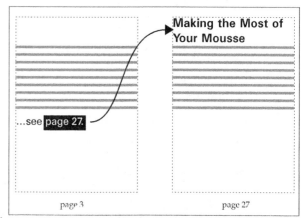

By default, cross-references are updated when you update a book file, open a file, save, or print. They are not, however, updated immediately when the file changes. For example, if you create a cross-reference to an item on page 17, then rearrange your file so that the referenced information moves to page 23, FrameMaker does *not* immediately update the reference to page 17. When you perform an update, the page number changes.

How Cross-References Work

When you create a cross-reference, FrameMaker inserts a marker at the source. The cross-reference itself points to the unique identifier (the marker) in the source paragraph.

Inserting Cross-References

When you insert a cross-reference, you must define the link's source and specify the formatting. Paragraph-based cross-references are the most common; another alternative is spot cross-references.

Creating paragraph-based cross-references

To create a paragraph-based cross-reference, follow these steps:

1. Open the file that contains the source paragraph. (Not necessary if you're creating a cross-reference within a single file.)
2. Locate the source paragraph that you want to point to. Make a note of its paragraph tag and what file it's in.
3. Position your cursor where you want to insert the cross-reference.
4. Select **Special > Cross-Reference**, or otherwise display the Cross-Reference pod.
5. In the Document drop-down list, click the file that you want to reference. Notice that only currently open files are available in this list. If your file isn't listed, return to step 1.
6. Select the Source Type drop-down, then select Paragraphs from the pop-up menu.
7. In the Paragraph Tags list, click the paragraph tag you want. The right side of the pod displays the first few words of each occurrence of that paragraph tag.
8. In the Paragraphs list, select the specific paragraph you want to point to. To help you locate the correct paragraph, an approximate page range for the items is displayed above the list.

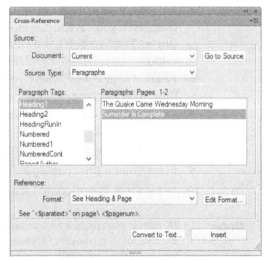

9. In the Format drop-down list, click a cross-reference format. If the format you want isn't available, you can create a new one. See "Setting Up Cross-Reference Formats" on page 179.
10. Click the Insert button to insert the cross-reference.

> **Note** If you selected some text or are modifying an existing cross-reference, you see a Replace button instead of an Insert button.

After inserting a cross-reference, you can double-click it to display the Cross-Reference pod and make changes (such as choosing a different format).

Creating spot cross-references

Paragraph references always point to the beginning of the paragraph, and when the paragraph is lengthy and splits across two pages, the information you're referencing can end up on a different page. Spot cross-references let you point to a particular location in a paragraph. To ensure that you point to the exact location where the information occurs, use a spot cross-reference. You create a marker and link the cross-reference to that marker.

To create a spot cross-reference, follow these steps:

1. Locate the spot that you want to link to and position your cursor there.
2. Select **Special > Marker**.
3. In the Marker Type drop-down list, click Cross-Ref.
4. In the Marker Text field, type a label for this cross-reference that is unique across the file. The label should be short and descriptive.
5. Click the New Marker button to save your changes. If you have text symbols turned on (**View > Text Symbols**), you will see a marker symbol.

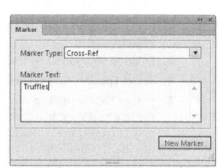

6. Position your cursor where you want to create the cross-reference.
7. Select **Special > Cross-Reference**, or otherwise display the Cross-Reference pod. (see page 177)
8. In the Document drop-down list, click the file that you are linking to.
9. In the Source Type pop-up menu, click Cross-Reference Markers.
10. In the Cross-Reference Markers list, select the marker you created.

Note   Whenever you create a paragraph-based cross-reference, FrameMaker automatically inserts Cross-Ref markers. However, those markers usually include the paragraph tag name and the text of the paragraph; yours is shorter.

11. In the Format drop-down list, click a cross-reference format.
12. Click the Insert button to insert the cross-reference.

Note   For spot cross-references, you usually want to select a format that displays just the page number and not the paragraph's text.

Setting Up Cross-Reference Formats

FrameMaker provides some default cross-reference formats, but you can also create your own new formats. To do so, you assemble a combination of text, character tags, and building blocks to create a cross-reference format. *Building blocks* are placeholders; they change for each cross-reference. For example, a cross-reference definition might read:

see page <$pagenum>

<$pagenum> is a building block; it is replaced with the page number of the source paragraph.

After you have created your cross-reference formats, you can import them into other documents or a book to make formats consistent across files.

To create a new cross-reference format, follow these steps:

1. Select **Special > Cross-Reference**.
2. Click the Edit Format button.
3. You can change an existing format or create a new one:

    - To change an existing format, click that format in the Formats list.
    - To create a new format, type the name for the new format in the Name field. The name is what you will select when creating a new cross-reference, so the name should be descriptive. If you're creating a new Chapter cross-reference format, you might call it:

    Chapter "title"

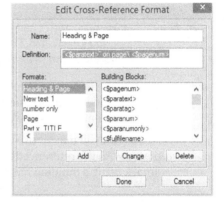

4. In the Definition field, modify the existing definition as needed or create a new definition.

    Cross-references usually have a combination of text (like "See also") and building blocks (like <$paratext>). You can select the building blocks from the list or type them in. You can also apply character formatting by inserting character tags, which are available at the bottom of the building block list. All building blocks begin with an angle bracket (<) and end with an angle bracket (>). Inside the angle brackets system variables are preceded by a "$", and character tags are not.

    Refer to Table 9-1 on page 182 for a list of available building blocks. For the new Chapter cross-reference, the definition could look like this if you type the special characters instead of using copy and paste:

    Chapter\ <$chapnum>, \'<$paratext>\'

Note A backslash followed by a space (\ ) creates a nonbreaking space. A backslash followed by a backtick (commonly found under the Escape key on U.S. keyboards) creates an opening curly quote. A backslash followed by a regular straight quote (\')creates a closing curly quote.

If you choose, you can copy and paste the curly quotes into the cross-reference definition instead; the resulting cross-reference will be the same:

Chapter\ <$chapnum>, "<$paratext>"

Note For more information about typing special characters in dialog boxes, see "Entering Special Characters in dialog boxes" on page 100.

5. To create a new format, click the Add button. To modify an existing format, click the Change button.

Caution If you select the Change button instead of Add, the original format is replaced with the one you just created. This means you renamed and redefined the cross-reference format you started with.

6. Click the Done button to return to the Cross-Reference pod. If the format you modified is already in use in your document, FrameMaker prompts you to update the existing cross-references. Click the OK button to update your cross-references.

Occasionally, you may notice that the building block is displayed in your text instead of being replaced with the referenced information. This is usually caused by a typo in your definition; for example, using a building block with a missing bracket:

$paratext>

To correct the preceding problem, add the opening bracket in the format's definition. The building block should then be replaced with the correct information.

Formatting cross-references

Inside the cross-reference definition, you can use character tags to create special formatting. Without character tags, the cross-reference will match the format of the surrounding paragraph. Though you could format with character tags applied to the cross-reference instance itself, doing so can lead to difficult consistency and maintenance issues.

One very common requirement is to create a cross-reference that is blue and underlined for hyperlinks. When you convert the FrameMaker file to a PDF file, users reading the PDF file can see where the links occur.

To create a blue, underlined cross-reference, follow these steps:

1. Create a character tag called hyperlink that applies blue and an underline. (For information on creating character tags, see page 143.)
2. Select **Special > Cross-Reference**. The Cross-Reference pod is displayed (see page 177).
3. Click the Edit Format button. The Edit Cross-Reference Format dialog box is displayed (see page 179).
4. In the Name field, type a name for the cross-reference format; for example:

   HeadingPDF

5. In the Definition field, insert the definition for the cross-reference format; for example:

   <hyperlink><$paratext><//>

   In this example, <hyperlink> is the character tag that formats the cross-reference and </> (which is the abbreviation for <Default Para Tag>) ends the character tag.

6. Click the Add button to add the new format.
7. Click the Done button.

Tip You can use the third-party tool FrameMaker-to-Acrobat TimeSavers from Microtype (www.microtype.com) to create formatting that is displayed in the onscreen version of the PDF file but not in the print version. For more information, see Appendix A, "Resources."

Cross-reference building blocks

FrameMaker provides several building blocks for creating cross-reference formats. A very basic cross-reference looks like this:

<$paratext> on page <$pagenum>

It tells FrameMaker to insert the contents of the source paragraph (<$paratext>) followed by a space, the words "on page", and another space. Then, it inserts the page number of the source paragraph. The result would look something like this:

Chocolate Mousse Pie on page 27

Table 9-1 lists the available building blocks. They are grouped by function; all of the building blocks that display paragraph text are in one group, all building blocks that display paragraph numbers are in another group, and so on.

Table 9-1. Cross-Reference Building Blocks by Function—Not in FrameMaker Order

Building Block	Description
Text	
<$paratext>	Displays the text of the linked paragraph.
<$paratext[paratag]>	Works from the linked paragraph toward the beginning of the document to locate a paragraph that uses the specified paragraph tag. Displays the text of that paragraph. If no paragraph in the document matches, nothing is displayed by this building block.
<$paratext[tag1,tag2, tag3]> *(not shown in the pod)*	Displays the first instance of paragraph tag1, tag2, or tag3, whichever occurs first. Useful when referencing headings that might use a different paragraph tag later, such as a chapter that you are considering making into an appendix. For example: <$paranumonly[ChapterTitle, AppendixTitle]>
<$paratext[+,paratag]> *(not shown in the pod)*	Searches for the last paragraph using the specified paragraph tag on the page instead of the first. (This building block is also available in running header/footer variables, where it's often used for dictionary or directory headers.)
Paragraph number	
<$paranum>	Displays the paragraph number of the linked paragraph.
<$paranum[paratag]>	Works from the linked paragraph toward the beginning of the document to locate a paragraph that uses the specified paragraph tag. Displays the paragraph number of that paragraph. If no paragraph in the document matches, nothing is displayed.
<$paranumonly[paratag]>	Works from the linked paragraph toward the beginning of the document to locate a paragraph that uses the specified paragraph tag. Displays the numeric portion of that paragraph's number. If no paragraph in the document matches, nothing is displayed.

Table 9-1. Cross-Reference Building Blocks by Function—Not in FrameMaker Order (Continued)

Building Block	Description
<$paranumonly>	Displays the only the numeric portion of the linked paragraph's paragraph number. For example, if the paragraph number is "Figure 17," this building block displays "17." Chapter and volume numbers are considered numeric. The <$paranumonly> building block includes everything from the first building block to the last in the autonumber definition, including any embedded text. If the autonumber definition is Figure\ <$chapnum>\+<n+>, this building block returns 1-4 (or whatever the current figure number is). Notice that the nonbreaking hyphen between the numbers is included in the definition. Also note that other punctuation that is not part of the paragraph numbering will be removed (such as an ending period). A complex numbering system including punctuation may produce unexpected results with this building block.
Page numbers	
<$pagenum>	Displays the page number of the linked paragraph.
<$pagenum[paratag]>	Works from the linked paragraph toward the beginning of the document to locate a paragraph that uses the specified paragraph tag. Displays the page number of that paragraph. If no paragraph in the document matches, nothing is displayed.
<$chapnum>	Displays the current chapter number.
<$volnum>	Displays the current volume number.
<$sectionnum>	Displays current section number as defined in book file.
<$subsectionnum>	Displays current subsection number as defined in book file.
Other	
<$paratag>	Displays the name of the linked paragraph tag.
<$filename>	Displays the file name of the linked file.
<$fullfilename>	Displays the path and file name of the linked file.
<character_tag>	Applies the specified character tag to the items that follow in the cross-reference definition. Note that this building block does not begin with a dollar sign ($) inside the angle brackets.
<Default Para Font> or </>	Removes any character formatting and returns to the regular paragraph formatting of the parent paragraph.

Cross-reference examples

Table 9-2 shows cross-reference formats you might need to create. Change the italicized names in the table to match your tag names. For more information about typing special characters in dialog boxes, see "Entering Special Characters in dialog boxes" on page 100.

Table 9-2. Cross-Reference Sample Definitions

Result	Definition
see page 3-17	see page\ <$chapnum>\+<$pagenum>
see Volume II, Chapter 7	see Volume\ <$volnum>, Chapter\ <$chapnum>
Chapter 7, "The Glories of Chocolate"	Chapter\ <$chapnum>, \'<$paratext>\'
see "Tempering chocolate" in Chapter 7, "The Glories of Chocolate"	see \'<$paratext>\' in Chapter\ <$chapnum>, \'<$paratext[*ChapterTitle*]>\'
see *Tempering chocolate* in Chapter 7, "The Glories of Chocolate"	see <*Emphasis*><$paratext></> in Chapter\ <$chapnum>, \'<$paratext[*ChapterTitle*]>\'
The file name is E:\mattrsullivan\data\projects\beach.fm.	The file name is <*Menu*><$fullfilename></>.

Deleting cross-reference formats

FrameMaker gives you the ability to delete cross-reference formats from your document. When you do this, any cross-references that use the format are converted to text (see page 189 for details).

To delete a cross-reference format, follow these steps:

1. Select **Special > Cross-Reference** to display the Cross-Reference pod (see page 177).
2. Click the Edit Format button to display the Edit Cross-Reference Format dialog. (see page 179).
3. In the list on the left, click the format you want to delete.
4. Click the Delete button.

Note You can delete multiple cross-reference formats by clicking each format in turn and clicking the Delete button (you cannot SHIFT+click to select several at once).

5. Click the Done button. If the cross-reference format is used in the document, FrameMaker prompts you to confirm that you want to convert the cross-references to text. You are prompted separately for each cross-reference format that's in use. Click the OK button to delete the cross-reference format and convert it to text; click the Cancel button to retain the cross-references that use the specified format.

Updating cross-references

FrameMaker updates your cross-references when any of the following actions occur:

- Opening a file
- Saving a file
- Printing a file
- Updating a book file (cross-references are updated for all files in the book)

If you need to update a single cross-reference, you can also double-click the reference to display the Cross-Reference pod, then click the Replace button. This works well for one or two references, but if you need to update all the references in a document (or book), the techniques described in this section are more efficient.

Updating Cross-References in a Book

If you are working in a document that's part of a book, whenever you update the book file, FrameMaker updates the cross-references for you. To update the book, go to the book window, then click the Update Book ( ) button at the top of the window.

This process is described in detail in "Updating a book" on page 329.

Forcing a cross-reference update

Occasionally, you'll want to force FrameMaker to perform a cross-reference update. To do this, follow these steps:

1. Select **Edit > Update References** to display the Update References dialog.
2. Check the All Cross-References checkbox, then click the Update button.

FrameMaker updates all of the cross-references in the current document.

Note   The update works only if FrameMaker can open all of the source documents.

Automatic updating problems

When FrameMaker performs an automatic update, it checks each cross-reference to ensure that the paragraph or marker specified in the cross-reference still exists. For references within a single file, this works perfectly. You can run into problems, though, when the reference is one file to another. To verify the cross-reference's source, FrameMaker opens the source file and looks for the source item. If FrameMaker cannot open the source file for some reason, the updating process fails.

Caution The most common causes of this failure are missing resource messages, like the Missing Fonts dialog. The "missing fonts" message that's displayed when you open the file causes the cross-reference verification process to fail. FrameMaker reports that you have unresolved cross-references when in fact you merely have a cross-reference to a file that FrameMaker couldn't open. To prevent this problem, open all of the files that the current file points to, update your cross-references, and see whether the message about unresolved cross-references is displayed.

Preventing automatic updates

In some cases, you may want to prevent FrameMaker from updating the cross-references automatically; for example, if your FrameMaker files are in a version control system, and you've only checked out one file.

To prevent FrameMaker from trying to update your cross-references inside a file, follow these steps:

1. Select **Edit > Update References** to display the Update References dialog. (see page 185)
2. Click the Commands button and select Suppress Automatic Reference Updating from the pop-up menu. This displays the Suppress Automatic Reference Updating dialog.
3. Check the Suppress Automatic Updating of All Cross-References checkbox.
4. Click the Set button.
5. In the Update References dialog box, click the Done button. (It is displayed in place of the Cancel button.)

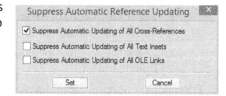

Note You can perform the same action for an entire book by selecting **Edit > Suppress Automatic Updating** when the book is the active window.

To turn automatic updating back on, repeat this procedure and uncheck the checkbox.

Renaming files without breaking cross-references

If you rename a file from inside the book file, FrameMaker will update the cross-references (also text insets and hypertext links) to that file.

Caution   FrameMaker checks only the files in the current book. If the file you're renaming is used in another book, references in that book are *not* updated.

To rename a file in a book file, follow these steps:

1. *(optional)* Open all the files in the book by selecting SHIFT+**File > Open All Files in Book**. This ensures that the update will work even if you have missing fonts in some of the files.
2. In the book window, click on the file you want to rename.
3. Click the file name again to make it editable. Type in the new name for the file, then press ENTER.
4. FrameMaker notifies you that it is going to update cross-references, hypertext links, and other items within the book file. Click the OK button to continue.

FrameMaker makes the needed updates and changes the file name.

Correcting unresolved (broken) cross-references

Cross-references can break or become *unresolved*. Most often, unresolved cross-references are caused by one of the following:

- Renaming or moving the file that the reference points to
- Deleting the paragraph that the reference is linked to
- Cross-referencing to text insets (for details on text insets, see Chapter 29, "Content reuse with text insets")
- Deleting the cross-reference marker at the source location

Unresolved cross-references are not flagged with any special formatting (though that would be nice!). When you open a FrameMaker file, cross-references are checked. If any are unresolved, FrameMaker

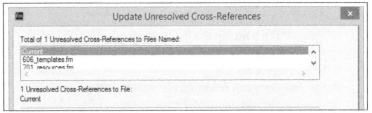

displays the Update Unresolved Cross References dialog. While it does tell you how many references are unresolved, you will likely need to use the Find/Change dialog in order to find them.

Note   Of course, if you suppress automatic updating, FrameMaker does not check for unresolved references when you open the file.

Cross-references are also updated when you update your book file. The book error report will list the unresolved cross-references. You can click each entry to go directly to that cross-reference so that you can fix it. (For details about the book error log, see "Troubleshooting book updates" on page 330.)

You can correct unresolved cross-references in two different ways:

- By changing the file that the reference points to
- By re-creating the cross-reference

Caution You may see "bogus" unresolved cross-reference messages. To verify references, FrameMaker peeks into the source file. If FrameMaker can't open the file, you get an unresolved cross-reference error. To ensure that your unresolved cross-references are not because of a file-opening error, open source files before opening the file in which the cross-reference occurs. If this approach eliminates the message, the cross-reference is alright. Missing fonts and other resources in the source file, and locked (or In Use) files on your system are common causes for this behavior.

Changing the referenced file

If the cross-reference is broken because the file it's pointing to has been renamed or moved, FrameMaker provides you with a very useful way of retargeting the references to the new file name. This actually works for all of the references that are using the old file name.

To update the file name that references are pointing to, follow these steps:

1. Select **Edit > Update References**.
2. Click the Commands button, then select Update Unresolved Cross-References from the pop-up menu.
3. At the top, this dialog box lists the total number of unresolved references in the file. Click on a file name in the list to see how many references to that file are unresolved.
4. For each renamed file, select the old name of the file at the top, then use the bottom half of the dialog box to locate the new file name. Click the file, then click the Open button.
5. When you have finished updating file locations, click the Close button to exit the dialog.

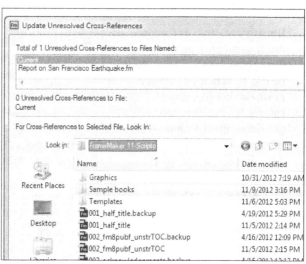

Note As you resolve the references, the numbers at the top of the dialog box are updated.

Replacing broken cross-references

If re-targeting the new file name doesn't work (or if you didn't move or rename the file), you'll need to rebuild the cross-reference. Locate each unresolved cross-reference (use **Edit > Find** and select Unresolved Cross-References in the Find drop-down list), double-click it to display the Cross-Reference pod, and recreate the reference.

Converting cross-references to text

When you convert cross-references to text, FrameMaker eliminates all of the special linking functions and changes the reference into ordinary editable text.

> **Caution** If you change your mind about converting to text, you will have to re-create the cross-references. The conversion to text cannot be undone.

Changing cross-references to text means that they are no longer live links and will not appear as such in any output. It destroys the link between the cross-reference and the source paragraph or marker. Cross-references to files outside your current book are an example of when you might benefit from a text-based cross-reference.

If you need to convert some or all of your cross-references to text, follow these steps:

1. Select **Special > Cross-Reference**.
2. Select the Convert to Text button.
3. Specify which cross-references you want to convert:

    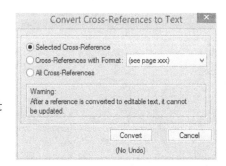

    - Selected Cross-Reference
    - Cross-References with Format:
      (All cross-references using the specified format will be converted to text.)
    - All Cross-References
4. Click the Convert button.

    FrameMaker converts the specified formats to regular, editable text.

A note regarding cross-references and text insets

FrameMaker will recognize Cross-Ref markers inside of text insets. To set a cross-reference to a marker like this, specify the "Source type" in the Cross-Reference pod to use Cross-Reference Markers.

However, you should refrain from inserting cross-references in your text insets. You may have difficulty updating the pagination, especially when the target of the cross-reference is outside of the file or outside of your current book.

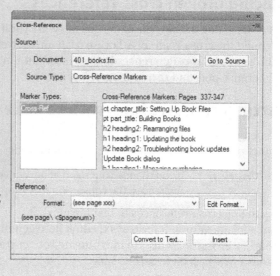

Chapter 10: Storing content in variables

Variables are small chunks of information that you can use throughout your document. By inserting variables, you save time typing and improve accuracy. More importantly, variables make it easier to update repetitive content your documents.

FrameMaker supports two types of variables:

- **System variables.** Display information updated by your computer. Many system variables relate to the time and date. The Current Date variable displays the current system date. Other system variables apply specifically to the current file. For example, there are four variables that display chapter number and other values that you set in the document numbering properties. Although you cannot create or delete system variables, you can edit how information from system variables is displayed.

- **User variables.** Allow reuse of small chunks of content, storage of text strings and formatting for easy access. A user variable might contain words that you use often, such as a book name. You can also store terms that are likely to change, such as a version number or product codename. User variables are static— you must edit a user variable definition yourself to change its value.

Here's a look at what's in this chapter:

```
Advantages of variables. . . . . . . . . . . . . . . . . . . . . . . . . . . . . . . . . .191
Inserting variables  . . . . . . . . . . . . . . . . . . . . . . . . . . . . . . . . . . . .192
    Using the Variables pod . . . . . . . . . . . . . . . . . . . . . . . . . . . . . .192
    Using Smart Insert for variables . . . . . . . . . . . . . . . . . . . . . . . .193
System variable definitions . . . . . . . . . . . . . . . . . . . . . . . . . . . . . .193
    Modifying system variables . . . . . . . . . . . . . . . . . . . . . . . . . . .193
    Valid locations for system variables  . . . . . . . . . . . . . . . . . . . .193
    Date and time variables  . . . . . . . . . . . . . . . . . . . . . . . . . . . . .194
    Numerical variables . . . . . . . . . . . . . . . . . . . . . . . . . . . . . . . . .195
    Filename variables . . . . . . . . . . . . . . . . . . . . . . . . . . . . . . . . . .196
    Table variables  . . . . . . . . . . . . . . . . . . . . . . . . . . . . . . . . . . . .197
    Running header/footer variables . . . . . . . . . . . . . . . . . . . . . .197
        Displaying text from a marker  . . . . . . . . . . . . . . . . . . . . .198
        Displaying dictionary-style headings . . . . . . . . . . . . . . . .200
        Displaying condition tags  . . . . . . . . . . . . . . . . . . . . . . . .200
    Updating System Variables . . . . . . . . . . . . . . . . . . . . . . . . . . .201
Creating user variables  . . . . . . . . . . . . . . . . . . . . . . . . . . . . . . . .201
    Modifying user variable definitions  . . . . . . . . . . . . . . . . . . .203
    Converting user variables to text . . . . . . . . . . . . . . . . . . . . . .204
    Deleting user variables . . . . . . . . . . . . . . . . . . . . . . . . . . . . . .205
```

Advantages of variables

Variables consist of building blocks (in system variables only), character tags, and text. *Building blocks* (pieces of code that display a specific value) do much of the work within system variables. For example, the <$paratext> building block displays text from a specific paragraph, such as a heading. If the heading changes, the <$paratext> building block displays the updated text.

Character tags let you format all or part of a variable. Including the formatting within the variable means that you don't need to apply a character tag to the variable after inserting it in the document. For instance, you can include an italicized character tag in a book title variable, and the book title will be italicized when you insert it.

Inserting variables

Access the Variables pod by selecting **Special > Variables**. System variables are marked with a red gear, and user variables are marked with a blue gear. The definition for a variable is displayed next to the name, along with the document in which the variable resides.

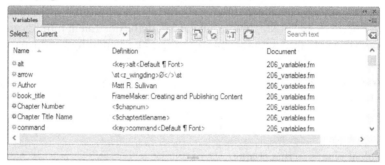

Like cross-references, variables look like standard text on the page until you try to click on them. Clicking selects the entire variable because variable information is not actually part of the text. Also,

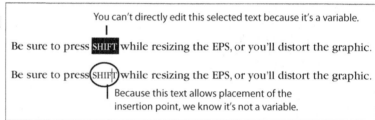

FrameMaker doesn't spell-check variables. Misspelled words must be corrected from the variable definition, which updates all instances of the variable in your document.

Some system variables are available only on the master pages. For example, the eighteen running header/footer variables can display information such as chapter and section headings in the header or footer, and must be inserted on the master pages. The Current Page Number variable inserts the page number on the master page. These variables are not available when you display the Variable pod on a body page.

FrameMaker provides two ways to insert a variable:
- Using the pod
- Using a Smart Insert for Variables

Using the Variables pod

To insert a variable using the Variables pod, follow these steps:
1. Position the cursor on the page.
2. Select **Special > Variable** or otherwise display the Variable pod.Creating Output.
3. Click the variable name, then click the Insert (🔁) button.

> Adding Variables to Documents
>
> Starting with FrameMaker 11, you can select the scope of variables viewed in the pod to Current, All Open Docs, or a specific document. This makes it easier to shuffle variables from one document to another.

Using Smart Insert for variables

To insert a variable using a keyboard shortcut, follow these steps:

1. Position the cursor on the page.
2. Press CTRL+0 to display Smart Insert for Variables.
3. Display the variable you want by doing any of the following:
   - Use the arrow keys or scroll bar to scroll through the list of available variables.
   - Type the first few letters of the variable name to uniquely identify the variable needed.
4. Press ENTER or double-click to insert the variable.

Note Some of the keys on your computer may be specially programmed. If so, one or more keyboard shortcuts may not work as described. Consult your system documentation for assistance.

System variable definitions

When you modify a system variable, you work with various building blocks. Different system variables use different building blocks. For example, FrameMaker provides several building blocks that return the value of the current month, such as <$monthname> and <$monthnum>. The date-related building blocks are valid only in variables that display dates. If you enter a date building block in a non-date variable such as the Filename variable, an error is displayed.

Modifying system variables

To modify a system variable, follow these steps:

1. Select **Special > Variables...** or otherwise view the Variables pod.
2. Do one of the following to display the Edit System Variable pod:
   - In the Variables pod, select the variable you wish to modify, and click the Edit (✎) button
   - Double-click the variable name you wish to modify
3. Modify the definition of the variable, adding formatting and changing codes as needed.
4. Click the Update button to save your changes.

Valid locations for system variables

System variables may not be available on all types of pages.

Table 10-1 shows the locations in which system variables are valid.

Table 10-1. Valid Locations for System Variables

Location	Available System Variables
Any text frame (body, master, or reference page)	Page Count Current Date, long or short Modification Date, long or short Creation Date, long or short Filename, long or short Volume Number Chapter Number Section Number Sub Section Number Chapter Title Name FM User Directory
Table title, heading, and footing rows	Table Continuation Table Sheet
Master pages only, in text frames with no flow tag	Current Page # Running H/F 1, 2, 3, 4,...,18

Date and time variables

FrameMaker provides system variables that display the month, day, and year in long and abbreviated formats. You can access the following date- and time-related variables in any text frame:

- **Creation Date:** Displays the month, date, and year when you created the file, such as November 14, 2001 or 11/14/01.
- **Modification Date:** Displays the month, date, year, and time the document was last opened or saved, such as September 29, 2015 10:02 am or 9/29/15. Updated when you open or save the file.
- **Current Date:** Displays the current month, date, and year, such as 10:52, September 30, 2015 or 9/30/15. Updated when you change the page magnification or display a different page.

These variables consist of building blocks that display different aspects of the date and time. You can change the building blocks in each variable, but you cannot change the function of the variable. For example, you can add seconds to the Modification Date variable by inserting the <$second> building block. However, you cannot force the Modification Date variable to display the *creation* date because FrameMaker predefines the function of system variables.

Note If you change the language for a paragraph that includes a date or time variable, the variable is updated to use the date or time format for that language.

Table 10-2 describes the building blocks used in date variables.

Table 10-2. Building Blocks for Date Variables

Building Block	Description	Example
<$ampm>	Lowercase morning or evening abbreviation	am or pm
<$AMPM>	Uppercase morning or evening abbreviation	AM or PM
<$dayname>	Name of the day	Monday
<$shortdayname>	Abbreviated name of the day	Mon
<$daynum>	Number of the day	20
<$daynum01>	Number of the day with leading zero	09
<$hour>	Hour	1
<$hour01>	Hour with leading zero	01
<$hour24>	Hour in military format	13
<$minute>	Minute	5
<$minute00>	Minute with leading zero	05
<$monthname>	Name of the month	September
<$shortmonthname>	Abbreviated name of the month	Sept
<$monthnum>	Number of the month	9
<$second>	Seconds	8
<$second00>	Seconds with leading zero	08
<$shortyear>	Abbreviated year	30
<$year>	Year	2030

There are additional building blocks available when your system supports typing asian and right-to-left text. For more information about these building blocks, see the FrameMaker online help.

Numerical variables

The numerical system variables display page, volume, and chapter numbers that are automatically updated when you open or save the file. You can insert a numerical variable in any text frame with the exception of the Current Page Number variable, which is available only on master pages.

The following are numerical variables:

- **Current Page Number:** Displays the number of the current page.
- **Page Count:** Displays the number of the last page in the file.
- **Volume Number:** Displays the file's volume number.

- **Chapter Number:** Displays the file's chapter number.

The formats of the page, volume, and chapter numbers depend on settings in the document or book file. You can specify uppercase and lowercase Roman, alphabetic, or text-based numerals, along with standard Arabic numerals. For details, see "Managing numbering" on page 332.

Note If you change the language for a paragraph that includes a numerical variable, the variable is updated to use the numerical format for that language.

Table 10-3 on page 196 describes the building blocks used in numerical variables.

Table 10-3. Building Blocks for Numerical Variables

Building Block	Description	Example
<$curpagenum>	Page number (only available on master pages)	735
<$lastpagenum>	Last page number in the document	758
<$paranum>	Value of autonumber field in paragraph tag	Step 2.
<$paranumonly>	Numerical value of the autonumber field (excludes punctuation and text)	2
<$chapnum>	Chapter number	10
<$volnum>	Volume number	II
<$sectionnum>	Section Number	10
<$subsectionnum>	Sub Section Number	10

Filename variables

Two system variables display your document's file name. You can use the following Filename variables in any text frame:

- **Filename (Long):** Displays the system path and file name.
- **Filename (Short):** Displays only the file name.

Table 10-4 describes the building blocks used in Filename variables.

Table 10-4. Building Blocks for Filename Variables

Building Block	Description	Example
<$fullfilename>	Displays the document's entire file name, including the path.	c:\Book\widget.fm
<$filename>	Displays the document's file name without the path.	widget.fm

Table variables

Starting with FrameMaker 2015, the Table Continuation and Table Sheet variables can be inserted via the Insert Table dialog. See "Inserting tables" on page 150 for more information.

The following system variables are available only in tables:

- **Table Continuation:** Displays the definition of the variable on the second and subsequent sheets of a table. You insert the variable into the title heading or footing row on the first sheet, but the value is not displayed on that sheet. By default, the Table Continuation variable displays "(Continued)," but you can change the text. The default definition of the Table Continuation variable contains a nonbreaking space ( ) which is displayed on the first sheet. You must have the document text symbols turned on to see the character.

    Table 1: Books for Technical Communicators — The nonbreaking space in this Table Contiuation variable displays a square bracket on the first sheet.

- **Table Sheet:** Displays the sheet number and the total number of sheets. By default, the variable displays "(Sheet # of #)," but you can edit the definition.

    Table 1: Books for Technical Communicators (Sheet 1 of 2)§

No building blocks are available in the Table Continuation variable except for character tags. Table 10-5 describes the building blocks used in the Table Sheet variable.

Table 10-5. Building Blocks for Table Variables

Building Block	Description	Example
<$tblsheetnum>	Displays the sheet number.	1
<$tblsheetcount>	Displays the total number of sheets in the table.	2

Running header/footer variables

Eighteen Running H/F system variables are available on the master pages of your document. You can insert these variables into headers and footers in your document. For example, they can display the text or autonumbering from a specific paragraph tag on the right pages of a document and the book title on the left pages. In this book, the left header consists of the Running H/F 1 variable. Here's the variable definition:

   Chapter <$paranumonly[Chapter_Number]>: <$paratext[Chapter_Title]>

In this example, the <$paranumonly> building block displays the value of the Chapter_Number paragraph tag, and the <$paratext> building block displays the value of the Chapter_Title tag (for example, Chapter 25: Cleaning Your Widget).

Running H/F variables often point to the first instance of a specified paragraph tag on the body page. If the variable references the Heading3 paragraph tag, and the body page has two Heading3 paragraphs, FrameMaker displays text from the first occurrence. If the body page lacks a Heading3, the running/header footer is determined by the previous Heading3 in the document. If no Heading3 occurs in the preceding pages in the document, the running header/footer is left blank. The running header/footer does not display text found on master pages.

You can also include an alternate paragraph tag in the variable definition, such as the Heading2 paragraph in the following example:

<$paratext[Heading3,Heading2]>

This variable looks first for a Heading3 paragraph on the page. In the absence of a Heading3 paragraph, text from the first Heading2 on the page will be displayed.

The following three running header/footer building blocks have special functions:

- **<$marker1> through <$marker8>:** Displays text stored in a marker instead of paragraph text. For details, see "Displaying text from a marker" on page 198.
- **<$paratext[+,paratag]>:** Displays the last instance of the specified paragraph tag. For details, see "Displaying dictionary-style headings" on page 200.
- **<$condtag[hitag,…,lotag,nomatch]>:** Searches for the specified condition tags on a page and displays the name of the first condition tag on this list that actually occurs on the page. For details, see "Displaying condition tags" on page 200.

Here are the building blocks used in Running H/F variables.

Table 10-6. Building Blocks for Running Header/Footers

Building Block	Description
<$paratext[paratag]>	Displays text from the specified paragraph tag.
<$paratext[+,paratag]>	Displays the last instance of the specified paragraph tag on the page. Often used to display dictionary-style headings.
<$paratag[paratag]>	Displays the paragraph tag name (rather than contents of the paragraph).
<$condtag[hitag,…lotag,nomatch]>	Displays the specified condition tag or alternate text, if the condition is not found.
<$marker1> through <$marker8>	Displays text from the Header/Footer $1 through Header/Footer $8 markers in the document.
<$paranum[paratag]>	Displays the autonumbering from the referenced paragraph, including punctuation.
<$paranumonly[paratag]>	Displays only the numeric portion of the referenced paragraph's autonumber.

Displaying text from a marker

Instead of using the Running H/F variables to display text from a paragraph, you can display text from a marker. This technique is useful for displaying abbreviated headings or information that is not available in a paragraph. In the following example, the heading on the body page won't fit in the running header/footer text frame. You can type an alternate heading in a special Header/Footer marker and refer to that marker in the Running H/F variable with the appropriate building block. The abbreviated heading from the marker is displayed instead of the long heading.

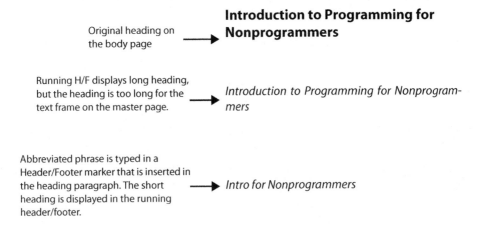

FrameMaker provides eight building blocks (<$marker1> through <$marker8>) and eight markers (Header/Footer $1 through Header/Footer $8) to display marker text in a running header/footer. The <$marker1> building block displays the definition of Header/Footer $1 markers; <$marker2> displays the definition of Header/Footer $2 markers; and so on.

To display text from a marker, follow these steps:

1. Position your cursor where you want the marker.
2. Create the marker using the Marker pod.
    a. Select a Header/Footer marker type.
    b. In the Marker Text field, type the alternate text that you want displayed. As for all markers, the limit is 1023 characters.
    c. Click the New Marker button. The marker is inserted in the paragraph.
3. View the document's master pages.

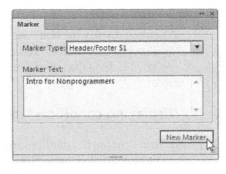

Caution Running H/F 3 and Running H/F 4 along with Running H/F 13–18 are the only variables that can use the Header/Footer building blocks.

4. Position your cursor in the header or footer as needed.
5. From the Variables pod, select the Running H/F variable that calls up the corresponding Header/Footer marker type and click insert.

When you display the body page, the marker text is displayed in the running header/footer. If the Running H/F variable does not find a marker on the page with the Running H/F variable, the running header or footer is left blank.

Displaying dictionary-style headings

By default, building blocks display the first instance of a paragraph tag on the page, but you can also use a variable to display the value of the last specified paragraph tag on the page. In the following example, the word at the top of the second column is also the last dictionary term on the page. The <$paratext[+,paratag]> building block in the running header refers to the Term paragraph tag.

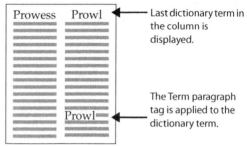

To create a dictionary-style running header/footer, follow these steps:

1. View the document's master pages.
2. From the Variables pod, select the Running H/F variable you want to change and click Edit ( ) to display the Edit System Variable pod.
3. Type **<$paratext[+,paratag]>** in the Definition field. (This building block is not displayed in the list of building blocks, so you must type it in.)
4. Replace **paratag** with the appropriate paragraph tag.
5. Click the Update button.
6. Click the Insert ( ) button in the Variables pod.

When you display the body page, text from the last occurrence of the paragraph tag you specified is displayed in the running header/footer.

Displaying condition tags

In your running header/footer, you can display a condition tag applied to text on the current page. To do so, you use the <$condtag[hitag,...lotag,nomatch]> building block. This building block includes one or more condition tags and, optionally, text that is displayed when the page has no condition tags, as in the following example:

Here, FrameMaker first searches the page for the Print condition tag. If found, "Print" is displayed in the running header/footer. If not found, FrameMaker looks for the Online condition tag and displays it's name, if found. If it is not found, the last value is displayed. This can be either a phrase, such as "No Conditional Text," or a blank space (if you use a nonbreaking space for the last value).

The ellipsis in the <$condtag[hitag,...lotag,nomatch]> building block indicates that you can provide additional condition tags, and FrameMaker will also search for them. It's essential, however, to indicate what FrameMaker should display if no condition tags are found. Without the nomatch value, FrameMaker displays the last condition tag in the building block ("Online," in this example), even if the condition tag isn't actually on the page.

Note This feature is ideal for tagging pages in classified material. The conditions might be called TopSecret, Secret, Confidential, and Unclassified. You can specify that the page header or footer display the highest classification tag that occurs on the current page. As described in the previous paragraph, it is critical to provide the Unclassified tag so that pages aren't flagged as classified when they are not.

To set up a condition tag variable, follow these steps:

1. Display the Variables pod. (see page 192)
2. Click the Running H/F variable you want to edit (if it's not already highlighted), then click the Edit (✎) button to display the Edit System Variable dialog box. (see page 200)
3. Click the <$condtag[hitag,...lotag,nomatch]> building block to add it to the variable definition, or type it in the Definition field.
4. In place of **hitag,...lotag**, type the condition tags you want to display, for example, "Print,Online."
5. In place of **nomatch**, type one of the following:
   - The text you want displayed if no match is found, such as "No Conditional Text."
   - A nonbreaking space to display nothing if no match is found. Type \ followed by a space.
6. Click the Edit button, then click the Insert button (or the Replace button if the variable was already on the master page).
7. Display the body page. The variable text is updated.

Updating System Variables

System variables (other than Running H/F variables) don't dynamically update their content; they must be updated by you or the system. For example, if you insert a current date variable that includes the hour, minute, and seconds, it will not be updated automatically as the time changes. The content is updated when you open or save the document. You can also force an update of the variables without saving the file.

To update, click the Update button in the Edit System Variables pod. Click the OK button in the confirmation dialog to update the variables.

Creating user variables

User variables provide a way to store common terms and update words that change often. Variables can also apply formatting, making them ideal for heavily formatted phrases like logotype. For example, during development, a software application may have a code name, such as Orion. By saving the code name in a Product_Name user variable, you can use the variable in your document. Once the product receives its final name, you can easily modify the variable. To update the term across an entire book, you import the variables from one document into all the documents in the book file.

In general, it is a good idea to name a variable for purpose, and not its content or formatting. For example, suppose you create a variable containing the name of a book's author—Shakespeare—and name the variable Shakespeare. If the author changes to Jack Smith, the variable name becomes both hard to remember and confusing. A more generic name, such as Author, would be a better choice.

However, in some cases, you might decide to make the variable name and definition similar. If you use variables to speed text entry, don't use generic names, such as Product1, Product2, and so on. The variables will be easier to identify if they're named after the product.

Caution FrameMaker doesn't spell-check variables, so make sure you spell the definition correctly.

To create a new user variable, follow these steps:

1. In the Variables pod, click the Create New User Variable button ( ) or otherwise display the Add/Edit Variable pod.

2. In the Name field, type a name for the new variable.

3. In the Definition field, type a definition for the variable. This text is displayed in the document. Along with regular text, you can use character tags in the variable definition by clicking the tags in the Character Formats list or by typing the tags in the Definition field. Variables cannot contain more than 1023 characters. Use the <Default ¶ Font> character tag (which can be abbreviated </> to save characters) where the character formatting should end, as shown in the following example:

   <SmallCaps>shift</> key

   This variable is used in the following sentence to format "shift" in small caps:

   Press the SHIFT key, then select **File > Save** All Open Files.

4. Click the Add button to add the variable to your list of user variables.

5. Click the Done button to close the Edit User Variable dialog box and return to the Variables pod.

6. Click the Done button to close the Variables pod. You can also insert the variable by clicking the Insert button.

Creating user variables

Modifying user variable definitions

After creating a user variable, you can modify the definition by changing the text or adding character tags.

To modify a user variable:

1. Select **Special > Variables...** or otherwise display the Variables pod.
2. Select the variable name in the Variables pod.
3. Click the Edit Variable ( ) button.

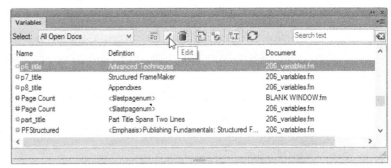

Note If you edit a system variable, the Edit System Variable pod is displayed (not the Add/Edit Variable pod). The Edit System Variable pod includes system variable building blocks not found in the Add/Edit Variable pod.

4. In the Add/Edit Variable pod:
   a. Make any changes needed.
   b. Click the Update button.
5. To add a building block or character style to the definition, position your cursor where you'd like the formatting to begin and enter the building block for the character style from the keyboard or click on it in the Character Formats list.
6. To format only part of a variable, use the <Default ¶ Font> building block to end the formatting (this can be abbreviated </> to save space).
7. Click the Update button to save your changes. The variable definition is modified, and each occurrence of the variable is updated in the current document.

Here is another example:

<Bold>Volume <$volnum>:</> Complete Classics

would display the following formatting:

Volume IX: Complete Classics

203

Controlling Time Display With Building Blocks

As mentioned in "Date and time variables" on page 194, you can use system variables to insert the creation, current, or modification date and time in a document.

FrameMaker provides three building blocks that display time in other system variables and in running header/footer variables:

- **<$creationtime>:** Displays the creation time.
- **<$currenttime>:** Displays the current time.
- **<$modificationtime>:** Displays the time the file was last opened or saved.

These building blocks don't do anything when used alone; instead, they work with the other time building blocks to display time. For example, the <$currenttime> building block is inserted with other time building blocks in the following running header/footer variable definition:

<$currenttime><$hour>:<$minute> <$AMPM>

This variable displays the following text:

11:53 AM

The three building blocks can be used in system variables that are unrelated to time, such as Filename, but they're mostly used in running header/footer variables to create a custom header or footer.

Converting user variables to text

You can convert user variables to standard text. For example, you might insert a term as a variable and decide later that the term doesn't need to be in a variable. When you convert the variable to text, character tags in the variable are applied to the converted term. For example, the Legacy Product variable is formatted by the Emphasis character tag. After converting the variable, FrameMaker applies the Emphasis character tag to the text on the page.

The converted variable can be edited and spell-checked; however, it can't be updated. For this reason, converting variables typically isn't a good idea unless you're cleaning up legacy content and see variables that shouldn't be there.

To convert a variable to text, follow these steps:

1. *(optional)* In the document, select the variable you want to convert.
2. In the Variables pod, click the Convert to Text () button to display the Convert Variables to Text dialog.
3. Do any of the following:
   - Click the Selected Variable radio button.
   - Click the Variables Named radio button, then click the name of a variable to convert.
   - Click the All Variables radio button to convert all variables in the document to text.

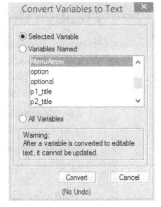

4. Click the Convert button. The variable definition is converted to text. If you click the word, it no longer highlights like a variable, and character tags used in the variable are applied to the converted text.

Deleting user variables

You can delete user variables (but not system variables) from your document. If the variable is in use, FrameMaker displays a warning, allowing conversion to standard text or canceling the action. If you're unsure whether a specific variable has been inserted in the document, search for the variable before you try to delete it. For details on searching for FrameMaker components, see "Searching and replacing" on page 38.

To delete a user variable, follow these steps: >

1. In the Variables pod, select the variable to delete, and click the Delete user variable () button.
2. A confirmation dialog box may display if the variable you want to delete is in the document. Click the OK button, and every instance of that variable in the document is converted to regular, editable text.

Creative Uses for Variables

User variables are great for frequently used words or phrases that often change. The variables may include character tags to format the text. Variables may contain a maximum of 1023 characters, compared to a 255-character limit for markers. Here are some popular uses of variables:

- **Software Release.** Type the software release number, as in "FrameMaker 13.0.0.331."
- **Product Name.** Type the name of the product. During development, your product might have an internal code name, such as "Pegasus." When the name is finalized, you update the variable definition. You could include the product name in the Software Release variable, but the Product Name variable can be used in several different ways with just the name.
- **Menu Item Separator.** Create a variable to represent the angle bracket with the leading and following space. By using the Smart Insert for Variables (CONTROL-0) you can now use the keyboard to rapidly enter this cumbersome string of characters.
- **Copyright.** Type a copyright statement. When the copyright year changes, you update the variable instead of searching for the year.
- **Version.** Type **First Draft** or **Second Draft** to indicate the version of your document.

Part III

Controlling Page Layout

Chapter 11: Understanding master pages

Page layout in FrameMaker is controlled by underlying pages called *master pages*. You assign master pages to the main content pages, which are *body pages*.

Here's a look at what's in this chapter:

Advantages of master pages	.209
Displaying master pages	.210
Assigning master pages to body pages	.210
Manually assign master pages	.210
Mapping paragraph tags to master pages	.211
Creating and managing master pages	.214
Creating default master pages	.215
Creating custom master pages	.215
Renaming master pages	.216
Rearranging master pages	.217
Modifying master pages	.217
Adding Text Flows	.218
Setting up headers and footers	.218
Setting up a watermark effect	.219
Creating a landscape master page	.221
Creating bleeding tabs	.225
Removing master page overrides	.229
Updating master pages and page layouts	.229
Deleting master pages	.231

Advantages of master pages

Master pages let you create a consistent look and feel for your content. For every body page, the assigned master page defines the location of headers and footers, background graphics, page numbers, and other items.

All body pages formatted by the Left master page, for example, will have the same basic layout (which you can override on the body page), and you can define a different layout for body pages formatted by the Right master page. For example, the Left master page might display the book title, and a left-aligned page number, whereas the Right master page displays the Heading1 text and a right-aligned page number.

Master pages may contain:

- **Body page text frames.** These usually contain the main text of the document. Many master pages have just one foreground object—the text frame in which

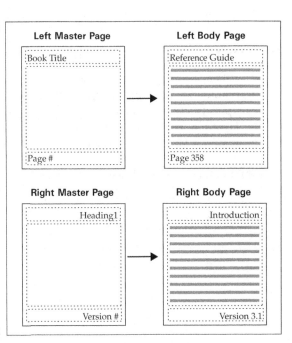

you insert text on the body pages. Chapter 12, "Text flows," describes how to modify foreground objects, such as adding columns and room for side heads.

- **Background text frames.** These are often header and footer text frames, containing static text, page numbers and other generated or referenced content.
- **Background objects.** These might be background graphics or other information that is repeated on the body pages

This chapter deals mainly with background text frames and background objects.

To create foreground and background objects, you place text frames, graphic frames, and graphics on the master pages. When you draw a text frame, you are prompted to specify whether the frame is for background objects or for body page information. Each foreground text frame has a flow tag, which controls how text fills the text frames. As you type text on the body pages, FrameMaker automatically adds new pages and allows the text to flow onto those pages. (See "Understanding text flows" on page 234.) When you place a graphic frame or a graphic on the master page, it automatically becomes a background object. Background objects are not part of a flow.

Note Many page layout properties, such as margin widths or the number of columns, can be modified on either the body or master page. If you update the master page, *all* body pages formatted by the master page are updated. If you update the body page, only that page changes, creating a Master Page override. Use caution when creating overrides, as reimporting Page Layouts from the template may revert to your earlier settings.

Displaying master pages

To see the master pages, select **View > Master Pages** or ALT+V M. To go to a particular master page, scroll down to find it or press CTRL-G and select the master page from the names in the list. To return to the body pages, select **View > Body Pages** or ALT+V D.

Assigning master pages to body pages

When you add content to a single-sided document FrameMaker document, the new pages are automatically set to the Right master page. In a double-sided document, pages will receive the appropriate Left and Right default page assignment. You can, however, manually change the layout by assigning a different master page to one or more body pages. You can also set up a Master Page Assignment table to allow automatic assignment of master pages. (see page 211)

Manually assign master pages

To assign a master page to one or more body pages, follow these steps:

1. Display the body page you want to format.

2. Select **Format > Page Layout > Master Page Usage** or ALT+O L M to display the Master Page Usage dialog.
3. In the Use Master Page section, click the master page you want to assign:

    - To assign default master pages, click the top button. (either Right or Right/Left, depending on pagination options)
    - To assign a master page you created, select that page in the Custom drop-down list.

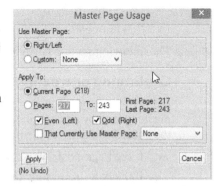

Caution Use the None master page with caution. Assigning the None master page will remove control over that page's formatting—it will no longer respond to updates to master pages. Setting the master page to None is useful for quickly removing all headers and footers on a page prior to assigning a different master page.

4. In the Apply To section, do one of the following:
    - To assign the master page to the current body page, click the Current Page radio button (the default setting). Verify that the page number displayed next to Current Page is correct. The Current Page value is based on the page being displayed, not on the cursor location. For example, if your cursor is on page 3 and the bottom of page 4 is displayed on the screen, the current page is page 4.
    - To assign the master page to a range of pages, type the page numbers in the Pages fields.
    - To assign the master page to even or odd pages within the page range you've specified, check the Even or Odd check box.
    - To assign the master page to all pages in the document, type the page range in the Pages section and check both the Even and Odd checkboxes. Note that FrameMaker provides the first and last page number in the current document to the right of the Pages fields.
    - To change all the assignments from one master page to another, specify the new master page in the Use Master Page section and the old master page in the That Currently Use Master Page drop-down list.
5. Click the Apply button. The master page is applied accordingly.

Mapping paragraph tags to master pages

FrameMaker also lets you map paragraph tags to master pages using settings stored on the reference pages. When reapplying master pages, FrameMaker scans for specified paragraph tags and assigns the correct master pages. For example, you might map the ChapterTitle paragraph tag to the First master page. Instead of manually applying the master page, you apply all master pages in the document with the Apply Master Pages menu choice. The First master page is instantly applied to the page containing the ChapterTitle paragraph tag. This feature also removes all custom master page assignments in the document at once.

Mapping paragraph tags to master pages requires these three steps:

- Confirm the existence of the Master Page Mapping table on the MasterPageMaps reference page.

- Set up the mappings.
- Tell FrameMaker to apply the mappings.

The following sections provide detailed instructions for each step.

Confirming or creating the master page mapping table

FrameMaker uses a mapping table on the reference page named MasterPageMaps to assign master pages by paragraph tag. If that page doesn't exist, you need to create it. The easiest way to confirm existence of the page is to select **Format > Page Layout > Assign Master Pages**.

- If the page exists, you'll get normal confirmation messages related to the process.
- If the page is missing, this command will ultimately create the needed page for you.

To set up the mappings, you'll need to enter paragraph tags and their corresponding master page names on the MasterPageMaps reference page. Once you're finished, master page assignments can be done from both the document and the book level. As a reminder, create the reference page in your template to ensure consistency across your book files. See "Importing settings from a template file" on page 91 for details.

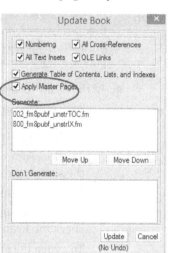

You can also apply master pages as part of the book update process. However, if you don't use the master page mapping feature, you don't want to select this option, as you will lose manual page mappings done with the Master Page Usage dialog.

Setting up your mapping table

After FrameMaker has created the mapping table, you will need to set up the mappings of paragraph tags to master pages.

To set up your mappings, follow these steps:

1. Select **View > Reference Pages** to view the reference pages.
2. Navigate to the MasterPageMaps reference page. On that page, there is a mapping table.

   UnstructMasterPageMaps
   Book Update (Yes or No): Yes

Paragraph Tag Name	Right-Handed Master Page (or Single-Sided Master Page)	Left-Handed Master Page	Range Indicator (Single, Span pages, Until changed)	Comments

3. FrameMaker provides the option to reapply all master pages from a book file. If you do not want this document updated when you reapply master pages from the book, type **No** in the Book Update field.

Note If the Book Update field above the mapping table is blank, FrameMaker uses the default value "Yes."

4. For each paragraph tag associated with a master page, type the tag name in the Paragraph Tag Name column, and the master page name in the Right-Handed Master Page column. If you also want the paragraph tag mapped to a different master page on left-handed pages, specify the master page name in the Left-Handed Master Page column. If you don't specify a master page for the left-hand body pages, the left-hand body pages use the setting for the right-hand body pages.
5. To assign master pages to a range of pages, type the range in the Range Indicator column. For example, the master page might be applied to a single page, a page range, or until you specify another master page. Table 11-1 lists the options available.

Table 11-1. Mapping Paragraph Tags to Master Pages

Field	Description
Required Fields	
Paragraph Tag Name	Paragraph that the master page is automatically applied to. Must be spelled and capitalized correctly, or FrameMaker ignores the setting and applies the default master page.
Right-Handed Master Page (or Single-Sided Master Page)	Master page applied to right body pages (in a double-sided document) or all body pages (in a single-sided document). If the text in this field is not spelled and capitalized correctly, FrameMaker displays an error when applying master pages and stops the process. If empty, the "None" master page is applied.
Optional Fields	
Left-Handed Master Page	Master page applied to left body pages. If this field is empty, FrameMaker applies the master page mapped to the right body pages.
Range Indicator	Pages to which the master page is applied. Options are as follows: • **Single:** Applied only to the page with the specified tag. • **Span pages:** Applied to a range of pages when each page contains the specified paragraph tag. • **Until changed:** Applied to all pages until FrameMaker finds another mapped paragraph tag. For instance, according to the next example, the RightSL and LeftSL master pages are applied until another mapped paragraph tag is found. If the setting is blank, FrameMaker applies the master page to a single page.
Comments	Notes on the mapping.

6. *(optional)* Type helpful notes in the Comments columns to document how the mappings are used.

Here are some sample mappings.

UnstructMasterPageMaps
Book Update (Yes or No): Yes

Paragraph Tag Name	Right-Handed Master Page (or Single-Sided Master Page)	Left-Handed Master Page	Range Indicator (Single, Span pages, Until changed)	Comments
Heading1	RightSL	LeftSL	Until changed	
ChapTitle	ChapTitle	ChapTitle	Single	
Intro	RightNonum	LeftNonum	Until changed	
Notes	NotesRight	NotesLeft	Span pages	

When there is more than one mapped paragraph tag on the same page, FrameMaker applies the master page for the first tag that occurs on the page.

Caution Make sure each row of the mapping table has at least one mapping; paragraphs without an associated master page receive a master page assignment of None when master pages are automatically applied. This removes the current or default master page assignment, which is probably not what you want.

Applying the master page mappings

To apply master pages you've mapped to paragraph tags, follow these steps:

1. Do one of the following:
   - (book file) Select **Format > Page Layout > Apply Master Pages**. Another method is to check the Apply Master Pages checkbox when you update the book.
   - (single file) Select **Format > Page Layout > Apply Master Pages**. The master pages are reapplied.

If applying the master page would cause the mapped heading to fall on the following page, FrameMaker ignores the mapping and applies the default Right or Left master page. See the console for messages about master pages that were not applied.

If you reapply master pages from your book file, FrameMaker displays an error if a file in the book doesn't have a MasterPageMaps reference page and mapping table. You must create a mapping table for one file in the book using the procedure in "Confirming or creating the master page mapping table" on page 212 and then import that file's reference pages into the other files in the book (see "Importing settings from a template file" on page 91) before reapplying master pages from the book file.

Creating and managing master pages

FrameMaker automatically creates default master pages, which control the layout of left and right body pages. You can also create custom master pages for specialized content. The following sections describe how FrameMaker creates default master pages and how you can create your own custom master pages, and how to perform tasks such as renaming or deleting master pages.

Creating default master pages

When you create a new document, FrameMaker creates default master pages for you—Right in a single-sided document, and Left and Right in a double-sided document. If you switch a single-sided document to double-sided, FrameMaker automatically creates the Left master page. You can modify the layout of default master pages, but you cannot rename, reorder, or delete them. For more information on changing the pagination of a document, see "Switching from single- to double-sided pages" on page 253.

Creating custom master pages

In a typical double-sided book, you use Left and Right for regular pages, and then you create custom master pages for cover, title, landscape, rotated, and other special page layouts. There is no practical limit to the number of master pages you can define (more than 100). All master pages in a document must be the same size; however, the pages can have different orientations. For example, you can mix 7-inch by 9-inch portrait pages with 9-inch by 7-inch landscape pages. You can change the paper size for an entire document, but to create a master page with different dimensions, you need to create a new document and define the size. See "Creating a blank document" on page 13 for details.

Tip   You can create a master page while you're viewing a body page, but you're limited to copying the master page applied to the current page. If you create a new master page while viewing the master pages, you have more design options, including copying the layout of any master page or creating an empty page.

To create a master page based on the current body page layout, follow these steps:

1. Display the body page that looks similar to the master page you want to create.
2. Select **Format > Page Layout > New Master Page** to display the New Master Page dialog.
3. Type the name of the new master page in the Master Page Name field, then click the Create button.
4. A confirmation is displayed, explaining how to apply and edit the new master page. Click the OK button to display the new master page.

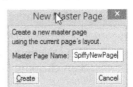

To copy the layout of any other master page or to create an empty master page, follow these steps:

1. Select **View > Master Pages**.
2. Select **Special > Add Master Page** to display the Add Master Page dialog.

3. In the Name field, type a name for the new master page.
4. Do one of the following:

   - Make a copy of an existing master page by clicking the page name in the Copy from Master Page drop-down list. Select this option if an existing master page resembles the page you want to create.
   - Create a blank page by clicking the Empty radio button. Select this option if none of the existing master pages resembles the page you want to create.

5. Click the Add button to display the new master page.

Note   If you copy an existing master page, the new master page will default to the same flow tag as the copied page. For more information about text flows, see Chapter 12, "Text flows."
If you added a blank master page, the new master page will use flow A. Make sure that the new master page you create uses the flow tag you need in your document.

Renaming master pages

When you rename a custom master page, the body pages that use that master page automatically use the renamed page. To rename a master page, follow these steps:

1. Display the master page you want to rename.
2. Click the name of the master page in the status bar to display the Master Page Name dialog.
3. Type the new master page name over the highlighted name, then click the Set button. The master page is renamed.

Note   You can't rename the Left or Right master pages.

Any body pages that used the master page with the original name now use the renamed page. You'll need to manually update any master page mapping assignments to the original name.

Caution   If the body pages have overrides to the master page formatting, those overrides remain. Whenever you move away from viewing the master pages, you will be prompted to keep the overrides on the body pages or remove them. Consider a workflow that minimizes overrides.

Rearranging master pages

Initially, master pages are displayed in the order you create them. You might prefer placing frequently used master pages at the beginning for easier access. You can rearrange custom master pages to display them in a different order in drop-down lists and in the master page view, as shown in the following example.

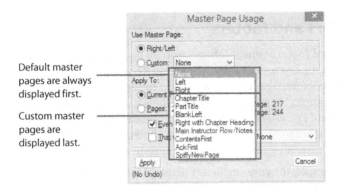

Default master pages are always displayed first.

Custom master pages are displayed last.

The order of the master pages can't be imported into other FrameMaker files; you'll need to rearrange them in each file. (However, new files based on the file with the reordered master pages retain the order.)

To rearrange the master pages, follow these steps:

1. Display the master pages.
2. Select **Format > Page Layout > Reorder Custom Master Pages** to display the Reorder Custom Master Pages dialog.
3. Click the master page you want to move, then click the Move Up or Move Down button.
4. Click the Set button. The custom master pages are rearranged.

Note The Left and Right master pages are not displayed in the list because you can't move them; they are always displayed first.

Modifying master pages

You can customize master pages to suit your document's requirements. For example, you can create multiple flows to separate text on the body pages, set up running headers and footers, and add a watermark effect. When you modify a master page that's already been assigned, the body pages assigned that master page layout are updated, and text reflows into the new layout, if necessary.

Adding Text Flows

Most master page text frames are associated with flow tags, which control how text fills text frames. In documents with more than one flow, text in each flow is displayed independently from the other flow. Newsletters often have multiple flows that continue a story from page 2, for example, to page 4. Many documents only have the default flow A, but you can add new flows. For details, see "Adding text flows to master pages" on page 236.

Setting up headers and footers

Running headers and footers repeat information on body pages, such as the page number, book title, modification date, or content from a heading. Because the information in the header or footer might change from page to page or section to section, you often use variables to provide this content. For example, the page number variable lets you format and display the page number on a master page, and FrameMaker updates it automatically as you work in your document. You might also set a variable to display heading text, allowing FrameMaker to automatically display the current section's heading text at the top or bottom of the page.

In the following example, the running footer displays the page number and the book title, both of which are provided by variables. For more information about using variables, see Chapter 10, "Storing content in variables."

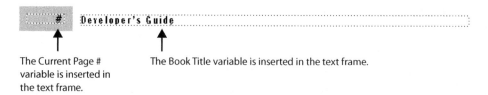

The Current Page # variable is inserted in the text frame.

The Book Title variable is inserted in the text frame.

FrameMaker provides eighteen variables specifically for running headers and footers, allowing display of text, paragraph number, or other info associated with a paragraph or marker. These running header/footer variables are updated automatically when the body text changes. For example, if the variable displays the first second-level heading on the page, and the heading changes, the new heading is immediately displayed in the running header/footer. For more information on running header/footer variables, see "Running header/footer variables" on page 197.

FrameMaker includes headers and footers in new documents by default. If you delete these text frames from the master pages, or add an empty master page, you might need to add a new running header/footer.

To create a Running Header/Footer variable on a master page, follow these steps:

1. Display the master page you want to change.
2. *(if necessary)* Choose **Graphics > Tools** to display the Graphics toolbar.
3. Select the Text Frame tool ( ) from the Graphics toolbar.

4. Draw the text frame where you want it on the master page to display the Add New Text Frame dialog.
5. Click the Background Text radio button, then click the Add button. The text frame is displayed on the master page.
6. Place your cursor in the text frame, then from the Variables pod, choose the running header/footer variable you want to insert and click the Insert ( ) button. The variable is displayed in the text frame.

Setting up a watermark effect

You can place text or graphics on the master pages to be displayed as a watermark on the body pages. Text watermarks can contain a rough draft statement or other messages.

You need to experiment with the watermark color and tint because optimum settings depend on your printer driver and paper. Most watermarks use a 10 percent to 25 percent gray tint. Print a page to verify the color before you print the whole document. If you plan to distribute a PDF file of the document, create a test PDF file to make sure the color isn't too dark on screen.

Note Some printer drivers also let you set up a watermark. You configure the watermark in the printer details and print the watermark only when needed.

Creating a text watermark

For text watermarks, you can use a paragraph tag's autonumber, a variable, or text typed on the master page. Paragraph tags and variables are updated more easily than if you type text on the page—you update the watermark in the tag or variable, and each instance of the watermark is updated automatically. You can't, however, apply a condition to the autonumbered watermark. Conditions only work with text you can select.

To create a text watermark in FrameMaker, follow these steps:

1. Display the master page you want to modify.
2. Select **Graphics > Tools**, then click the Text Frame button ( ).
3. Draw a background text frame where you want the watermark.
4. In the Add New Text Frame dialog, choose the Background Text button and press OK. The text frame is displayed on the page.
5. Position your cursor in the text frame and do one of the following:
   - To store the watermark in a paragraph tag, create a paragraph tag or display an existing tag in the Paragraph Designer, then type the watermark text in the Autonumbering Format field of the Numbering sheet. For a new paragraph tag, you could create a paragraph tag that formats the watermark with a 15 percent black sans-serif font set to 48 points, as shown in the following example. Print a page to make sure the color prints dark enough without reducing document readability.

- To store the watermark in a variable, create a variable, type the watermark text in the variable, and insert the variable.
- To type the watermark on the page, click in the text frame and type the watermark text. The watermark is displayed on the master page.

6. *(optional)* Resize the text frame to remove white space between the watermark and text frame.

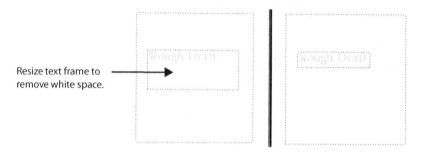

Resize text frame to remove white space.

Next, you can rotate and align the watermark. See "Rotating" on page 315 for details.

You can also create and apply a condition tag to the watermark text. When you need to hide the watermark, change the Show/Hide settings for the condition tag, and the condition is updated on all body and master pages. For details, see Chapter 25, "Setting up conditional text."

Creating a watermark graphic

For graphic watermarks, you create the graphic and import it onto the master page. If you have a specific watermark design that's difficult to set up in FrameMaker, the graphic watermark is a good alternative to text.

When you create a watermark graphic, you should make sure the color of the watermark is light enough to be displayed behind the main content without reducing readability. You might start with 15 percent to 25 percent black tint. To angle the graphic on the page, rotate the watermark in your graphics program or within FrameMaker.

To place a graphic watermark, follow these steps:

1. Display the master page you want to modify.
2. Click in the margin outside the master page's main text frame so your cursor isn't in the text frame, then select **File > Import > File**. (If the cursor is in the text frame, the graphic will import inside an anchored frame in the text frame and won't appear on body pages.) The Import dialog box is displayed.
3. Import the watermark graphic as you would other graphics. The graphic is displayed in the main text frame.

   You can embed a watermark graphic onto a master page or import the file by reference. If you import the graphic by reference, the file size doesn't increase by the size of the graphic. For more information on embedding and importing graphics, see "Importing a graphic" on page 259.

Modifying master pages

Tip If the graphic isn't displayed, select the main text frame (by CONTROL+CLICKING) and select **Graphics > Send to Back**. This moves the text frame behind the graphic.

Creating a landscape master page

You can create pages that have portrait and landscape orientation in a single file. (You can't use different page sizes in a single file but you can have books containing files with different sizes.)

Landscape master pages can contain headers and footers that are portrait- or landscape-style.

The following illustration shows a landscape page with portrait-style footers:

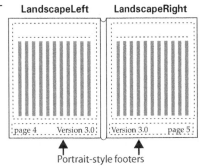

Portrait-style footers

In contrast, this illustration shows landscape-style footers:

To decide which direction the headers and footers on landscape pages should face, consider the following factors:

- **Portrait-style header/footer.** The headers and footers on landscape pages match those found on portrait-style pages so the reader can see them more easily while scanning the book. Typically, this option makes more sense in terms of document readability.
- **Landscape-style header/footer.** The header and footer information is harder to read from the vertical perspective, and placement must take any binding method into account to avoid problems with visibility or trimming of the pages. However, this style is useful if the reader will navigate the landscape sections based on the footer content.

Landscape-style footers

Creating a landscape master page with portrait headers and footers

To create a landscape master page with portrait headers and footers, follow these steps:

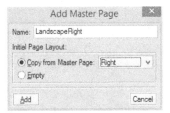

1. Create a new master page based on the Left or Right master page with a name such as LandscapeLeft or LandscapeRight.

221

2. On the new master page, select the main text frame, then select **Graphics > Object Properties** to display the Text Frame Properties dialog.

3. In the Dimensions properties, set the rotation of the frame (90 degrees Clockwise in our example).

4. Swap the Width and Height values.

5. Click the Apply button to return to your modified master page.

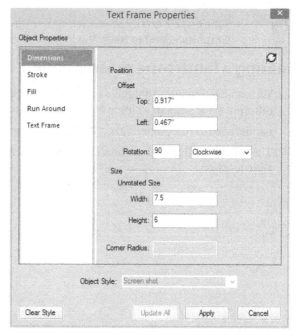

If text symbols are visible, your master page should look something like this:

6. Next, select **Format > Customize Layout > Rotate Page Clockwise**. The entire master page is rotated so that authors can type text from left to right (as on a portrait page) instead of typing sideways from bottom to top.

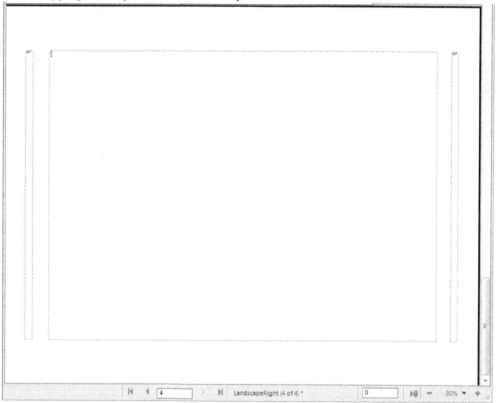

You now have a frame that allows the editor to create content in the same way that it will be read by the end user.

Creating a landscape master page with landscape headers and footers

To create a landscape master page with landscape headers and footers, most of the steps are similar to steps for creating a page with portrait headers and footers:

1. Repeat step 1 and 2 in the preceding procedure for the header and footer text frames to get the information about the height of those frames and their margins. Also, make a note of the variables used in the frames.
2. Delete the header and footer frames.
3. Select the main text frame, then rotate by following steps 3 and 4 in the preceding procedure.
4. Rotate the page (see step 6 in the previous section).
5. Redraw the header and footer text frames; use the information from step 2 to calculate the necessary settings.

6. Reinsert any variables in the previous. You might need to adjust tab settings for the header and footer paragraph tags (or create new tags with modified tab settings) to accommodate the longer frames.

Creating bleeding tabs

Thumb tabs help a reader find information in a book. They can contain the chapter heading, chapter number, or other descriptive text. *Bleeding thumb tabs* are printed on the edge of the page so that the reader can easily identify section breaks while flipping through the book. For simplicity, here's an explanation of single-sided thumb tabs. In this example, all thumb tabs are stored on the Right master page. During the production edit, the unnecessary tabs can be deleted, and the variable definition updated. For example, in Part 1, all thumb tabs were deleted except for the first one. In Part 2, all thumb tabs were deleted except for the second one. To create double-sided thumb tabs, you will repeat this process on the Left master page as well.

The following figure shows the page layout before and after removing tabs and updating the variable. The thumb tabs appear to be one long graphic, but they're actually several overlapping graphics. The overlap helps prevent gaps due to tolerances inherent in the binding process.

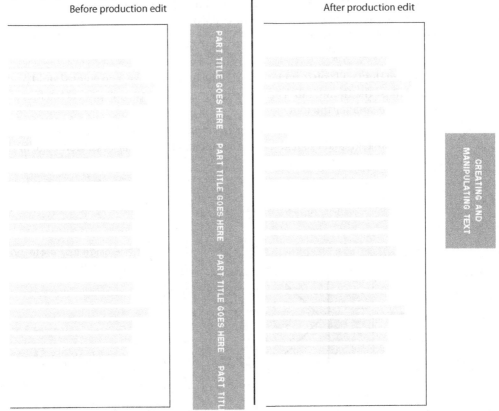

Thumb tabs have four components:
- **Shaded rectangle:** Drawn on the tabs as rectangles with a solid fill.
- **Text frame:** Drawn on each tab to hold the content.
- **Variable:** Inserted into the text frames to display the content.
- **Paragraph tag:** Created to format the content in a color that displays clearly against the solid background. The tag may also center-align the text.

Consider the following factors when you plan thumb tabs:
- **Size:** How much text will be displayed on the tabs? For the chapter number and title, you'll need to create a longer tab and text frame than for just the chapter number.

Tip If your commercial printer is capable of printing all the way to the edge of the page, and you want content to extend to that page edge, you'll need to extend your image(s) 1/8" beyond the page area. (referred to as "extending to the bleed size")
Typically your print vendor will trim the paper to the final book size. If so, you extend images to bleed size on all four sides to avoid possible errors in printing and trimming, which can result in unsightly white gaps at the edge of the page.
If your printer isn't capable of printing all the way to the edge of the page edge, you'll have a gap anyway, and don't need to worry about "bleeding" your images.

- **Formatting:** How will you format the text? The font face, size, and color should be displayed clearly against the thumb tab. The amount of content on the tab affects the font size. The more text, the smaller the font.
- **Placement:** Will the paper be trimmed? To create bleeding tabs, the tabs must be large enough for the text, and the text frames must be placed closer to the left edge of the tab so the content isn't trimmed off.

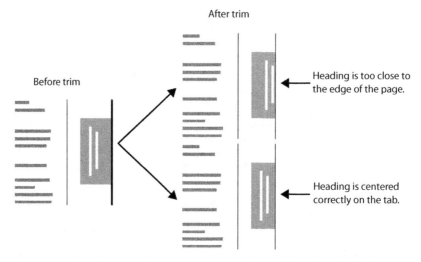

- **Master page setup:** Some users create one page for all thumb tabs, while others prefer creating a master page for each part or chapter (depending on your thumb tab setup) with the thumb tab in the correct position for that section. Creating separate thumb tab master

Modifying master pages

pages saves you the step of deleting extra tabs in every section, but it also multiplies the number of master pages in your template.

The following procedure describes how to add thumb tabs to a right master page.

Drawing the thumb tab

Creating thumb tabs involves several stages. First, you need to draw a rectangle for the tab:

1. Display the master page you want to change.
2. Select **Graphics > Tools**, then click the Rectangle icon ( ).
3. Draw a rectangle against the right edge of the page. The blank rectangle is displayed. Don't worry about the position. You'll calculate the bleed and distribute the tabs later.
4. With the rectangle selected, click the Fill Pattern icon ( ), then click the shading pattern (such as *solid*) from the Fill pop-up menu. The pattern is displayed on the tab.

The shaded thumb tab is displayed on the master page.

Adding the text frame

After you draw the thumb tab, you need to add the text frame. To do so, follow these steps:

1. Draw a small background text frame over the thumb tab. It shouldn't be the size of the final text frame. You're going to rotate the frame, so it will change dimensions. The background text frame is displayed on the shaded tab.
2. With the text frame selected, rotate it 90 degrees clockwise. See "Rotating" on page 315 for details.
3. Lengthen the text frame and move it closer to the left edge of the tab. This helps prevent the text from being trimmed off. You can either drag the borders of the text frame to resize it or change the size in the text frame properties.

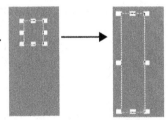

4. Insert a variable in the text frame and apply a paragraph tag. The variable is displayed on the tab, and the paragraph tag formatting is applied.

Tip   To shorten long headings so they fit on the tab, you can insert a Header/Footer marker in the heading paragraph, type the shortened heading in the marker, and reference that marker in the variable. For details, see "Displaying text from a marker" on page 198.

5. Select **View > Body Pages** to display the body pages.

6. Apply the thumb tab master page to the correct body pages, and check the position of the thumb tab text. If the paragraph is too long, you can resize the text frame or change the font size.

Creating additional thumb tabs

To place additional thumb tabs down the edge of the master page and align them, follow these steps:

1. Select **View > Master Pages** and page up or down to find the thumb tab page.
2. Select the thumb tab rectangle and text frame, then select **Graphics > Group**. The text frame and thumb tab are grouped.
3. Copy the tab, then paste it down the edge of the page.
4. Select all tabs, then left align them. See "Aligning" on page 310 for details.

Caution You should wait to delete extra thumb tabs and modify any thumb tab variables until production. If you reimport the page layout properties and variable definitions from your template, the original thumb tabs are displayed, and you'll have to modify the thumb tabs and variables again.

Creating the bleed

To make sure that the thumb tab prints past the edge of the page, follow these steps:

1. Select one of the thumb tabs, then select **Graphics > Object Properties** to display the Object Properties dialog. The Offset From section is shown here.
2. Change the offset values, which are different for the first and last tab. If the top or bottom edge will not be trimmed, you need to follow only step b.
   a. *(first tab only)* Type a value between -0.125 and -0.25 inches in the Offset From Top field.
   b. *(all tabs)* To calculate the Offset From Left value, subtract the tab width from the page width and add between 0.125 and 0.25 inches. Type the value in the Offset From Left field.
   c. *(last tab only)* To calculate the Offset From Top value, subtract the tab height from the page height, add between 0.125 and 0.25 inches. Type the value in the Offset From Top field.

Tip To check the page size, select **Format > Page Layout > Page Size**. The page width and height are displayed.

3. Click the Set button to save your changes. You'll need to experiment with the offset values and then trim the page to find the best location. After changing the offset, you can widen the tab to make room for text.
4. After you place all tabs on the page, you need to distribute them evenly. See "Distributing" on page 312 for details.

After you set up the tabs on the master page, you can import the page layouts into other files, apply the thumb tab master page, and delete the unnecessary tabs.

Here is an example of tab settings for a 2-inch by 0.5-inch tab, set to bleed off of the top and right corners:

This example is for a standard US Letter sized page.

Removing master page overrides

One way to create overrides to the master page is by modifying text frame properties on the body page. For instance, when you adjust the size of the text frame to copyfit a body page, you create an override. FrameMaker warns you about the overrides when you switch from the master page to body page view. If you remove the override, FrameMaker removes all master page overrides—not just the overrides on the current page—so that the body pages match the master pages again. If you keep the override, the changed objects retain their current settings.

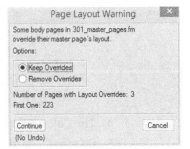

Updating master pages and page layouts

After you modify the text frame on a body page, you can update the master page applied to that page, or you can update the entire flow, as shown in the procedures that follow. When you update the master pages, all body pages to which you've applied the master page are reformatted. For example, suppose you experiment with different side head widths on a body page to see how text is displayed in the new layout. If you like the new size and want to modify all pages in the flow to match the current body page, you update the column layout, and FrameMaker changes the master pages in the flow according to your specifications.

Caution   Before updating master pages, make sure that the change you want is correct for all the master pages in the flow. For example, make sure the master pages that will be changed are using symmetrical layouts; if they are not and you make this change, you might get unintended results.

Removing all layout overrides

To remove all layout overrides by displaying the master pages, see "Removing master page overrides" on page 229.

Updating master pages and body pages simultaneously

To update default master pages and body pages in the same flow all at once, follow these steps:

1. Display either the master pages or the body pages.
2. Position your cursor in the text flow.
3. Select **Format > Page Layout > Column Layout**.

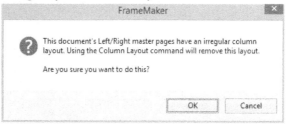

If the left and right master pages have different layouts, a warning is displayed. Click the OK button.

If you have rotated pages, a warning is displayed.

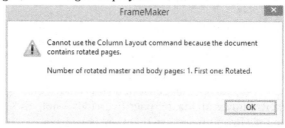

4. The Column Layout dialog box is displayed.
5. Change the settings you want to update. For example, to remove the side head area, uncheck the Room for Side Heads checkbox.
6. Click the Update Entire Flow button. The master page and body page layouts are updated.

   Note Only the default master pages are updated. You'll need to update custom master pages individually.

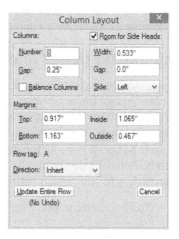

Updating master page via body page changes

If you create master page overrides on body changes, you may decide to update the change to the master page and across all similar pages. To transfer overrides on the body pages back to the applied master page, follow these steps:

1. Display the customized body page, then select **Format > Page Layout > Update Column Layout** to display the Update Column Layout dialog.

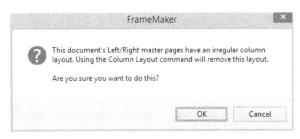

2. Click the Update button. The master page applied to the current body page is updated.

Deleting master pages

Although you can delete several body pages at once, you must delete master pages individually. You can delete only the master pages you create; the default master pages (Left and Right in a double-sided document and Right in a single-sided document) cannot be removed.

Before you delete a master page that has been applied to a body page, you must apply another master page to all pages that use that master page. FrameMaker prevents you from deleting the master page if the master page is in use.

See "Assigning master pages to body pages" on page 210 for details.

To delete a master page, follow these steps:

1. Display the master page you want to delete.
2. Select **Special > Delete Page "master page name."** If the master page is applied to a body page, the error above is displayed. You'll need to display the body page, apply a different master page in its place, and then repeat this procedure.
3. If a confirmation dialog box is displayed, click the OK button. The master page is no longer in the document.

Ideas for custom master pages

Print publishing can require custom layouts to indicate the type of information on a given page. Here are some situations that may require a custom master page for print documents:

- Rotated table. If using the Apply Master Pages feature, your rotated table title paragraph style will need to be applied to the paragraph containing the table anchor. FrameMaker will not recognize the paragraph style used in the table title itself for mapping master pages. Because of this, remember to remove the Table Title from your rotated table's table format.
- Full-page graphic. This page might allow a larger body page text frame by eliminating header and footer content.
- Chapter title page. Chapter title pages typically have less header and footer information. Assigning the chapter title style to the appropriate master page ensures the extra information is not displayed.
- This page intentionally left blank. Used by many industries where safety is a concern, this convention can be implemented in a number of ways.

  One easy way: Create a paragraph named TPILB, and map it to a master page with no headers or footers. The paragraph style can start at the top of next left page. It can also have autonumbering properties for the text itself, or a frame below paragraph to automatically insert a reference page graphic frame. The corresponding TPILB master page needs only a single text frame, large enough to display any autonumbering properties or Frame Below properties.

Custom master pages can be assigned manually (**Format > Page Layout > Master Page Usage**) or automatically (mapping paragraph styles to master pages via the MasterPageMaps reference page).

Chapter 12: Text flows

In FrameMaker, you produce content by placing text frames and graphics on pages. Content on the body pages usually flows through the document according to text frames that are defined on the master pages. For example, if you create a two-column text frame on the master page, text that flows onto a body page based on that master page is displayed in two columns.

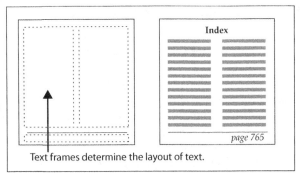
Text frames determine the layout of text.

Note You can also create new text frames on the body pages, but text inside them does not automatically flow from frame to frame.

Most text frames are associated with *flow tags*. Flow tags control how text fills the frames; in other words, how the frames are connected. If frames are set to *autoconnect*, FrameMaker automatically creates new pages when the last page is full. This means you generally don't need to insert new pages manually.

Here's a look at what's in this chapter:

Text flow considerations 233
Understanding text flows 234
   Text frame overflows 235
   Understanding text frame properties 235
   Adding text flows to master pages 236
   Drawing text frames on the body page 237
   Selecting text frames 238
   Resizing text frames 238
   Moving and copying text frames 239
   Deleting text frames 240
Customizing text flows 240
   Connecting text flows 241
   Creating room for side heads 242
   Aligning text across columns 246
   Splitting text frames 251
   Disconnecting text frames 251
   Changing the flow tag 252
   Switching from single- to double-sided pages .253

Text flow considerations

You can define more than one flow in a document to create multiflow documents like newspapers or magazines. Text in each flow is controlled independently throughout the document. Take, for example, a newsletter in which the front-page article continues on page three. For this layout, you create a text flow that begins on page one and continues on page three. The article will flow automatically through from the flow's frame on page one to a corresponding frame on page three, skipping page two, which might contain another flow or other content.

Understanding text flows

You define text flows by drawing text frames (usually on the master pages) and assigning a flow tag. (For information about creating master pages, see Chapter 11, "Understanding master pages.") In a default FrameMaker document you will see the Flow A used automatically. As you type text, FrameMaker automatically adds new pages and allows the text to flow onto those pages. The End of Flow symbol (§) marks the end of text in the container, whether it's a text frame, table title, or table cell.

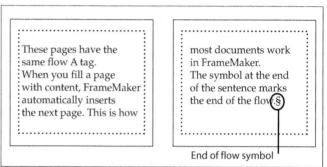

The document shown in the following example has two text flows. The text with flow tag A goes into the A columns, and the B text flow goes into the B columns. One common use for multiple flows is in creating newsletters. You can "jump" a story from the front page to an inside page by creating custom text flows and connecting them.

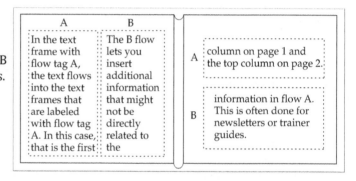

The flow tag for a text frame is displayed in the status bar next to the paragraph tag name. For example, the First master page displayed in the following example is in Flow A.

Note Reference pages also have text flows. These special flows help generate the index, table of contents, list of tables, and list of figures. They also contain master page mappings, graphics, and so on. See Chapter 21, "Creating Other Generated Files," for details. Custom flows in the reference pages can be used as separate text insets. See "Content reuse with text insets" on page 513.

Text frame overflows

Manually created text frames and other text frames without Autoconnect set to On will display a solid line along the bottom edge of the text frame, indicating there is more text in the frame than can be displayed. To fix this, either connect the frame using the **Format > Customize Layout > Connect Text Frames** command, edit the text, or resize the frame.

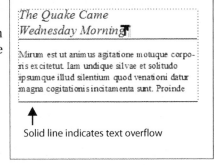

Solid line indicates text overflow

Understanding text frame properties

The characteristics of each text frame in a document are displayed in the text frame properties. Select the frame and then select **Graphics > Object Properties** to display the properties. (You can select **Format > Customize Page Layout > Text Frame Properties** to display the same properties, but using the Graphics menu is quicker.) You can modify the flow tag, text frame size, side head space, columns, and other items in the text frame properties.

If you change the text frame properties on a body page, only the selected text frame is modified. On the master page, editing a text frame updates all body pages that use that master page.

You can more rapidly assign common sets of properties to text frames and other objects by defining and applying object styles to them. The Text Frame Options dialog box will show the object style assignment (if any) at the bottom of the dialog. See Chapter 14, "Object styles" for more information.

Table 12-1 describes the categories available in the Text Frame Properties dialog.

Table 12-1. Text Frame Properties dialog box

Categories	Description
Dimensions	Sets the position, rotation, size and corner radius feature (corner radius introduced in FrameMaker 11).
Stroke	Sets the width, style (straight, dashed, double), and pattern. The arrow style feature is not available for text frame objects.

Table 12-1. Text Frame Properties dialog box (Continued)

Categories	Description
Fill	Sets the color, pattern, tint (percentage of base color, as a pattern of dots) and overprint.
Run Around	Sets the text runaround for the frame. This only impacts text frames that overlap other text frames.
Text Frame	Sets the columns, flow tag and side head options for the frame.

Adding text flows to master pages

On master pages, text frames provide layout for foreground and background items. Text frames for foreground items have an associated flow tag, identifying which text frames are related. If a text frame has a flow tag, any content in the frame on the master page is not displayed on the body pages. Instead, you provide content in those frames on the body pages. Text frames for background items, such as headers and footers, have no flow tag. Text placed in these frames on master pages is displayed on body pages and cannot be edited on body pages.

When you draw a body page text frame on the master page, you connect it to an existing flow on the page, or you assign a new flow tag. If the text frame is connected to an existing flow, text on the body page will flow into the new text frame. If you create a new flow, existing text will not be displayed in the new flow; you need to type new text on the body page. See "Connecting text flows" on page 241 for more information.

Some text frames on the master page aren't part of a flow. For example, the header and footer text frames, which contain repeated information, are considered background items. They don't have flow tags because they're not connected and don't create additional pages for content as foreground text frames usually do. Adding a flow tag to a background text frame on a master page will convert that text frame into a body page text frame.

To draw a text frame on the master page, follow these steps:

1. Select **View > Master Pages** to display the master pages, then page up or down to find the page you want to modify.
2. *(if necessary)* Select **Graphics > Tools**, then click the Text Frame icon (▤).
3. Click and drag the cursor across the page to draw a text frame. (To draw a square, press SHIFT while you draw the frame.) The Add New Text Frame dialog box is displayed.
4. Do one of the following:
   - To create background text, click the Background Text radio button.
   - To create a frame for the body page, click the Template for Body Page Text Frame radio button and assign a flow tag. Flow A is appropriate for most text frames, but you can type another tag name (one or more characters) in the Flow Tag field or change the tag later if the text frame belongs in a new flow.

5. *(optional)* To create a multicolumn text frame, type the number of columns in the Number field, then type the measurement between the columns in the Gap field. See "Adding columns" on page 245 for more information.
6. Click the Add button to display the new text frame.
7. If you added a background text frame, then add content as needed. If you added a body page text frame, leave the frame empty.
8. Select **View > Body Pages**. Your changes are displayed on the page formatted by the modified master page.

Note   If you added text to a background text frame, the text displays on the body page. If you created a body page text frame and added text, you will not see that content when viewing the body pages.

If you want to change the dimensions of the text frame, you can resize the text frame manually, using the FrameMaker rulers as a guide. (Unlike programs such as InDesign, FrameMaker does not let you drag guides from the top or side of the window to help you align objects.) You can also edit the text frame properties to assign precise dimensions. For details, see "Understanding text flows" on page 234.

If the document has multiple flows, text frames within each flow are connected. You add to that flow by creating the text frame first and then changing the assigned flow tag. For details, see "Changing the flow tag" on page 252.

Drawing text frames on the body page

Text frames added to the body pages don't belong to a flow. You can connect the text frames to a flow later, and content will fill the first frame and then flow to the second frame. If you want to reuse the layout on another page, however, you should modify the master page instead of the body page.

Tip   One common use of text frames on the body page is to contain callouts for graphics in anchored frames.

To draw a text frame on the body page, follow these steps:

1. Display the body page you want to modify.
2. *(if necessary)* Select **Graphics > Tools** to display the Tools panel, then click the Text Frame ( ) icon.
3. Drag the cursor across the page to draw a text frame to display the Create New Text Frame dialog.
   (To draw a square, press SHIFT while you draw the frame)
4. Type the number of columns. For multiple columns, type the space between each column in the Gap field.

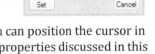

5. Click the Set button. The text frame is displayed on the page. You can position the cursor in the frame, add content, apply a paragraph tag, and modify other properties discussed in this chapter.

Note   In FrameMaker, you can specify the unit of measure used throughout the document—in paragraph tags, text frame properties, table tags, and so on. To change the unit of measure, select **View > Options** and choose an option in the Display Unit drop-down list.

Selecting text frames

You can select a text frame by CTRL+CLICKING the edge of the frame or by selecting the Select Object tool in the Graphics Tools ( ). Handles are displayed on the edge of the frame, indicating that the text frame is selected. you'll need to select the Smart Select ( ) tool before placing your cursor in the text again.

There are three ways to select several frames:

- Drag the cursor across the frames as if you're drawing a box (called drawing a marquee box). An outline is displayed around the frames as you hold down the mouse button. When you release the mouse button, the outline disappears, and the frames are selected.

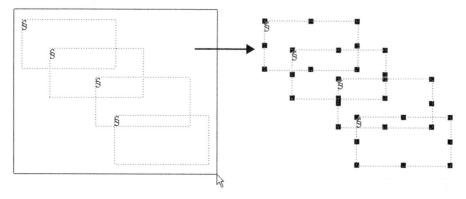

- CTRL+CLICK the first frame, then the second frame, and so on.
- Click outside the main text frame and press CTRL+A. This selects all objects on the page, and you can CTRL+CLICK the objects you don't want selected. Typically, you use this option if the text frames are in an anchored frame or on a page with few objects. (If your cursor is inside the text frame when you press CTRL-A, the text in the flow is highlighted instead.)

Resizing text frames

You may want to force pagination of a line of text or a particular paragraph by resizing the text frame until the text is displayed on the next page. On a body page, you can resize a specific text frame, and only that body page is updated (creating a master page override); the text frame on the master page remains unchanged. To save the values on your body page to the master page, See "Updating master page via body page changes" on page 231.

When you resize text frames on the master pages, the body pages formatted by those master pages are updated (unless there are master page overrides), and text reflows to fit the new frames.

Note Adjusting a text frame on the body page (instead of on the master page) creates an override. You'll get a message when switching from the master page to body page, but you can select the Keep Overrides radio button to preserve your changes. Generally, you should avoid overrides. Adjusting text frames to copyfit text, however, may be acceptable during final document production. (that's what I've done in this book) See "Removing master page overrides" on page 229 for more information.

There are two ways to resize a text frame. You can select the text frame and then drag the handles, or you can change the dimensions in the text frame properties. Typically, you manually resize text frames when copyfitting a page or modifying simple items, such as figure callouts. If you need to place the text frame at a specific location on the page, you modify the height and width in the text frame properties.

To resize a text frame manually, follow these steps:

1. Display the body page or master page containing the text frame you want to modify, then select the frame. The handles are displayed on the edges of the text frame.
2. Do one of the following to resize the frame:
    - To change the height, drag one of the middle handles on the top or bottom edge.
    - To change the width, drag one of the middle handles on the left or right edge.
    - To change the width and height at the same time, drag a corner handle.

You can use the FrameMaker ruler or the dimensions in the left corner of the status bar as a guide. As you drag the edge of the text frame, FrameMaker updates the dimensions displayed in the status bar, and the ruler marking moves.

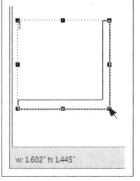

To modify the height and width of the text frame in the text frame properties, follow these steps:

1. Display the body page or master page containing the text frame you want to modify, then select the frame. The handles are displayed on the edges of the text frame.
2. Select **Graphics > Object Properties** to display the text frame properties (see graphic on page 235 for example), then change the measurements in the Height and Width fields.
3. Click the Set button. The text frame is resized.

    Notice that the top and left offset don't change when you resize using the lower right corner handle or when changing only the Height and width in the text frame properties dialog.

Moving and copying text frames

You can move and copy a text frame by dragging it on the page. To move the frame, CTRL+CLICK the edge of the text frame and then drag and drop the frame. To copy the frame, select the frame and press CTRL while you drag and drop the frame. Of course, you can copy and paste a text frame as well.

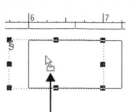
Moving the frame using
the ruler as a guide

Box displays beneath cursor in
to show you're moving the
frame.

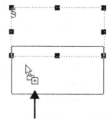
Copying the frame

Two boxes display with a plus
sign to indicate you're copying
the frame.

Note   Don't select the text frame handles when you move or copy a text frame by dragging. The handles are for resizing the frame. If you try to drag the handle, you'll resize the text frame.

With FrameMaker's snap feature turned on, the edge of the text frame moves to the nearest imaginary line in the snap grid (see "Working with grids" on page 297). In the preceding example, the text frame snapped to the 6-inch ruler marking. If you turn snap off, you can move the text frame between grid lines. For details, see "Working with grids" on page 297.

To move a text frame to a specific place on the page, you edit the top and left offset in the text frame properties. For some text frames, such as a callout, it's easier to drag the frame where you need it.

Deleting text frames

After you select one or more text frames, you can delete the frames by pressing the DELETE key or selecting **Edit > Clear**. If the frame you are trying to delete is autoconnected to another frame, the content of the deleted frame moves into the next frame in the connected sequence, and FrameMaker warns you that a new page may be generated to hold the displaced content. If the frame is not autoconnected, deleting the frame also deletes any content it contains.

Customizing text flows

You can customize text flows by adding columns, setting side head space on body pages, splitting text frames to create multiple flows, and disconnecting flows. When you modify a flow on the master page, text on the body pages formatted by that master page reflows into the new layout. You can also fine-tune the display of text in columns by balancing columns, feathering text across columns, and aligning baselines. These fine-tuning features come in handy when formatting documents such as newsletters. The following sections describe these methods.

Customizing text flows

Connecting text flows

A master page can have text frames that are in different flows. The text frames either have different flow tags, or one of them may have no flow tag. You can connect text frames to put them in the same flow. To connect more than one frame, you must connect the first text frame you want in the flow to the second frame, then connect the second frame to the third frame, and so on.

When you run out of room for text in an untagged frame, text disappears below a solid line, indicating that an overflow has occurred, and the end of flow symbol is not displayed, as shown in the illustration.

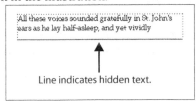

Line indicates hidden text.

In this example, the line on the bottom edge of the frame indicates that the frame should be resized to make room for the hidden text. To highlight everything in the flow, position the cursor in the top frame, and select **Edit > Select All in Flow**.

A disconnected flow can be connected to the end of an existing flow. For example, suppose flow A in the illustration is displayed on pages 1 and 2, and the disconnected flow is on page 1. When you connect the disconnected frame to flow A, the autoconnect feature is turned off throughout flow A. This means that FrameMaker doesn't generate new pages when the existing frames fill with content. After you connect the frames, you will likely want to turn on autoconnect back on for the flow. See the following illustration.

241

To connect text flows, follow these steps:

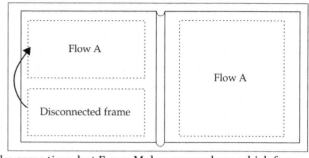

1. Display the master page or body page containing the flow you want to modify.
2. Select the first frame you want in the flow, then select the second frame. If the frames are on different pages, the frames aren't displayed as selected at the same time, but FrameMaker remembers which frame you selected first and makes it the first in the flow.
3. Select **Format > Customize Layout > Connect Text Frames**. One of the following occurs:
   - If you're connecting an untagged flow to a tagged flow, FrameMaker connects the frames.
   - If you're connecting two different flows, a confirmation dialog box is displayed. Click the OK button to connect the frames.
   - If one of the frames isn't visible—for example, it's on a page that's not displayed—a confirmation dialog box is displayed. Click the OK button to connect the frames.

Tip Reapply the master page to return the page layout to its original design. Text is displayed according to the master page layout; it will not be deleted.

Creating room for side heads

A side head area provides an area of white space on the side of your document. You set up side head space for a flow, and the side head is displayed on all body pages in that flow.

The side head space may be on the left side, right side, side closer to binding, or side farther from binding. The following graphic shows the four side head positions.

After you set up side head space, you can create tags for

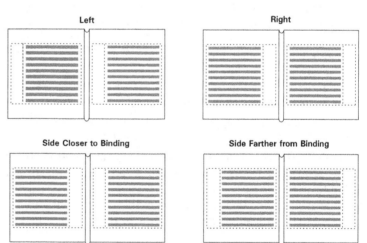

paragraphs that stay within the main column, stay within the side head, or flow from the side head into the main column. The following illustrations shows examples of all three types.

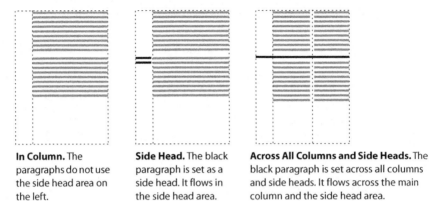

In Column. The paragraphs do not use the side head area on the left.

Side Head. The black paragraph is set as a side head. It flows in the side head area.

Across All Columns and Side Heads. The black paragraph is set across all columns and side heads. It flows across the main column and the side head area.

For more information on changing the pagination settings in a paragraph tag, see "Pagination sheet" on page 119.

Headings can paginate across the side head space and the main column. The side head borders are displayed only for paragraphs other than side heads, and when the document borders are turned off, you won't see any borders.

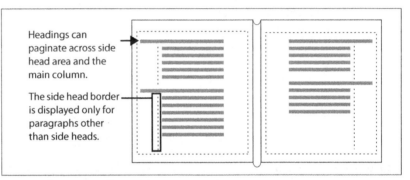

Headings can paginate across side head area and the main column.

The side head border is displayed only for paragraphs other than side heads.

Creating side heads for regular page layouts

If your pages use text frames that are the same size and have the same number of columns, your document has regular page layouts. If your document has irregular page layouts, follow the procedure in "Creating side heads for irregular page layouts" on page 244.

To create space for side heads in a document with regular page layouts (the text frames on all pages are the same width and have the same number of columns), follow these steps:

1. Back up your file. Side heads can be tricky, and you can't undo your changes.
2. On either a body or master page, position your cursor anywhere within the text flow.
3. Select **Format > Page Layout > Column Layout** to display the Column Layout dialog.

4. Check the Room for Side Heads checkbox.
5. In the Width field, type the width of the side head area.
6. In the Side drop-down list, click the location of the side head (see page 242):
    - **Left:** Places the side head area on the left side of the page.
    - **Right:** Places the side head area on the right side of the page.
    - **Side Closer to Binding:** Sets the side head closest to the binding or spine of the document. On a single sided document, the side head is on the left side of the page.
    - **Side Farther from Binding:** Sets the side head away from the binding or spine of the document. On a single sided document, the side head is on the left side of the page.
7. In the Gap field, type the width between the side head area and the main text.
8. Click the Update Entire Flow button. Side heads are added to all pages within the flow.

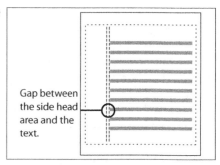

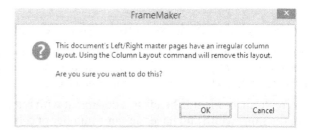

Gap between the side head area and the text.

Caution In documents with irregular column layouts, FrameMaker displays this warning: If you don't want FrameMaker to change the layouts of other pages, click Cancel. Instead, see "Creating side heads for irregular page layouts" on page 244.

Creating side heads for irregular page layouts

If your document or template has pages with irregular page layouts (for example, some pages have one column while others have two, or some pages have narrower columns), follow these steps to add room for side heads without changing the size of the text frame or number of columns on other pages:

Customizing text flows

1. Back up your file.
2. View a master page needing an irregular layout.
3. Select the text frame, then select **Graphics > Object Properties** to display the Text Frame Properties dialog.
4. Navigate to the Text Frame section.
5. In the Flow section, check the Room for Side Heads checkbox.
6. Follow step 5 through step 6 on page 244 to define the side head.
7. Click the Apply button.
8. Repeat this procedure on all master pages containing the flow.

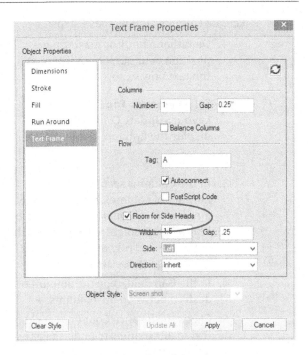

Note Master pages in the same flow can have different sized sideheads, but the sidehead option must be either on for all or off for all those pages in the same flow. Creating a side head of minimal width is one way to work around this.

9. View the body pages.

Tip If side head area is not displaying as you expect in a document, reapplying the master pages usually fixes the problem.

Adding columns

You can add columns to a text flow and specify the gap between each column. You can set the columns to be balanced so that text ends on or near the same line across all columns on the page. For details, see "Balancing columns" on page 246.

In a document with columns, you may want paragraph tags for headings to run across columns or to stay in column. See "Pagination sheet" on page 119 for details.

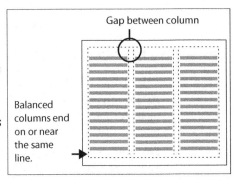

To create columns, follow these steps:

1. On either a body or master page, position your cursor anywhere in the text flow.

2. Select **Format > Page Layout > Update Column Layout** to display the Column Layout dialog.

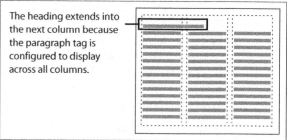

The heading extends into the next column because the paragraph tag is configured to display across all columns.

3. In the Columns section, type the number of columns. If you try to set the number of columns to a large number, FrameMaker might warn you that the columns will be too narrow, and if that is the case, it will not change the setting.

4. In the Gap field, type the amount of space desired between the columns.

5. Check the Balance Columns checkbox for the text in all columns to end as near as possible on the same horizontal line.

Tip If you set up balanced columns on the master page, all pages formatted by the master page are affected. This gives you less control over layout on specific pages, and the results can be unattractive. Instead of balancing columns on the master page, consider balancing columns only on specific body pages. See "Balancing columns" on page 246 for more information.

6. Click the Update Entire Flow button. The master page and body pages formatted by the master page are updated.

Aligning text across columns

There are three settings for adjusting the vertical space in columns:

- **Balancing columns:** Moves the baseline of each column on the page so that text ends on or near the same horizontal axis. This option does not change the amount of space between lines of text.
- **Feathering text:** Adds space between lines of text and paragraphs so that text in a column ends at the bottom edge of the frame. You set the maximum amount of space to be added.
- **Synchronizing baselines:** Lines up text across columns. You specify the grid spacing, and you can limit how headings are aligned at the top of columns.

Balancing columns

When enabled, Balance Columns allows FrameMaker to adjust the text flow so that the columns have approximately the same amount of content. If you balance columns on a specific body page, you modify only the selected page.

To balance columns on a body page, follow these steps:

1. Display the body page you want to modify, then select the text frame.

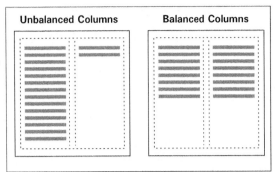

2. Select **Graphics > Object Properties** to display the Text Frame Properties dialog.
3. Check the Balance Columns checkbox, then click the Apply button. Text on the body page reflows to balance across both columns.

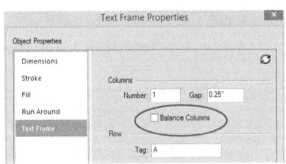

Feathering text

When you feather text, FrameMaker adds space, or padding, between lines of text and paragraphs to fill each page in the flow with text. You specify the maximum amount of space FrameMaker can add. There are two types of padding:

- **Interline:** FrameMaker changes the amount of line spacing on the page.
- **Interparagraph:** FrameMaker adjusts the space above and below paragraphs.

Feathering overrides the line spacing and space above and below settings from the paragraph tags. Feathering can improve the presentation of the content, but also adjusts the density (referred to as the "color") of the text on any given page.

In Figure 12-1, the original unfeathered page is at the top. The bottom-left page shows interline padding. The bottom-right page shows interparagraph padding. Notice that adding space between lines creates more even spacing than adding space between paragraphs. In the last column, the extra padding between the bulleted paragraphs looks awkward.

Figure 12-1. Feathering text

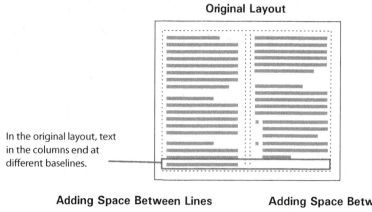

In the original layout, text in the columns end at different baselines.

Adding Space Between Lines

Adding Space Between Paragraphs

Caution Unlike balancing columns, which changes only the selected body page, feathering text affects all pages in the selected flow.

You can combine interline and interparagraph padding to get the look you want, but you should add as little space as possible so that paragraph tags control the spacing. It's also best to feather text after you're sure that the line spacing and space above and below settings in the paragraph tags won't change. For example, suppose you configure interline padding and then decrease line spacing in the Body paragraph tag. The text will move up on the page, and you might need to change the interline padding again.

Customizing text flows

Before modifying the Body paragraph tag, the feather settings control line spacing.

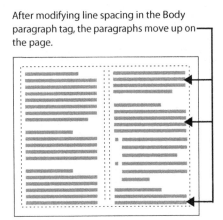

After modifying line spacing in the Body paragraph tag, the paragraphs move up on the page.

To feather text, follow these steps:

1. Display a body page in the flow you want to modify, then select the text frame.
2. Select **Format > Page Layout > Line Layout** to display the Line Layout dialog.
3. Check the Feather checkbox.
4. Do one or both of the following:
   - Add padding between lines using the Maximum Interline Padding field.
   - Add padding between paragraphs using the Maximum Inter-Pgf Padding field.

   If you specify a value for both options, FrameMaker feathers the text between paragraphs first and then feathers between lines if necessary.

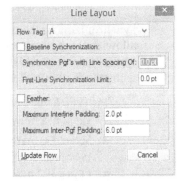

5. Click the Update Flow button. The spacing in the flow is updated. If the spacing doesn't change on the page, experiment with increasing the padding.

Tip To experiment with settings, turn on feathering, then select **Edit > Undo**.

Synchronizing baselines (align to baseline grid)

To align baselines across columns, you specify the line spacing of the paragraphs you want to align, and FrameMaker snaps the text to an imaginary grid. In the following example, the body paragraphs have 12-point line spacing, so the paragraph synchronization is set to 12 points. Paragraphs that have different line spacing, such as the headings, are not aligned.

You can also limit the alignment of headings when they appear at the top of the column. For example, you might want all headings up to 12 points aligned. The headings are actually displayed above the text column, as shown in the next illustration. If using baseline synchronization, leading and any space above/below paragraphs should all reflect the baseline grid value.

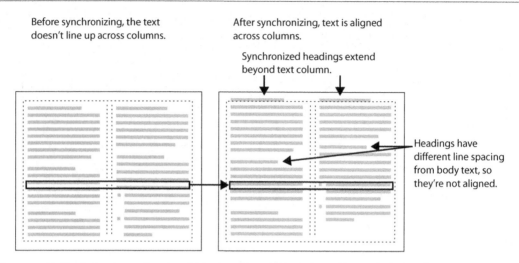

Caution Unlike balancing columns, which changes only the selected body page, synchronizing baselines affects all pages in the selected flow.

Synchronizing baselines is useful when you're working with newsletters or magazine articles, allowing the paragraphs to line up and form a pleasing layout. In technical documentation, you typically don't use this feature unless you're working with a highly customized layout. If you are using the feathering feature, you will need to turn if off for synchronizing to work correctly.

To synchronize baselines, follow these steps:

1. Display a body page in the flow you want to modify, then select the text frame.
2. Select **Format > Page Layout > Line Layout**. The Line Layout dialog box is displayed (see page 249).
3. Check the Baseline Synchronization checkbox.
4. Uncheck Feather if it's checked, or synchronization will not work.
5. Do one or both of the following:
   - To specify the line spacing for paragraphs, type a value in the Synchronize Pgf's with Line Spacing Of field.
   - To specify the limit for synchronizing headings at the top of a column, type the heading size in the First-Line Synchronization Limit field.
6. Click the Update Flow button. The text in the flow is updated.

Tip To compare the synchronized and unsynchronized layouts, turn on synchronization, then select **Edit > Undo**.

Splitting text frames

You can split a text frame on the master or body pages. Content continues from the first text frame to the second frame because they're in the same flow. The insertion point determines where FrameMaker splits the frame. In the following example, the cursor is at the end of the first paragraph, so the frame is split below that line.

On a page with two text frames, you can make the last text frame the first in the flow by deleting the top frame and then splitting and resizing the remaining frame.

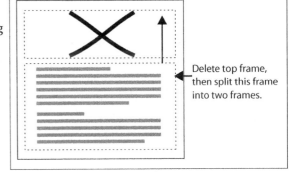

> **Tip** If you split text frames on the body page and decide you don't need the split layout, you can reapply the master page, and the page will be reformatted by the master page.

The position of the insertion point determines the initial size of the split frames. After you split a text frame, you can adjust its dimensions in the text frame properties.

To split a text frame, follow these steps:

1. Display the body or master page you want to modify.
2. Make a text selection above the line where you want the frame to split.
3. Select **Format > Customize Layout > Split Text Frames**. The text frame is divided into two frames. You can resize the frames to get the layout you want. See "Resizing text frames" on page 238 for details.

Disconnecting text frames

You disconnect text frames to separate content from the rest of the document. If the frames are on the same page, FrameMaker converts them to two untagged flows. If the frames are on different pages, the flow tag is preserved but the autoconnect setting is turned off in the frame on the first page. In the second text frame, autoconnect is still on, so FrameMaker adds a new page when the second page is filled.

> **Caution** Disconnected text flows will autonumber within a flow but not across flows.

Existing text remains unchanged when you disconnect text frames. You can cut and paste the text into a different text frame, but when you add new text to the disconnected frame, the text disappears under the bottom edge of the frame. To make room for the text, you can resize the bottom edge of the frame. See "Resizing text frames" on page 238 for details.

To find disconnected text frames in a document, place your cursor in a text frame and select **Edit > Select All in Flow**. Text in the same flow is highlighted. You might need to decrease the document magnification to 25 percent or 50 percent to see the entire flow.

To disconnect text frames, follow these steps:

1. Display the body page you want to modify, then select one of the text frames you want to disconnect.
2. Select from the choices available under **Format > Customize Layout**:
   - Disconnect Previous.
   - Disconnect Next.
   - Disconnect Both.

If the text frames are on the same page, a confirmation dialog box is displayed. Click OK to disconnect the text frames and remove them from the flow.

If the text frames are on different pages, FrameMaker disconnects the frames without warning and turns off autoconnect on the first disconnected flow.

Changing the flow tag

After you create a flow, you can change the flow tag. FrameMaker lets you assign a flow tag to the current frame only or to all frames in the current flow. Text on the body pages will reflow according to your changes.

To change the flow tag, follow these steps:

1. Select **View > Master Pages** to display the master pages, then page up or down to find the page you want to modify. (You could also change the flow on body pages, if you needed to.)
2. Select the text frame, then select **Graphics > Object Properties**. The Text Frame Properties dialog box is displayed (see page 235).

3. In the Tag field, type in a new flow tag.

   While you can use A, B, C, and so on for flow tags, you can also create meaningful flow tag names. For example, a front-page newsletter article may be in the LeadStory flow and the organizational logo in the Masthead flow. You also want to keep the flow tag name short because a long flow tag name takes up more room in the status bar.

4. If you want FrameMaker to automatically create new pages when necessary, check the Autoconnect checkbox.

5. Click the Apply button.
   If the text frame previously had no flow tag, the master page is updated, and you can skip the remaining steps.

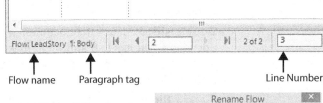

Flow name   Paragraph tag   Line Number

6. If you're changing the tag assigned to the text frame, the Rename Flow dialog box is displayed.

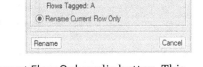

7. Do one of the following:
   - To rename all flows assigned to the current flow tag, click the Rename All Body and Master Page radio button. This replaces the original named flow.
   - To rename just the current flow, click the Rename Current Flow Only radio button. This creates a new separate flow in the document.

8. Click the Rename button to rename the text flow.

Switching from single- to double-sided pages

Documents can be set to either single- or double-sided. For a single-sided document, FrameMaker creates a Right master page and applies it to every page. For a double-sided document, FrameMaker creates default Left and Right master pages. The Left master page is applied to the even pages, and the Right master page is applied to the odd pages by default. You can create additional master pages and apply them to create different layouts. See Chapter 11, "Understanding master pages," for details.

You may decide to switch a document from single- to double-sided or vice versa. Instead of requiring you to create and apply master pages manually, FrameMaker instantly adjusts the page layouts. You can also manage the page count in a document or book. For instance, in a double-sided book, you can force an even page count so that chapters always begin on the right page. FrameMaker can also be set to delete empty pages automatically. This is particularly handy for Portable Document Format (PDF) output if you don't want to convert the empty pages in your FrameMaker book.

Note   When part of a book file, set file pagination properties on a file from within the book file. FrameMaker will use the book's pagination settings (not the settings in the individual file) when performing an Update Book function.

To modify pagination properties, follow these steps:

1. Do one of the following:
    - Display the book file, then CTRL+click to highlight each file you want to modify.
    - Display the document you want to modify.
2. Select **Format > Page Layout > Pagination**. The Pagination dialog box is displayed.

3. Set the pagination. The choices are as follows:
    - **Single-sided:** Makes the document a single-sided document that has a master page called Right and applies it to all pages by default.
    - **Double-sided:** Makes the document a double-sided document that has Left and Right master pages and applies them to the even and odd pages, respectively. You also specify the side of the document's first page:
        - **Right:** The first page is a right page (unless you've applied a custom master page to the first page).
        - **Left:** The first page is a left page (unless you've applied a custom master page to the first page).
        - **Read from File:** Begins document with the page specified in the file. The front matter of a book would be a common use for this feature. Use with caution, as this setting can cause pagination problems.
        - **Next Available:** Begins document with the next available page and deletes the empty pages. If a chapter ends on a left page, the next chapter begins on the right, and vice versa.
4. To set up the page count, select one of the following choices from the Before Saving & Printing list:
    - **Make Page Count Even:** Adds an empty page to files that have an odd page count. In a book file, for example, you may want to force an even page count so that all chapters begin on the right page and end on the left page.
    - **Make Page Count Odd:** Adds an empty page to files that have an even page count.
    - **Delete Empty Pages:** Automatically removes empty pages from the end of each file. You might want to do this before converting a book to PDF output. The printed book, for instance, can have an even page count. When you make a PDF file of the book, you can change the pagination to prevent empty pages from printing at the end of chapters. This applies only to left or right pages; pages formatted by custom master pages are unchanged.
    - **Don't Change Page Count:** Doesn't change the page count. If the document has several blank pages at the end, those pages are not deleted.
5. Click the Set button. The page layout is updated.

For more information on setting up numbering in book files, see "Managing numbering" on page 332.

Chapter 13: Importing graphic content

FrameMaker supports vector graphics (line art) and bitmap images (images composed of pixels, such as screen captures and photographs) in many formats. Most graphics are displayed in frames. On body pages, a graphic is usually in an anchored frame, which attaches the graphic to a paragraph. When the paragraph moves up or down on the page, the anchored frame moves with the anchoring paragraph. This lets you control the graphic alignment and position.

For example, here is an illustration showing a graphic anchored to the previous paragraph.

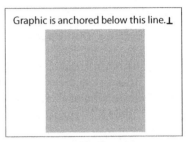

Graphics may also be placed in a graphic frame. However, graphic frames do not move with the text of your document. Because of this, we recommend graphic frames mostly for placing images onto master pages. You will also use graphic frames on your reference pages to store frequently used objects, such as icons or art that is called upon with the Frame Above and Frame Below features.

You create graphic frames using the Graphic Frame ( ) tool in the **Graphics > Tools** toolbar. You can crop graphics with both anchored and graphic frames by sizing the frame around the part of the graphic you need to crop.

Here's a look at what's in this chapter:

    Advantages of using external graphics 255
    Anchoring Graphics 256
        Inserting an anchored frame 256
        Positioning anchored frames 259
        Importing a graphic 259
        Shrink-wrapping an anchored frame 265
        Relinking missing imported graphics 266
        Setting anchored frame object properties 267
    Graphic Formats 270
        Choosing the best graphics format 270
        Image facets 273
        Cross-platform images 273
        Transparency 274
    Placing graphics on the reference pages 275
        Placing graphics above and below the paragraph 275
        Placing graphics "beside" a paragraph 277
        Changing the name of a graphic frame 277
        Importing graphics on a master page 278

Advantages of using external graphics

FrameMaker also supports interactive content, such as Adobe Captivate files, video and 3D objects (U3D format) files. You can set many properties for these objects, such as a default view in FrameMaker. When you save your FrameMaker file to an electronic format that supports multimedia content (such as PDF, SWF, EPUB, or WebHelp), the interactive content is generally preserved.

Anchoring graphics

⊥ When you import a graphic into a text flow, the graphic is automatically inserted into an anchored frame. The anchor, which looks like an upside-down T, is displayed on the line where you positioned your cursor. Alternatively, you can insert the anchored frame first, with the alignment and position you want, and then import the graphic into the frame. Inserting the frame first lets you select the alignment and position. However, importing the graphic directly into the document saves you the step of inserting the empty frame.

Tip   Define an object style to quickly apply attributes to anchored frames and other objects. See "Object styles" on page 279 for more information.

There are seven anchored frame positions. Each position has specific alignment options. Some frames can display outside the text column or text frame like the icon in the margin at the beginning of this section. Other frames are displayed in different positions inside a text frame. Table 13-1 on page 257 describes the anchored frame options.

Inserting an anchored frame

You can insert an anchored frame in a document, then import a graphic or place objects in the frame. Although this adds a step to the import process, it lets you position and align the frame first.

To insert an empty anchored frame, follow these steps:

1. Select **Special > Anchored Frame** to display the Anchored Frame pod.
2. Click the anchored frame position in the Anchoring Position drop-down list.
3. Click the anchored frame alignment in the Alignment drop-down list. The available choices depend on the anchoring position. See Table 13-1 for details.
4. Complete additional fields that are displayed for the alignment you selected. Table 13-1 also describes these fields.
5. Type the size of the anchored frame in the Width and Height fields.
6. Click the New Frame button. The empty frame is inserted on the page. You can click it to import a graphic, or you can draw objects inside the frame. For details on drawing objects, see "Drawing basic shapes" on page 294.

Table 13-1. Anchored Frame Options

Position	Description
Below Current Line 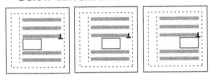	Places the anchored frame below the line containing the insertion point. Select the position of the frame: • **Left:** Places the frame on the left margin of the paragraph containing the anchor. • **Center:** Centers the frame in the text column or across the columns and side head. • **Right:** Places the frame on the right margin of the paragraph containing the anchor. • **Side Closer to Binding:** In a single-sided document, aligns the frame with the left side of all pages; in a double-sided document, aligns the frame with the right side of even-numbered pages and the left side of odd-numbered pages. • **Side Farther from Binding:** In a single-sided document, aligns the frame on the right side of all pages; in a double-sided document, aligns the frame with the left side of even-numbered pages or the right side of odd-numbered pages. Closer to Binding / Farther from Binding  Side head area crops the frame. / Frame floats where it fits. 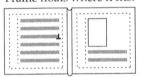 You can apply two additional properties: • **Cropped:** Fits the anchored frame within the text column (or side head area, if one exists). Cropped anchored frames are cut off by the borders of the text frame or the side head area. Uncropped anchored frames extend into the side head or margins. • **Floating:** When there is not enough space for the anchored frame in the current column, moves the anchored frame to the top of the next column without moving the anchor.
At Top of Column	Places the anchored frame at the top of the column containing the insertion point. Has the same alignment and options as the Below Current Line setting.
At Bottom of Column	Places the anchored frame at the bottom of the column containing the insertion point. Has the same alignment and options as the Below Current Line setting.

Table 13-1. Anchored Frame Options (Continued)

Position	Description
At Insertion Point	Aligns the anchored frame vertically with the bottom of the current line. The anchor is not displayed. The Distance Above Baseline option adds space between the anchored frame and the current line. For example, the illustration on the left shows a graphic placed -0.5 inches from the baseline.
Outside Column	Places the anchored frame in the side head space outside the text column. Select where the frame is displayed: • **Left, Right, Side Closer to Binding,** or **Side Farther from Binding:** These options are the same as for "Below Current Line" on page 257. • **Side Closer to Page Edge:** Aligns the frame closer to the edge of the page in single-sided documents. Same as Side Farther from Binding in double-sided documents. • **Side Farther from Page Edge:** Aligns the frame farther from the edge of the page in single-sided documents. Same as Side Closer to Binding in double-sided documents. You can also change the vertical and horizontal spacing: • **Distance above Baseline:** Space above or below the anchored frame. A negative setting extends the anchored frame below the baseline. • **Distance from Text Column:** Space to the left or right of the anchored frame.
Outside Text Frame	Places the anchored frame in the margin of the document outside the main text column. In a single-column document, shown in the illustration on the left, this position is the same as the Outside Column position. These options are the same as for "Outside Column."
Run into Paragraph	Places the anchored frame in the text column with the paragraph. Text is displayed around one side of the anchored frame. To adjust the space between the text and frame, you modify the gap setting in the anchored frame properties. These options are the same as for "Below Current Line" on page 257, except the center alignment is unavailable.

Tip You can format anchored frames as you do other objects in FrameMaker. If you need a border around the imported graphic, for instance, you can add a border to the anchored frame. See "Modifying objects" on page 299.

Positioning anchored frames

Many users have strong personal preferences regarding placement of anchored frames. Some like to put frames on their own line with the At Insertion Point option, while others prefer to place anchored frames in a title, displaying them Below Current Baseline. Another option is to set the anchored frame to run into the paragraph, which is what you'll often find in this book.

The following table describes the options available when placing anchored frames.

Table 13-2. Pros and Cons of Anchored Frame Placement

	Advantages	Disadvantages
At Insertion Point	When placed on a separate paragraph, page breaks occur predictably. This method allows the frame to remain connected to the next paragraph; useful for figure titles which must follow the graphic, as may be required by your style guide.	Requires management of extra paragraph tag(s). Often these tags are defined as 2-point text and/or with a color of white, which make them difficult to see and find in the document. In electronic formats, links to these graphics point to the bottom (or the paragraph below) the graphic. This results in unexpected link behavior.
Below Current Baseline	When figure titles can be placed above anchored frames, this is the fastest way to place graphics, and results in titles and figures staying together. In online formats, links to the figure title work as expected.	If graphics do not have a figure title, the anchored frame exists in the paragraph above, and thus can cause strange page breaks due to the widow/orphan settings of the paragraph holding the anchor.
Run Into Paragraph	This is the only anchored frame position that allows for text runaround. As a result, you may save significant amounts of space in your document relative to figures that have whitespace along the edges. Run Into Paragraph can also result in instructions within the text having a more favorable position next to the graphic being discussed.	Anchors for frames of this type can be placed anywhere in the paragraph. This means that upon export to different formats, the graphic will appear above, in the middle of, or below the content, depending on where the anchor was placed. Use caution to place your anchor in the appropriate position, and definitely avoid placing the anchor in the middle of a word.

Importing a graphic

When you import a graphic into a text flow, the graphic is automatically inserted into an anchored frame. You can either copy the graphic directly into the document or import the file by reference, which creates a link to the source graphic. You and your coworkers need to weigh your options before deciding which method to use. Table 13-3 describes the pros and cons of importing graphics by reference.

Table 13-3. Pros and Cons of Importing Graphics by Reference

	Advantages	**Disadvantages**
Organizing	• You maintain a central repository of graphics for your documentation, which lets coworkers insert the same graphic in multiple documents. • Because you can reuse the graphic, you save storage space. • If your FrameMaker files become corrupted, you still have the original graphics. • You can generate a list of graphics that shows you the location of each imported graphic and the resolution at which it was imported. This list makes it easy to navigate to graphics, count the number of graphics in your book, find which folders graphics are stored in, and so on. See Chapter 21, "Creating Other Generated Files." • If a FrameMaker file gets corrupted, you can reimport the original graphic from disk.	• You must organize graphics clearly so that others can find what they need. • Anyone who opens a document with referenced graphics must have access to the graphics in addition to the FrameMaker file. If the graphics are on your hard drive and the drive isn't shared, the graphics are not displayed. • The link to the referenced graphic will break if the graphic is renamed, moved to another directory, or deleted. • If cross-platform compatibility is an issue, you need to follow file naming and graphic format guidelines. See "Advice on organizing graphic files" on page 268 for details.
Updating	• To modify a graphic, you edit the file and save it with the original name. All FrameMaker files that reference the graphic are updated when you open or save the files. • The name of the referenced graphic is displayed in the object properties. The file name isn't displayed with copied graphics.	You and a coworker might try to update the same graphic in different documents simultaneously.
Archiving	Your archives are smaller because the same graphic isn't placed in multiple documents.	You must be careful archiving a document with graphics that are used in other documents. To avoid breaking links in current documentation, generate a report of imported graphics in FrameMaker to see which graphics need to be archived.
Minimizing File Size	Files are smaller when they contain links to graphics and not the graphics themselves. Large FrameMaker files can cause memory problems, even on the most powerful computers.	None!

The graphics you import can reside on your local machine, a networked computer, the Internet, or a company intranet. If you import a graphic from the Internet or the company intranet by reference, a copy of the file is placed on your machine in your local Temporary Internet Files folder. By default, if the file on the Internet or intranet is updated, your local file is not updated until you delete the temporary file and import the graphic again.

Tip   If you want FrameMaker to always update the graphic from the Internet or intranet site, edit the maker.ini file in a text editor and set the AlwaysDownloadURL flag to On. See Appendix D, "Customizing maker.ini."

FrameMaker supports both bitmap images (such as screen captures and photographs) and vector graphics (line art created with drawing tools). When you import bitmap images, which consist of pixels, you must specify the DPI (dots per inch) setting. The DPI indicates the number of pixels per inch in the imported graphic. The higher the DPI, the smaller the dimensions of the graphic. The DPI setting is crucial because it controls the image resolution. If you resize a bitmap image by dragging one of its corners, the DPI no longer determines the resolution, and the image may become distorted.

Note   You can resize images proportionately by pressing SHIFT while dragging a corner handle or by using the **Graphics > Scale** command.

For consistency's sake, you usually decide on one or two DPI settings to use in a single book. If you use different DPIs when importing screen captures of dialog boxes, for instance, the reader can't judge the true size of the images.

Most vector graphics don't have DPI settings because they consist of lines, not pixels. You do, however, need to set the DPI when you import Scalable Vector Graphics (SVGs). FrameMaker converts SVGs to bitmap images (or rasterizes them) for display in your document. The rasterization doesn't change printing or converting the graphic; it's only for FrameMaker's display of the SVG. You can even import an animated SVG, and the animation will be displayed when you export the file to HTML or XML. Although PDF files can display animated SVG if you have the appropriate plug-in, embedded SVG from FrameMaker is just a static image. The PDF specification is capable of displaying SVG directly, but there is no feature in Acrobat for doing this.

Note   To display SVG files in a browser, you may need to download an SVG viewer plug-in. Some browsers give you this option; others may require you to download and install the viewer yourself.

FrameMaker also lets you import a range of multimedia files into documents. For more information, see Chapter 15, "Placing rich media."

Importing a vector graphic

To import a vector graphic (other than an SVG file), follow these steps:

1. Position the insertion point in the paragraph where you want to display a graphic. (If you have already inserted an anchored frame, select the anchored frame.)

2. Select **File>Import>File** to display the Import dialog.

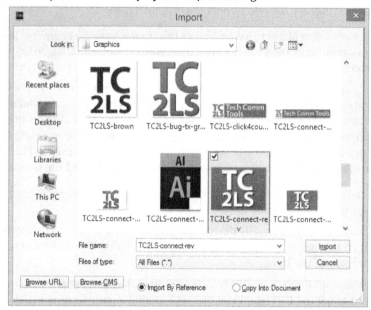

3. Browse to find the file you want to import, then click the file name. If applicable, you can select images using the Browse URL and Browse CMS buttons to access images stored on the web or in a content management system, respectively.

4. Do one of the following:
   - Click the Import By Reference radio button to create a link to the graphic.
   - Click the Copy Into Document radio button to embed the graphic in the FrameMaker file.

5. Click the Import button. The file is displayed in the document within an anchored frame.

If you have specified an HTTP file path and the file is not available, FrameMaker displays a warning:

Importing an SVG file

To import an SVG file, follow these steps:

1. Follow the previous procedure for importing a vector graphic. Click the Import button, to display the Import SVG dialog.

2. Do one of the following:
   - To set the DPI, type a number in the Raster Quality drop-down list. This resizes the graphic proportionately. (The current DPI for the graphic is displayed in this list by default.)

Anchoring graphics

- To change the specific dimensions, click a width in the Width drop-down list and a height in the Height drop-down list. This option may distort the image if you're not sizing the image proportionately.

Caution For the best results, resize SVGs in a graphics program. For the SVG to resize correctly when imported into FrameMaker, the SVG (which is an XML file) must have a viewBox attribute in the top-level XML element, and the units of measure must be specified in the file. Check your graphics program documentation for more information.

3. Click the Set button. The image is converted to a bitmap image and displayed in the document.

Note When you use FrameMaker to save a document to HTML or XML, you have the choice of converting SVG files to GIF, JPEG, or PNG—or you can pass them through unchanged in their native format.

Importing a bitmap image

To import a bitmap image, follow these steps:

1. Follow the steps for importing a vector graphic. (See "Importing a vector graphic" on page 261.) When you click the Import button, one of the following results occurs:
   - If FrameMaker can't detect the graphic format, the Unknown File Type dialog box is displayed. Select the format that matches the graphic and click the Convert button to display the Imported Graphic Scaling dialog.
   - If FrameMaker automatically selects a graphics filter, the Imported Graphic Scaling dialog box is displayed.

2. Select the size of the graphic:

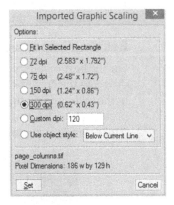

   - To use the DPI from the graphic, leave the selection in the Custom dpi field. The number that is displayed is the DPI of the imported graphic.
   - To change the size of the graphic, click a different radio button for the DPI, or type another number in the Custom dpi field.
   - The projected dimensions of the graphic appear for the standard resolutions. To change units, select **View > Options**.

Tip To protect the quality of the image, avoid the Fit in Selected Rectangle option. This setting resizes the object to fit inside the selected anchored or graphic frame, setting the DPI to Unknown and usually distorting the graphic in the process.

3. Click the Set button. The graphic is displayed in an anchored frame. (The following anchored frame has a border, so you can see the space between the graphic and frame.)

After you import a graphic, you can remove all white space between the graphic and anchored frame by shrink-wrapping the frame. See "Shrink-wrapping an anchored frame" on page 265 for details.

Setting hotspot properties

Starting with FrameMaker 11 you can directly add a hyperlink to a placed graphic. To do so, right-click on the graphic and choose the **Hotspot Properties...** option. For more detail see "Using hotspot properties to create links" on page 491.

Understanding graphic filters

During the FrameMaker installation, you install filters by default. Some filters convert text; other filters are for importing graphics. When you import a graphic, FrameMaker either detects the file type automatically or requires you to pick the correct filter. For most graphics, however, FrameMaker chooses the filter automatically. All you do is set the DPI.

If you select a filter and an error is displayed, the file might be corrupt, or your computer might not have the correct filter installed. For more information about graphics filters provided with FrameMaker, check the *Using Filters* online manual.

Note The *Using Filters* online manual had not been updated since FrameMaker 7 at the time this book was written. Check the Adobe web site for the latest information.

Resizing imported graphics

For the best results, you resize an imported bitmap image by changing the DPI. You should avoid dragging the corner of a bitmap image to resize it. Doing so often distorts the graphic, as shown in the following example. Although you can resize objects and vector graphics proportionately by pressing SHIFT while dragging a corner, using this method on bitmap images resets the DPI to "Unknown." As a result, the graphic may not print or export as expected.

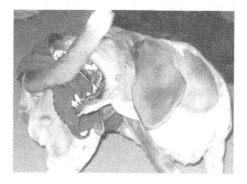
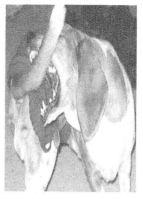

72-DPI graphic was imported at 150 DPI to decrease the dimensions of the displayed image.

150-DPI graphic was resized by dragging a corner.

The DPI of the imported graphic is displayed in the object properties, which you display by selecting **Graphics > Object Properties**. Next to the DPI, the scaling value shows the percentage by which the graphic was scaled when imported. For example, a 72-DPI graphic imported at 150 DPI is scaled to 48 percent.

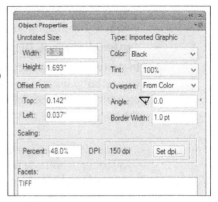

To resize an imported vector graphic, you can either scale the graphic (select **Graphics > Scale** and type a percentage) or press SHIFT and drag a corner of the graphic. Be sure to press SHIFT while resizing the graphic manually to maintain proportions.

Shrink-wrapping an anchored frame

You can shrink-wrap an anchored frame to quickly remove space between the imported graphic and its frame. Shrink-wrapping makes the anchored frame just large enough for its contents, changes the anchoring position to At Insertion Point, and displays the frame 0 points above the baseline of the text. If the anchored frame is on the same line as text, the 0-point baseline causes a problem—the graphic covers the text on the preceding lines.

Example 1

This graphic's anchored frame is set to At Insertion Point and overlaps the text.

To prevent the problem, you may want to import a graphic on a blank line. The shrink-wrapped object doesn't cover text on blank lines. The following illustration shows the difference between inserting and shrink-wrapping an anchored frame in text and on a line by itself.

Example 2

Here's my paragraph. The anchored frame is on a blank line.

Placing the same anchored frame in its own paragraph solves the problem.

When you import a graphic on a blank line, you can customize the paragraph tag applied to the line. The graphics in this book are inserted on a blank line formatted by the Figure paragraph tag. This tag has the following properties:

- **Two-point font.** Works with the space above and below the paragraph to control vertical spacing. You use the smallest supported font in FrameMaker so that the space above and below settings actually position the figure. The font family doesn't matter.
- **Eight points above and below the paragraph.** Separates the figure from the preceding and following paragraphs.

- **Line spacing must not be fixed.** Allows the paragraph spacing to expand vertically to the height of the graphic. Uncheck the Fixed checkbox in the Basic tab.
- **Center alignment.** Displays the figure in the center of the text column.
- **In-column pagination.** Displays the figure in the text column so that the figure is aligned consistently with in-column text. As a result, anchored frames that are wider than the column overlap the right margin, except for cropped anchored frames, which are clipped by the borders of the text frame or the side head area.

Tip If you often have wide figures, create a separate paragraph tag that displays content across all columns and side heads and apply the tag to the paragraph containing the figure. See "Pagination sheet" on page 119 for details.

To shrink-wrap an anchored frame in an empty paragraph, follow these steps:

1. Click the anchored frame or the contents of the frame.
2. *(optional)* If the anchored frame contains several objects, select **Graphics > Group** to group them.
3. Press ESC M P. The anchored frame is shrunk or expanded to fit the contents and is positioned according to the paragraph pagination settings.
4. If you want to unwrap a shrink-wrapped graphic, press ESC M E.

Note There is not a menu choice or mouse action for shrink-wrapping graphics.

Relinking missing imported graphics

When FrameMaker can't locate an imported graphic, it displays a gray box. The link to the graphic can break if the graphic is renamed, moved, or deleted. When you open a document with a missing graphic, FrameMaker prompts you to find the graphic. You can either browse to find the file on your computer, or you can skip the search and fix the link later. (If the graphic was copied into the FrameMaker file rather than imported by reference, you may not be able to recover the graphic if the FrameMaker file becomes corrupted. Look for earlier copies of the FrameMaker file that may still show the graphic.)

The name of the missing graphic is also displayed in the object properties. You can use this information to reimport the graphic if you find gray boxes while working on a document.

Note Gray boxes can indicate missing graphics, but they are also displayed when the computer platform doesn't support the graphic and preview formats. You can tell the difference by looking at the facets in the object properties. The Facets field shows the graphic format and preview format (for graphics not supported on the platform). If there are no facets, the graphic is missing. If there are facets, the formats are not supported on the computer. For details, see "Image facets" on page 273. Gray boxes can also indicate a corrupted imported file.

Anchoring graphics

To relink a missing graphic when opening a document, follow these steps:

1. After you open the document with the missing graphic, the Missing File dialog box is displayed.

2. Do one of the following:
   - To relink the graphic, browse to find the file, and then select the file from the list of files. Make sure the Update Document To Use New Path radio button is selected.
   - If the name has changed, type the new name in the File Name field.
   - To skip the graphic, click the Skip This File radio button.
   - To skip all missing graphics, click the Ignore All Missing Files radio button.
3. Click the Continue button. The document is displayed along with the missing graphic (if you chose to relink the graphic).

Setting anchored frame object properties

In the anchored frame object properties, you can describe the contents of the frame by including the following information:

- **Alternate text.** Displayed when you place your cursor over a graphic in a tagged PDF file (or in XML produced from structured FrameMaker). Screen readers (such as the one built into Adobe Reader) will speak the alternate text.

Advice on organizing graphic files

It's important to plan how you'll name and store graphics that are imported by reference. Lack of planning can cause organizational and computer resource problems. You might also need to follow cross-platform guidelines. Consider the following tips:

- **Decide how to organize your files.** You might group screen captures by the product or feature they depict. It's generally not a good idea to store graphics in folders named after particular chapters in a book—if you rearrange the text and move content into a different chapter, you'll have to move the graphic in the file system and reimport the graphic. To resolve missing graphics that are in different directories, you'll have to select the directory for each graphic, which is time-consuming. If graphics are in one directory, you point FrameMaker to that directory for the first graphic you're relinking, then the rest of the graphics in the directory are automatically relinked.

- **Consider a document management system.** Large corporations often use a document management system to organize files. Such systems not only track the location of files, but they also can track who modifies the files.

- **Use short folder and file names.** The Object Properties pod truncates long path names, so the file name of the imported graphic is sometimes not displayed. Short file names are also easier to remember.

- **Make file names platform-independent.** If your graphics files are used on multiple platforms, file names should conform to the following conventions:
    - Include file name extensions. Windows identifies files by their extensions.
    - Keep file names short.
    - Use lowercase file names. UNIX is case sensitive.
    - Exclude spaces, asterisks, quotes, slashes, question marks, colons, and other special characters from file names. Many special characters are reserved for specific functions on the Windows, UNIX, and Mac platforms. You can, however, use alphanumeric characters, underscores, and hyphens.

For details on cross-platform graphic formats, see "Cross-platform images" on page 273.

- **Actual text.** Describes characters in the anchored frame (as opposed to graphics); exported only to tagged PDF files. For tagged PDF files saved from unstructured FrameMaker, screen readers treat alternate text and actual text the same. No text is displayed when you place the cursor over the graphic in the tagged PDF file. (This setting is more useful for structured FrameMaker.)

See "Threading articles" on page 405 for information about setting up tagged PDF files.

Note   If you define both alternate and actual text for an anchored frame, only the alternate text is converted.

Attributes. The attributes you define in the anchored frame object properties are not exported to XML or tagged PDF files. Other applications may process the anchored frame attributes, or you may use them for your own benefit to store notes or other information.

Note   FrameMaker will drop the object attributes when using **File > Save as… > HTML**. Using RoboHelp to process the file will preserve the alternate text, but not any custom attributes defined for the frame.

To define anchored frame object properties, follow these steps:

1. Select the anchored frame.
2. From the Anchored Frame pod, select the Object Attributes button to display the Object Attributes dialog.
3. To describe the anchored frame contents for display in tagged PDF files (or in XML produced from structured FrameMaker), type text in the Alternate field.
4. To describe characters in the anchored frame for display in tagged PDF files, type text in the Actual Text field.

Remember   Not all object attributes are preserved to all available output formats. Check on your workflow prior to spending much time setting attributes.

5. To add and modify attributes, do any of the following:
   - To create an attribute to describe the object, type the attribute in the Name field, type the attribute definition in the Definition field, then click the Add button.
   - To change the attribute, type the attribute in the Name field of the New or Changed Attribute section, modify the definition in the Definition field, then click the Change button.
   - To delete an attribute, click the attribute in the Defined Attributes section, then click the Delete button.
6. Click the Set button to return to the Object Properties pod.

When you export the document, the alternate text or actual text is converted. For more information on creating PDF files, see Chapter 22, "Print, PDF output, and package."

Graphic Formats

There are three types of graphic images: vector, which consist of lines; bitmap, which consist of pixels; and meta, which may contain both vector and bitmap images. FrameMaker supports many file formats in each category. The format you choose depends largely on the final output of your document and your computer platform.

Choosing the best graphics format

Your computer platform and the final output of your document determine which graphic formats you can use. Some image formats are more compatible with printed documents, while others work well in both printed and online documentation. For example, if you plan to print the document or convert a FrameMaker file to PDF, Encapsulated PostScript (EPS) graphics are a good choice because the vector images in EPS files reproduce at the maximum resolution of your output device. You can also import PDF files as graphics in FrameMaker; like EPS files, vectors in PDF files also have crisp edges. For HTML or XML, screen captures are best displayed as Portable Network Graphic (PNG) files, Graphics Interchange Format (GIF) files, or SVG files. For HTMLHelp, GIF files work well for screen captures, and JPEG files work well for photographs.

Table 13-4 lists the supported graphic formats for different kinds of output.

Table 13-4. Choosing Graphic Formats

Graphic Format	Optimal for Print	Optimal for PDF	HTML, XML, and Web Help	HTMLHelp	DotNetHelp	WinHelp	RTF	EPUB3	Kindle E Ink
Bitmap (BMP)				•	•	•	•		•
CorelDRAW (CDR)	•	•							
Computer Graphics Metafile (CGM)	•	•	•						
Enhanced Metafile (EMF)							•		
Encapsulated PostScript (EPS)	•	•							
Graphics Interchange Format (GIF)	•	•	•	•	•	•	•	•	•
Joint Photographic Experts Group (JPEG or JPG)	•	•	•	•	•	•	•	•	•
MathML	•	•	•	•				•	
Portable Document Format (PDF)	•	•							
Portable Network Graphic (PNG)	•	•	•	•	•		•	•	•

Table 13-4. Choosing Graphic Formats (Continued)

Graphic Format	Optimal for Print	Optimal for PDF	HTML, XML, and Web Help	HTMLHelp	DotNetHelp	WinHelp	RTF	EPUB3	Kindle E Ink
Scalable Vector Graphic (SVG)	•	•	•	•	•			•	•
Tagged Image File Format (TIFF or TIF)	•	•					•		
Windows Metafile (WMF)						•	•		

Note Third-party help authoring tools usually can convert the graphic format you are using to a supported format for online help, but you may get better results by avoiding conversion when using a graphic format that is supported in your final output.

Vector graphics

Best practice: Use Adobe Illustrator as your vector editing tool. Along with FrameMaker's native support of .ai file placement, you'll benefit from a faster workflow by right-clicking on the image when you need to edit your placed files.

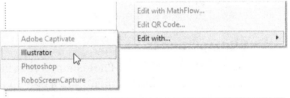

You also can launch Illustrator and other Adobe applications directly from the **File > Launch** menu item.

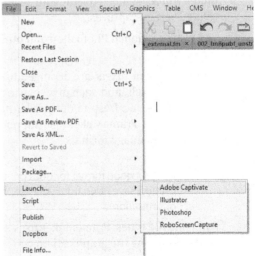

A vector graphic consists of mathematical formulas that describe the drawing. EPS, PDF, and SVG are popular vector formats. If you open a vector graphic in a text editor such as TextPad or EditPlus, you'll see code similar to this illustration.

```
%AI3_DocumentPreview: PC_TIFF
%AI5_ArtSize: 424.8 792
%AI5_RulerUnits: 0
%AI5_ArtFlags: 0 0 0 1 0 0 1 0 0
%AI5_TargetResolution: 800
%AI5_NumLayers: 1
```

You can decipher some of the code, but most of it is incomprehensible by any entity other than a printer. The main advantage of vector graphics is their resolution—they can be scaled drastically and maintain sharp edges. For line drawings, the files are also smaller than bitmap images. The following illustration shows an original vector graphic and another version scaled 300 percent, which looks just as crisp as the original.

Original vector graphic

Graphic was scaled 300 percent without losing resolution.

Some vector graphics have bitmap image previews that are saved with the graphic in their native application. If you use PDF files for your vector images, the preview has sharp edges instead of a blocky bitmap image.

Bitmap images

Best practice: Use Adobe Photoshop as your bitmap editing tool. Along with FrameMaker's native support of .psd file placement, you'll benefit from a faster workflow by right-clicking on the image when you need to edit your placed files. See "Importing a vector graphic" on page 261 for related screen captures.

You also can launch Photoshop and other Adobe applications directly from the **File > Launch** menu item.

Colored dots, or pixels, are arranged to create bitmap images. Bitmap images have a fixed number of pixels, and each pixel expands as you enlarge the image. The result can be a fuzzy, distorted graphic. Despite the similar name, bitmaps (BMP files) are just one possible type of bitmap image. FrameMaker also supports GIF, JPEG, EMF, PNG, TIFF, PDF, WMF, and several other bitmap image formats.

Tip PNG is a good format for both print and online output. To avoid bloating FrameMaker's color catalog, save PNGs from your graphics program with all possible colors (no subsetting or optimization).

For some file types, FrameMaker uses a filter to import the file. You select the appropriate filter when importing the graphic, then you set the DPI. See "Understanding graphic filters" on page 264 for more information.

> **What are those gray boxes?**
>
> No, not *this* gray box…the ones in your FrameMaker document!
>
> Imported graphics may be displayed as gray boxes for several reasons:
>
> - The image is missing. The referenced file might have been renamed, moved, or deleted. When you open a file with a missing image, FrameMaker prompts you to find the correct path, skip the file, or skip all missing files. Skipping the file causes the gray box to be displayed. See "Relinking missing imported graphics" on page 266 for details.
> - The preview format may not be supported on the current platform, or there may be no preview at all. Windows machines cannot display an EPS file with a Mac preview, for instance. EPS graphics with PC previews, however, are displayed on the Windows platform.
> - The file format isn't supported on your platform.
> - The graphic file might be corrupted. Open it in a graphics editor to verify that the file is not corrupted.
>
> To prevent the gray boxes, you should follow guidelines for cross-platform graphics (if applicable) and make sure referenced graphics are not moved, renamed, or deleted.

Image facets

Many graphics include more than one image, although only one image is displayed. FrameMaker considers each image a *facet*, or a separate format of the same image. Facets make graphics formats compatible across computer platforms and monitors. With an EPS graphic, the TIFF facet is displayed as a preview of the graphic. The facets are listed in the imported graphic's object properties.

Cross-platform images

FrameMaker supports cross-platform graphics. This feature lets you view graphics on more than one platform. Some graphic formats, such as WMF and PICT, are displayed only on their native platforms. To view the image on non-native platforms, you copy or import the graphics into your document and set the FrameMaker preferences to create FrameImage facets (for bitmap images) and FrameVector facets (for vector graphics) when you save the document. The facet is displayed when a user, for example, views a document containing a WMF graphic on the UNIX platform. Without the facet, the WMF is displayed as a gray box on the UNIX platform.

FrameImages are saved along with the copied graphic, so the file size increases, especially with graphics-intensive documents. As a result, you might run into the following problems with FrameImages:

- If your computer doesn't have enough free memory, FrameMaker may not be able to save the document.
- Lack of free memory slows down the display of images, so scrolling through the document can take longer.
- The inflated file size can cause printing errors and problems converting the document to another format.

FrameImages are primarily for viewing the document. If you plan to do anything other than view the document, you're better off converting the graphics to a compatible format and importing them by reference.

To save FrameImage or FrameVector facets, follow these steps:

1. Select **Edit > Preferences… > General** to display the Preferences dialog.
2. In the Compatibility Preferences section, check the Save FrameImage with Imported Graphics checkbox.
3. Click the OK button. The FrameImage or FrameVector will be saved in all documents that contain copied graphics until you change the preferences again.

Importing using OLE

You can use object linking and embedding (OLE) to embed objects while maintaining the object's association with its native program. When you double-click a Visio graphic, for instance, the Visio interface is displayed inside FrameMaker. You can then edit the graphic inside the FrameMaker window instead of leaving FrameMaker and opening Visio. You can also create a number of different objects inside FrameMaker without opening the appropriate program or saving the object as a separate file.

OLE has some major drawbacks. OLE objects come in as WMF/EMF formats in FM, which sometimes do not render as cleanly to print or PDF as from the native application.

Documents are large because the OLE items are embedded, not referenced. Those who open your document must have the same version of software you used to create an OLE graphic. It's difficult to resize an OLE object—you can't set the DPI, and scaling the object often results in a distorted or pixelated image.

For best results, create graphic objects in a graphics program and import the files. For more information, see "Importing a graphic" on page 259 and "Cross-platform images" on page 273. In particular, I recommend PSD (Photoshop) and AI (Illustrator) files, because FrameMaker has native support for these formats, allowing better workflow for editing placed images.

Transparency

FrameMaker will recognize transparency settings in imported files, but only relative to the imported file itself. For example, an imported PDF with transparency is internally converted to EPS and is flattened for output. The resulting image appears transparent relative to other parts of the image, but not transparent relative to the FrameMaker document.

Note Flattened files have been simplified without reducing quality to improve the likelihood of successful printing. Flattened files are not as easily edited as their layered counterparts.

In other file formats like TIFF or PNG, the outer transparent area is actually a clipping path that allows FrameMaker to apply a fill of None to the frame. Although the image appears to have transparent regions, it's not really transparent; If you try to move a graphic with transparency over another object in a PDF generated from FrameMaker (using a tool like Pitstop Pro or Acrobat XI Pro), you'll see that it's not really transparent.

Placing graphics on the reference pages

You can store frequently used objects, such as logos or borders, on the reference pages. Some FrameMaker users maintain a library of objects on the reference pages so they can copy an object on the reference page and paste it on a body page when they need it.

New FrameMaker documents contain four graphic frames on the reference pages—Footnote, TableFootnote, SingleLine, and DoubleLine. You can use display these frames above or below paragraphs using the Advanced properties in the Paragraph Designer.

Each graphic frame has a name. You can either copy graphics or import them by reference into the graphic frame. If you copy graphics, keep in mind that the graphics increase the size of the file. Most of the time, you'll import the graphics by reference to make updating the graphics throughout the book easier. See "Importing a graphic" on page 259 for more details.

Note Reference pages also store definitions for the table of contents and other generated files, as well as HTML conversion information. See "Creating paragraph-based lists" on page 380 for details.

Placing graphics above and below the paragraph

You can set up a paragraph that always has a line or other graphics above or below it. FrameMaker provides a few graphics on the reference page named Reference, or you can create your own.

You might also insert notes, tips, and caution graphics using a similar technique. By creating paragraph styles that use Frame Above settings in a two-column, one-row table, you can quickly insert an icon in one cell, with content in the other. In the graphic above, a TipGraphic paragraph tag (applied in the left column) inserts the Tip icon above the paragraph.

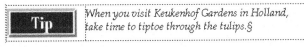

The Tip icon is stored on the reference page. The label *tip_graphic* describes the image; the name of the actual graphic frame (displayed in the status bar when selected), may differ from the label, or the label may not exist at all.

Graphic frame names are listed in the Advanced tab of the Paragraph Designer, in the Frame Above Pgf and Frame Below Pgf drop-down lists.

Graphic frames can contain more than one object, in case you need more sophisticated graphics.

Label describes the graphic; the graphic frame name may differ. → tip_graphic

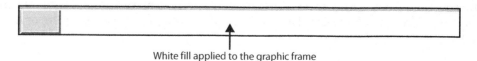

White fill applied to the graphic frame

Note The graphic frame may appear to crop a graphic on the reference page, but the full dimensions of the graphic are displayed on master and body pages. To crop a graphic in a reference frame, edit the object properties on the reference page.

To set up a custom graphic to precede or follow your paragraph, follow these steps:

1. Select **View > Reference Pages** to display the reference pages. The default FrameMaker graphics are included on the first reference page, which is labeled "Reference."
2. Click the Graphic Frame icon in the Graphics toolbar.
3. Click and drag to create a new frame. The location doesn't matter. When you release the cursor, FrameMaker prompts you to name the frame.
4. In the Name field, type a name for the frame, then click the Set button. This is the name you will select later in the Paragraph Designer.
5. Put the graphic objects you want inside the graphic frame. You can use FrameMaker's drawing tools, import an external graphic, or both.

Note Unlike other programs, FrameMaker graphic and anchored frames can contain multiple graphics and text frames. You can use this as a method of grouping objects as a unit.

6. Above the frame, insert a text label (you can copy, paste, and then edit one of the existing ones) with the name of the graphic frame to help you keep track of your graphic frames.
7. Select **View > Body Pages** to return to the body pages.

8. Display the Paragraph Designer, then click the tag that you want to add a graphic to.
9. Click the Advanced tab.
10. In the Frame Above Pgf or Frame Below Pgf drop-down list, click the frame you want to use.
11. Click the Update All button to save your changes. The new graphic is displayed in each paragraph formatted by the tag.

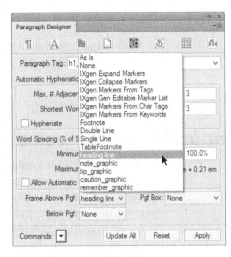

Placing graphics "beside" a paragraph

Typically, the white space in the graphic frame is displayed when you apply the paragraph tag. However, if you need to display text on the same line as the frame object, you can create another tag for the heading text and set the space above that paragraph tag to a negative number. In this example, set the Heading1 paragraph tag to have -36 points spacing above the paragraph. You'll need to adjust the spacing based on the size of the font and of the image called by the graphic paragraph tag.

Another way to create graphics that appear next to text is by using the FrameAbove setting to place a graphic, then using a negative value in the Space Above setting to move the paragraph up into the graphic frame area. Use the left margin settings control the space between the graphic and the text.

Changing the name of a graphic frame

When you add a graphic frame to the reference pages, you name the frame. You can change this name later, and the new graphic frame name will be displayed in the Frame Above Pgf and Frame Below Pgf paragraph tags.

Caution Don't rename graphic frames that are already referenced in paragraph tags, or you'll remove the object from the paragraph tag definition. FrameMaker doesn't provide a warning, so you must verify that the graphic frame isn't being used by checking your paragraph tags.

To change the graphic frame name, follow these steps:

1. Select **View > Reference Pages** to display the reference pages.
2. Select the graphic frame.

Chapter 13: Importing graphic content

3. Click the frame's name in the status bar to display the Frame Name dialog.

4. Modify the text in the Name field, then click the Set button. The new graphic frame name is displayed in the status bar, object properties, and paragraph tags.

5. *(recommended)* Update the text label above the reference frame to the new name.

Importing graphics on a master page

Graphics can be placed on master pages to create customized page layouts. Anchored frames are less common for master page graphics because there's typically no text to displace the graphic. Also, if you import a graphic into a text flow on a master page, the graphic isn't displayed on the body pages. It is common to place the graphic onto the page outside the main text flow, using a graphic frame to crop the graphic, if necessary.

Caution   Run-around properties of the graphic on a master page will affect text on the Body pages. This can create unexpected effects within your document.

In this book, the title page of each section has a graphic. The graphic was imported by reference into a graphic frame on the SectionTitle master page.

Note   Objects on the master pages (along with any other objects outside the body page text flow) aren't exported to HTML or XML.

To import a graphic on the master page, follow these steps:

1. Select **View > Master Pages** to display the master pages.
2. Scroll to the master page you want to edit, or select **Special > Add Master Page** to create a new one.
3. Do one of the following:
   - To import the graphic into a graphic frame, click the Graphic Frame button in the Tools panel and draw the graphic frame, then click the frame and import the graphic. You can also copy and paste the object (from another master page, for instance) into the frame.
   - To import the graphic into a background text frame, place your cursor in the text frame and import the graphic. You can shrink-wrap the anchored frame to remove extra space.
   - If you don't need a frame for the graphic, click outside the text frame to deselect it, then import the graphic. The graphic is centered on the page. If you need to select the graphic, but the graphic is displayed behind the text frame, click the text frame, then select **Graphics > Send to Back**.

After you import the graphic, you can adjust the frame or offsets to create the design you need. For details on cropping images with graphic frames, see "Cropping and Masking Graphics" on page 309.

Chapter 14: Object styles

Starting with FrameMaker 11 you can use *object styles* to quickly format graphics, frames, and text. You can define object styles using the Object Style Designer and then apply the styles just like paragraph and character styles using the Object Catalog.

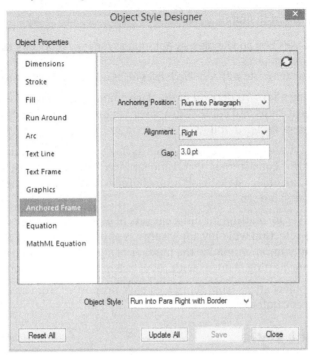

Note While you can apply Object Styles via the Object Styles catalog, as of FrameMaker 2015, there's no Smart Insert, as with Paragraph and Character Styles.

For example, in the FrameMaker 8 version of this book, graphics were handled mostly by placing anchored frames Below Current Baseline or At Insertion Point. This made the graphics easy to control, but resulted in a great deal of unused space, and in many cases, descriptions were not on the same page as the referenced graphic.

Here's a look at what's in this chapter:

    A case for using object styles 280
    Create an object style 280
        Create an object style from an object 281
        Create an object style from scratch 281
    Display the Object Style pod with a shortcut 282
    Using object styles in a regular workflow 282

A case for using object styles

Using the Object Style Designer, I quickly set up object styles for my most commonly used anchored frame positions, and was able to manage graphics more rapidly and with greater accuracy. The screen capture above shows one of my most commonly used object styles. Applying the object style reduces a number of custom formatting steps to a single click, and allows me to control all frames of the same style from a template, further simplifying the publishing of content to mobile devices.

Note Formats like EPUB and Kindle may not support runaround options, requiring more basic handling of graphics.
A separate template with modified styles helps output to these formats.

Along with saving about 10 percent of the page count of the book, these styles allows most steps and content to appear next to the related graphic, making the content easier to follow.

You can apply styles to various objects, such as images, anchored frames, and text frames, for consistent size and appearance. For example, you can create and apply an object style to all the anchored frames in a document to make them of the same size.

Object styles cover formatting for different sets of objects, so applying an object style to a selection will only apply settings appropriate to that object type. In this way, you can minimize the number of styles needed, and also apply multiple formatting changes with savvy use of the As Is... feature.

You can define object styles from scratch, or save properties from an existing object as a named style in the object catalog.

Create an object style

The Object Style Designer gives you control over object formatting in the same way the Paragraph Designer gives you control over paragraph formatting.

Since different objects have different properties, the Object Designer stores a superset of properties in the Object Style, ignoring properties that don't exist for a given object.

Note Properties exist for many types of objects. Thus, a style might have different results if applied to both an image and a text frame.

Create an object style from an object

Formatted objects allow you to quickly create reusable Object Styles within your document. For properties that will vary across formatted objects, select As Is so that the object style will not apply formatting to those features. Get ready to create a new object style by doing the following:

1. Select one or more objects.
2. Select **Graphics > Object Style Designer**. The Object Style Designer displays the properties relevant to the selected object and populates the values of these properties from the selected object. If you select multiple objects and then launch Object Style Designer, the properties and values are relevant to the object you selected first.
3. In the Object Designer dialog, type a name for the object style and click Save.
4. Edit the property values, as needed and click Update All.

Create an object style from scratch

You may prefer to create styles only using values you specifically set. To do this and apply to an object, perform the following steps:

1. Click outside the text frame to deselect all objects.
2. Select **Graphics > Object Style Designer.**
   The Object Style Designer displays with all the property values blank.
3. Type a name in the Object Style field and choose Save.
4. Edit the property values as necessary and choose Update All.
5. Select the object you intend to format with a graphic style.
6. If not currently visible, select **Graphics > Object Catalog.**
7. Select the object style from the catalog.

Display the Object Style pod with a shortcut

The new Object Style Catalog provides graphic formatting similar to the formatting for paragraphs, characters, and tables. The Smart Insert feature is not available for object styles, but you can use ESC, G O to display the object style catalog.

1. If the Object Style Catalog is not visible, select **Graphics > Object Style Catalog**.
2. Select one or more objects onto which you want to apply the style.
3. In the Object Style Catalog pod, click the name of an object style to apply it.

Using object styles in a regular workflow

You can import object styles from other FrameMaker documents.

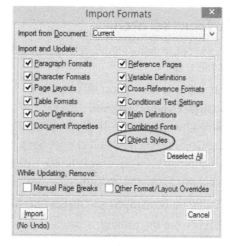

Note Object styles were introduced with FrameMaker 11, so this option is not available in earlier versions of FrameMaker.

Chapter 15: Placing rich media

As publishing moves toward electronic formats on multiple devices, you will likely place video, interactive content, and 3D models into your documents. FrameMaker supports a surprising number of these formats, and allows you to place controls for them that are available to the end user.

Here's a look at what's in this chapter:

Supported formats	.283
Placing a multimedia file	.284
Full-motion video	.285
Set movie poster	.285
Set graphic name	.286
Create video player controls	.286
Create cue points	.287
Activate SWF and FLV by default	.288
Interactive simulations	.288
3D objects	.288
Inserting a multimedia links table	.289
Link 3D part to text	.289
Activate 3D By default	.290
Display 3D and multimedia in pop-up windows	.290
Other 3D control options	.290
3D object support for JavaScript	.291
QR codes	.292

Supported formats

FrameMaker supports over 20 audio and video formats, including:

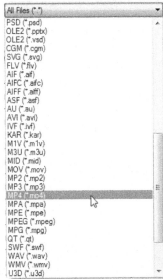

- FLV (Flash video)
- SWF (Flash animation)
- AVI (video format)
- MOV (QuickTime video)
- MP3 (compressed audio)
- MP4 (compressed video)
- MPEG (compressed video)
- QT (QuickTime video)
- WMV (Windows video format)
- U3D (Universal 3D)
- WAV (audio format)

For a full list of supported formats, see the FrameMaker online help.

Within FrameMaker, you can double-click on an image to play the file in a supported application, or right-click to choose from available editors.

If you want to save a PDF that contains internal copies of your rich media content, select **Edit > Preferences… > Global > General**.

In the Preferences dialog box, make sure that the Embed in PDF options are set appropriately, then click Set to close the dialog. (If these options are left unchecked, the multimedia objects will display as images in the PDF, but will no longer include the animation or video.)

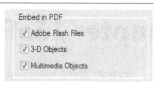

Remember As with all placed files, for online delivery, you can specify a valid HTTP file path to the file. Click the Set button. If you have specified an HTTP file path and the file is not available, FrameMaker displays a warning.

Placing a multimedia file

Best practice: If you have the option to use Adobe Captivate for creating and editing your multimedia files, I recommend doing so. Captivate has direct integration with the more recent versions of FrameMaker that can make placing and editing multimedia files substantially easier. The **File>Launch** command and the ability to right-click on placed Captivate files are two examples of this additional integration. See "Importing a vector graphic" on page 261 for more information.

Since multimedia files will by definition be used in electronic output, you may want to consider placing them underneath a figure title paragraph, or other textual description. This approach will help you control pagination, and will help users jump to the title in a paragraph above an image, rather than a jumping to the point just past the media file. This example shows a video placed above the figure title, and the PDF created from the file. When navigating to the title from a hyperlink, the text is displayed at the top of the screen, meaning that the video is hidden from view until you scroll backward in the file. Other electronic formats produce similar results when the content is placed above the text, rather than below. If you are placing multimedia (SWF) and would like the link to be to the graphic itself, and not the title, see "Set graphic name" on page 286.

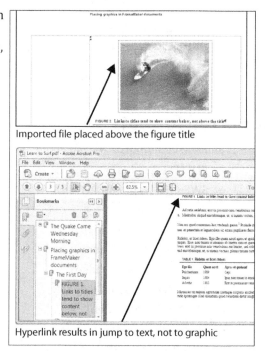

Imported file placed above the figure title

Hyperlink results in jump to text, not to graphic

Full-motion video

You can place movie files just as you would any graphic. To do so, follow these steps:

1. Position your cursor in the document, or select an anchored frame.
2. Use the **File > Import > File** menu item to display the Import File dialog.
3. Select your video file and click OK.
4. If an Unknown File Type message is displayed, choose the appropriate file type in the Convert From: area and click the Convert button.
5. Set your resolution from the available choices and click the Set button.

Tip   Rich media file sizes can be large, so try to produce the files at their final published resolution to keep files sizes for both the working files and the published files manageable.

Set movie poster

Video files are imported with a generic icon to represent the media type. You can replace this generic image with a *poster* image. The poster image is a static representation of the file, analogous to the posters in front of a movie theater. This poster image is used when you print the content. If you prefer not to display the video for print, you can hide the content using the conditional text features found in Chapter 25, "Setting up conditional text."

Set poster from image file

To use an existing graphic as a poster image for a placed movie, follow these steps:

1. Right-click on a placed video file and click on the Set Poster... option.

2. Click the Image File radio button, and click the Select Poster File ( ) button on the right to specify an image.

If you don't have an existing poster image, use the Media File radio button.

Set poster from media file

To create a graphic to be used as a poster image from the movie itself, follow these steps:

1. Right-click on a placed video file and click on the Set Poster… option.
2. Click the Media File radio button, and click the Play button.
3. When the desired frame is displayed, click the Capture Frame button and save the file. The file will save in the DIB, or device independent bitmap format.

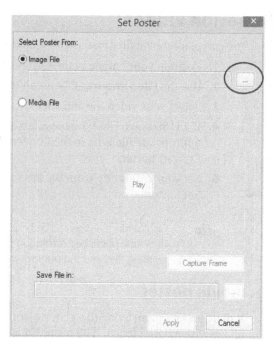

Caution   You may find that the low-resolution image created is a bit rough. If so, create a higher-resolution image manually in the originating application, and use the Image File option instead of the Media File option in the Set Poster dialog.

Set graphic name

Some multimedia formats (like SWF, FLV, and U3D) allow you to assign a graphic name to the image. You can then use the label to more easily create hyperlinks to the image and to assign controls for playback. To set the graphic name on an FLV file, follow these steps:

1. Right-click on your placed multimedia file and select Set Graphic name… to display the Set Graphic Name dialog.
2. *(optional)* Change the unique ID provided automatically by FrameMaker, and click the OK button.

Caution   FrameMaker doesn't support the Set Graphic Name option for MP4 and some other rich media formats.

Create video player controls

Once you've placed your rich media file, you can insert playback controls for the video via the Create Link to Graphic… command. To do so, follow these steps:

1. Type the description for the player control needed within the body of your document.

2. Highlight the description and select the **Graphics > Create Link to Graphic...** menu item to display the Create Multimedia Link dialog.
3. From the drop-down menus, identify the object type and the object for which you would like to create a control.
4. Choose from the available Link Type commands and click the Insert button.

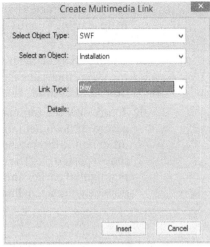

Note The *seek* command is used for cue points, and is covered in the following section.

5. To preview the command, save your document to PDF or other online formats and open the exported file. The text you highlighted earlier should now be your active link.

Create cue points

FrameMaker will recognize existing *cue points* within FLV, or Flash Video. Cue points are like bookmarks you can jump to within your video file. If you don't yet have cue points that you need, set them by following these steps:

1. Highlight the text you want to use as a cue point control. Select the **Graphics > Create Link to Graphic...** menu item. The Create Multimedia Link dialog box is displayed (see page 287).
2. From the drop-down menus, identify the object type and the object for which you would like to create a control.
3. From the Link Type: drop-down, select the seek option.

Tip Existing cue points will be displayed by selecting the Defined Cue Points radio button.

4. If you don't have the required cue point defined, choose Play to display the Create Cue Point dialog.
5. Choose an available frame in the video, then click the Create Cue Point button.
6. To preview the command, save your document to PDF or other online formats and open that exported file. The text you highlighted earlier should now be your active link.

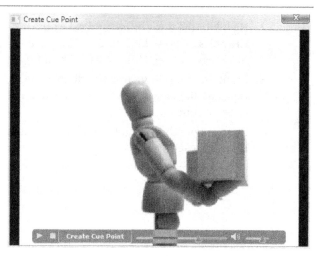

Activate SWF and FLV by default

Rich media are not set to play automatically by default. You can set each of your rich media files to start up immediately when viewed in PDF. To set automatic activation on a SWF or FLV object, follow these steps:

1. Right-click the object.
2. Select the **PDF View Options> Activate by Default**.

Caution If you do this, you are a horrible human being. Nobody likes autoplay.

Interactive simulations

You can place SWF files (like those created with Adobe Captivate) in your document, to include audio and video in your electronic output. If your SWF was not created in Captivate, treat the file as any other placed multimedia. However, for SWF files created in Captivate, FrameMaker 11 provides three additional commands in the File menu:

- Launch Captivate
- Insert Captivate Demo
- Edit with Captivate

These commands give you more direct access to placing and editing the Captivate content than you have from the **File > Import > File** menu item.

3D objects

FrameMaker supports placement of U3D files. Support for the ECMA U3D 3.0 standard began with FrameMaker 11. Using U3D, you can place fully articulating 3D models into your FrameMaker documents. When viewing these models in PDF output, they are complete with zoom, animations, and drill-down capabilities. FrameMaker can generate a table containing hyperlinks to all existing links found within a model. For more detail, read the following sections.

As with other rich media files, you may want to set the graphic name and poster image to make other tasks easier.

Inserting a multimedia links table

3D artists often choose to define views, animations and other controls within these complex files. To generate a table of links embedded in a U3D file, follow these steps:

1. Place a U3D file in your document. Set poster and graphic name options as needed.
2. Position your cursor where you would like the table to appear.
3. Select the **Graphics > Create Link Table to Graphic...** menu item to display the Insert Multimedia Link Table dialog.
4. Select the appropriate object type for your placed file.
5. Select the object for which you would like a table to be generated.
6. Choose a table format if desired.
7. Select view, parts, or animation from the Select Table Type: drop-down menu.
8. Save the file to PDF or other electronic formats to see the working links.

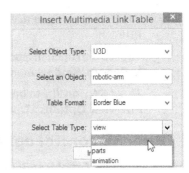

Link 3D part to text

If your placed 3D and vector files contain named parts you can use these named parts as links to destinations in your FrameMaker documents. To set up a link from a multimedia part to a location in a FrameMaker document, follow these steps:

1. Position your cursor in the desired destination location.
2. Show the Hypertext pod (**Special > Hypertext**), and insert a newlink marker using the Specify Named Destination command (see page 488).
3. Right-click on the object with a named part, and select the **3D Menu > Link to Text...** menu item to display the Link to Text dialog.
4. Select from the named parts in your object.
5. Match with the appropriate marker in your FrameMaker document.
6. Click the OK button.

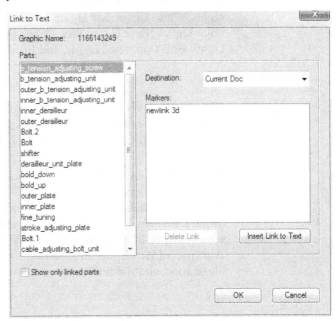

Activate 3D By default

Rich media are not set to play automatically by default. You can set each of your rich media files to start up immediately when viewed in PDF. To set automatic activation on a 3D object, follow these steps:

1. Right-click the 3D object.
2. Select the **3D Menu > Activate 3D by Default**

Display 3D and multimedia in pop-up windows

You have the option to display rich media files in their own pop-up windows when playing in PDF.

To set 3D to open in a pop-up window, right-click on the object and choose the **3D Menu > Open in Pop-up Window** menu item.

To set SWF or FLV to open in a pop-up window, right-click on the object and choose the **PDF View Options > Open in Pop-up Window** menu item.

Other 3D control options

From the **Graphics > 3D Menu** list, you can also change any of the following:

- **Background color.** Select a background color from the Color palette. You can define a custom background color. The background color is not added to the color catalog.
- **Lighting.** Try out the lighting choices until you get the effect you want. The choices are Lights From File, No Lights, White Lights, Day Lights, Bright Lights, Primary Color Lights, Night Lights, Blue Lights, Red Lights, Cube Lights, CAD Optimized Lights, and Headlamp.
- **Show Existing Views.** The person who created the 3D object may have saved views of the graphic in the U3D file. You can select these predefined views, but you can't set up new ones.
- **Render Mode.** By default, the 3D object is rendered as a solid object. The available render modes are Bounding Box, Transparent Bounding Box, Transparent Bounding Box Outline, Vertices, Shaded Vertices, Wireframe, Shaded Wireframe, Solid, Transparent, Solid Wireframe, Transparent Wireframe, Illustration, Solid Outline, Shaded Illustration, and Hidden Wireframe.

- **Properties.** The **Graphics > 3D Menu > Properties** dialog aggregates most of the properties available as discreet choices from the 3D menu.

3D object support for JavaScript

For actions not directly supported by FrameMaker features, you can now attach JavaScript to a U3D object. To do so, right-click on the object and choose the **3D Menu > Select JavaScript** menu item.

QR codes

While you've always been able to embed QR code graphics into a FrameMaker document, starting with FrameMaker 12 you can generate QR codes directly from within FrameMaker. Your user (or you) can scan the printed or digital QR code to access the information (for example, the training.techcommtools.com web address) directly on their device.

To generate and edit Quick Response (QR) codes using FrameMaker.

1. Select **Special > Generate QR Code**.
2. Choose the type of content you wish to create.
   - SMS
   - URL
   - phone
   - e-mail
   - text
3. Set remaining options as needed, and click Insert.

Chapter 16: FrameMaker's graphics tools

FrameMaker includes drawing tools similar to those found in basic graphics programs. FrameMaker's tools let you draw different shapes and combine them to create objects, such as those shown in the following examples. You can change the colors and borders, apply shading, resize, rotate, layer, align, and so forth. The tools aren't appropriate for designing sophisticated magazine layouts or for detailed technical drawings, but they provide most of the features you need for illustrations in reference guides, newsletters, training manuals, or brochures.

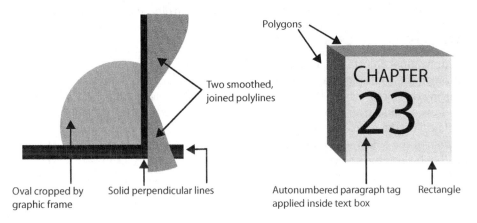

Here's a look at what's in this chapter:

Advantages of FrameMaker graphics tools	294
Drawing basic shapes	294
Working with grids	297
Selecting objects	298
Deleting objects	299
Modifying objects	299
Resizing	303
Reshaping	304
Smoothing corners	306
Changing the number of sides	306
Joining lines	307
Cropping and Masking Graphics	309
Rearranging objects	310
Aligning	310
Distributing	312
Grouping	313
Running text around an object	313
Layering objects	314
Changing the orientation	315
Rotating	315
Flipping horizontally or vertically	317
Transparency in FrameMaker graphics	318
Image hotspots	318

Advantages of FrameMaker graphics tools

You can draw objects directly on the page, or you can place them in a frame. An *anchored frame*, which is a container for graphics and text, attaches the object to a paragraph and can crop or mask images. When the paragraph moves, the anchored frame moves as well. See "Anchoring Graphics" on page 256 for details about inserting anchored frames. *Graphic frames* can crop or mask objects. They also store frequently used items on the reference pages.

Note You can place imported multiple graphics, objects and frames in the same anchored or graphic frame. In this book, for example, we imported the graphics, drew small text frames inside the anchored frame for callout text, and then drew the callout arrows.

Drawing basic shapes

You draw most objects by clicking a button on the Tools panel and then dragging the cursor across the page. As you draw an object, an outline is displayed on the page. This outline is the *path* of the object—an imaginary line that bisects the object border. When you let go of the cursor, the object border is displayed on the page, and the object is selected.

FrameMaker's drawing tools are in the Tools panel. You can format objects with a color, pattern, border, and other properties. To display the tools, select **Graphics > Tools**.

Table 16-1 summarizes the features on the Tools panel.

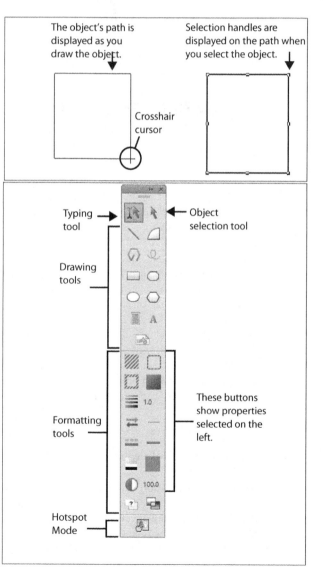

Table 16-1. Tools panel Buttons

Button	Description
	Smart Select. For selecting text or objects. Press CTRL to force selection of an object.
	Select Object. For selecting an object but not text.
	Draw a Line. For drawing a straight line. To draw a line at a 45-degree angle, press SHIFT when drawing the line.
	Draw an Arc. For drawing a curved line. To draw one-fourth of a circle, press SHIFT when drawing the segment.
	Draw a Polyline. For drawing a line with several segments. To draw a polyline, click at each segment point and then **double-click** at the end of the line.
	Draw a Freehand Curve. For drawing a smoothed polyline. To draw a freehand curve, click and drag the cursor.
	Draw a Rectangle. For drawing a rectangle. To draw a square, press SHIFT when drawing the shape.
	Draw a Rounded Rectangle. For drawing a rectangle with rounded corners. To draw a square with rounded corners, press SHIFT when drawing the shape.
	Draw an Oval. For drawing ovals and circles. To draw a circle, press SHIFT when drawing the shape.
	Draw a Polygon. For drawing a closed shape composed of several line segments. To draw a polygon, click at each corner and then **double-click** at the end point.
	Place a Text Frame. For drawing a text frame on any page in a document, including master and reference pages.
	Draw a Text Line. For typing lines of text. To type several lines of text, click where the text begins, type the text, and press ENTER to type another line. (Text lines don't wrap, so you must press ENTER where you want the line to break.) Each text line is a separate object that you click and move on the page.
	The previous cursor location determines the default font properties. For example, if your cursor was previously in a paragraph with a 24-point blue font, the text line has the same properties. You can change the text line font properties using the Format menu, object properties, character tags, or the Color button in the Tools panel. You cannot apply a paragraph tag to text lines.
	Place a Graphic Frame. For drawing a frame to crop or mask objects. Also used on the reference pages to hold frequently used objects, such as note icons and borders.

Table 16-1. Tools panel Buttons (Continued)

Button	Description
	Set Fill Pattern. For choosing a fill pattern for the selected object. Some striped fill and pen patterns don't display accurately in PDF files. If you plan to convert your document to PDF, you should create a test PDF file before using a particular pattern. (If you click the fill pattern, pen pattern, or line width buttons when your cursor is in a table, the Custom Ruling and Shading dialog box is displayed.)
	Set Pen Pattern. For choosing a pen pattern, or border style, for the selected object. As is the case with fill patterns, you should create a test PDF file before using a specific pen pattern if you plan to distribute the document as a PDF file.
	Set Line Width. For choosing from four preset line widths or changing the default widths.
	Set Line End Style. For selecting from four plain or arrowhead line ends. You can also choose from eight additional arrow styles, specify a custom arrow style, and select the shape for line ends.
	Set Dashed Line Pattern. For selecting plain or dashed line patterns. You can also choose from eight additional line patterns.
	Set Color. For setting the color of selected objects. Select from the predefined colors, plus any custom colors that you have added. See "Managing color definitions" on page 432 for details.
	Set Tint. For setting the tint of a previously selected color in 1-percent increments from 0 percent to 100 percent. You can also type in decimal values between 0 and 100 percent. The tint is reset to 100 percent if you apply a different color to the object.
	Set Overprint. Indicates how FrameMaker prints the colors of an overlapping text frame and object. When you change the color of an object, the overprint is reset to From Color. • **Knock Out:** FrameMaker prints only the color of the top object. • **Overprint:** FrameMaker prints both colors from the overlapped portion of the objects. • **From Color:** FrameMaker prints the color using the overprint setting from the color definition. The overprint setting matters only for process printing and color separations. See Chapter 24, "Color output," for more information on color.
	Hotspot Mode: When turned on, Hotspot Mode automatically adds image hotspots to any shape created with an area tool. (Straight lines cannot create a hotspot.)

Drawing basic shapes

To draw an object, follow these steps:

1. *(optional)* Insert one of the following:
   - An anchored frame to attach the graphic to a specific paragraph.
   - A graphic frame to crop the graphic or to draw the graphic on the reference page. (See "Placing graphics on the reference pages" on page 275 for more information.)
2. Select **Graphics > Tools**. The Tools panel is displayed (see page 294).
3. Click the button for the object you want to draw.
4. Click where you want the object.
5. Draw the object according to descriptions in Table 16-1. For example, to draw a freehand curve, click at the starting point and drag your cursor to form the line. To draw a rectangle, rounded rectangle, or oval, click at the starting point and drag to the opposite corner or edge. To draw a polygon, click at the first point, click at the midpoints, and then double-click at the end point.

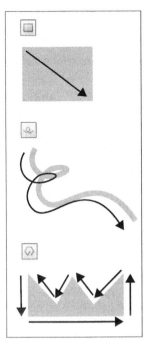

Tip   To judge the size of the object as you draw, use the FrameMaker rulers or status bar dimensions as guides. The ruler markings show the size of the object as you draw, as do the dimensions displayed in the status bar. Select **View > Rulers** to toggle the ruler display.

For more information on anchored frames, see "Anchoring Graphics" on page 256. See "Cropping and Masking Graphics" on page 309 for details on graphic frames.

Working with grids

Along with the baseline grid, there are two other grids available in FrameMaker.

- **Snap grid.** Lets you specify spacing for an imaginary grid. Typically, you can draw, move, copy, or resize an object and drop the edge of the object at any point on the page. When snap is turned on, the edge of the object "snaps" to the imaginary grid. Snap also applies when you reshape, rotate, and align objects. To toggle the snap grid, select **Graphics > Snap**.

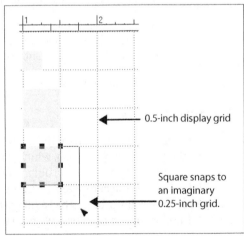

0.5-inch display grid

Square snaps to an imaginary 0.25-inch grid.

- **Grid Lines.** Displays grid lines on the page, which may have different dimensions from the snap grid. The following example shows 0.5-inch grid lines, and the snap grid is set to 0.25 inches. When the square is resized, it snaps to the closest 0.25-inch mark on the ruler.

To toggle the grid lines, select **View > Grid Lines**. To change the spacing for snap grid and grid lines, follow these steps:

1. Select **View > Options** to display the View Options dialog.
2. To change the snap grid spacing, type a number in the Grid Spacing field.

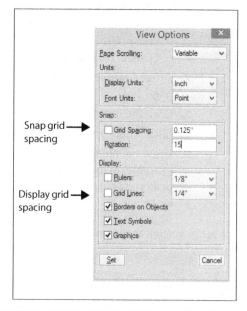

3. *(optional)* To snap rotation of objects to a specific number of degrees, type that value in the Rotation field..
4. To change the display grid spacing, click a measurement in the Grid Lines drop-down list.

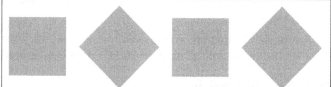

Tip After you change the grid line spacing, the Grid Lines checkbox is automatically checked. This means the grid lines will be displayed when you save your changes. You can uncheck the checkbox if you don't want to display the grid lines yet.

5. Click the Set button to save your changes.

Selecting objects

You can select objects with the Smart Select pointer ( ), which is displayed when you edit text, or the Object Pointer ( ), which you click in the Tools panel to select objects. Using the Smart Select pointer, you select an object by CTRL+clicking the object or border. Using the Object Pointer, you click the object to select it. To return to text mode, click the Smart Select pointer button in the Tools panel. If the object has a fill pattern ( ), you can click anywhere in the object to select it; for hollow objects (objects without a fill pattern), you must click the border.

When an object is selected, square selection handles are displayed on the path of the object. You drag a handle to manually resize the object. See "Resizing" on page 303 for details.

To select several objects at once, press either CTRL or SHIFT as you click each object. To select objects with a selection rectangle, press SHIFT and ALT as you drag. You can also select all objects inside an anchored or graphic frame by clicking the frame and pressing CTRL-A.

Caution Pressing CTRL-A with your cursor in the main flow selects all content in the flow instead of the objects.

Deleting objects

To delete one object, select the object and press the DELETE key. To delete several objects, do one of the following:

- Select only the objects you want to delete by pressing either CTRL or SHIFT as you click each one.
- Select all objects, press either CTRL or SHIFT, and click the objects you *don't* want to delete to deselect them. Another way is to hold down SHIFT and ALT while you drag a selection rectangle around selected objects to deselect them.

Press the DELETE key to remove the selected objects.

Modifying objects

You change basic properties such as color, tint, pattern, and border attributes in the Object Properties dialog or the Tools panel. For a single object, you can use either method. To format several objects, modify those properties in the Tools panel. Although the Tools panel lets you change basic properties, other settings—the object dimensions, offsets from the left and top of the page, and angle—must be changed in the object properties.

Tip To speed up repetitive formatting of objects in FrameMaker, see Chapter 14, "Object styles."

The following example shows the color, tint, and border width properties that are displayed in both the object properties and the Tools panel.

After you select a property in the Tools panel, that property is saved as a default until you change it or close FrameMaker. The property is also saved as a default when no objects are selected. To apply the same border to other objects, select the objects, then click the Line Width button.

Tip To display the properties of any object in the Tools panel, select the object, press SHIFT, and then select **Graphics > Pick up Object Properties**. The formatting of the selected object is displayed on the buttons.

To format an object using the Tools panel, follow these steps:

1. Select the object or objects you want to format, then select **Graphics > Tools**. The Tools panel is displayed (see the previous illustration).

2. Click the button for the property you need to change, then click an item in the pop-up menu. The properties are described in Table 16-1 on page 295. Your changes are displayed on the object, and the buttons on the Tools panel show the selected properties.

Consider the following rectangular graphic:

To format this item through the object properties, follow these steps:

1. With the object selected, choose **Graphics > Object Properties** to display the Properties dialog.
2. Set any of the following:
   - **Dimensions.** Controls offset, rotation, size, and corner radius.
   - **Stroke.** Set line width, line style, pen pattern, and arrow options.
   - **Fill.** Set color, pattern, tint, and overprint options.
   - **Runaround.** Control runaround options and the gap between graphic and text.
3. When finished, click the Apply button, then close the dialog.

Note The options in the Object Properties box changes based on the context of your selection. You will get varying properties, depending on the type of object selected. You may apply an object style to any and all objects, but the object style properties may not be valid for a given object type. If this occurs, the settings will be ignored.

Redefining line widths

In the Tools panel, you can change the four predefined line widths. The new line widths may be applied to objects, but they're not automatically updated in existing objects. For example, if you change the thinnest default line to 1 point, the 0.5-point lines in your document aren't updated. You have to apply the new line width to objects manually.

Unlike other properties, custom line widths are saved until you change them, even if you close FrameMaker. The chosen line width, however, defaults to the thinnest value when you close and reopen FrameMaker. So if you want the default line width to be 1 point, modify the predefined line widths to include a 1-point line as the thinnest line.

Modifying objects

To change the default line widths, follow these steps:

1. Select **Graphics > Tools** to display the Tools panel. (see page 294).
2. Click the Line Width button to display the Line Widths pop-up menu.

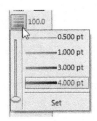

3. Click Set to display the Line Width Options dialog.
4. Type new point values in the fields, then click the Set button. The values are mapped to lines in the Line Widths pop-up menu.

Tip Using the Object Properties dialog box (see page 300), you can specify a line width in the Stroke sheet that isn't in the default line widths shown in the pop-up menu in the Tools panel.

Changing the default arrowhead

FrameMaker provides eight predefined arrowheads. You can also creat custom arrowheads by specifying the base angle, tip angle, and length. The following example shows the settings for a custom arrowhead. Though the Line End Style stores only eight options, in FrameMaker 11, you can create any number of arrowhead styles via the **Object Designer > Stroke > Arrow Style > Edit Arrow Style** and save them to the Object Catalog, effectively extending your options. These new styles can then be exported to other documents in a normal template workflow.

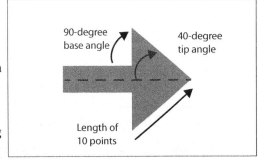

You can select a predefined or custom arrow style as the default. FrameMaker's default arrow style looks like the following:

When you close and reopen FrameMaker, the arrow style reverts to FrameMaker's default style. If you use a different arrow style in your documentation, you'll need to select it again in the Tools panel after reopening FrameMaker.

To change the default arrow style in a document, follow these steps:

1. Select **Graphics > Tools**. The Tools panel is displayed (see page 294).

2. Click the Line End Style button to display the Line Ends pop-up menu.

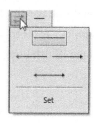

3. Click Set to display the Line End Options dialog.
4. Do one of the following:
   - Click a radio button for the arrow you like.
   - Click the Custom radio button, then type the base and tip angle in degrees and the length in points. Click the arrowhead style in the Style drop-down list.

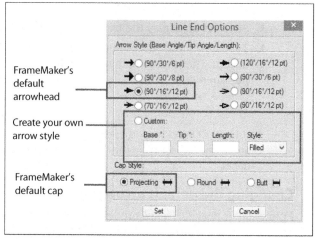

Note For arrow styles, the tip angle must be at least 5 degrees smaller than the base angle.

5. To change the end of the arrowhead, click one of the radio buttons in the Cap Style section. The default is a projecting cap.
6. Click the Set button. The default arrow style is updated and can be applied to new and existing objects. (Existing objects are not automatically updated.)

Modifying the default dashed line

FrameMaker provides eight default dashed lines. You can set one as the default; however, when you close and reopen FrameMaker, the line style reverts to FrameMaker's default style. For more permanent changes to default behavior, edit the maker.ini file. See Appendix D, "Customizing maker.ini" for more information.

To modify the default dashed line, follow these steps:

1. Select **Graphics > Tools**. The Tools panel is displayed (see page 294).
2. Click the Dashed Line Pattern button to display the Line Styles pop-up menu.

3. Click Set to display the Dashed Line Options dialog.

4. Click the radio button for the line you like, then click the Set button. The default dashed line is updated and can be applied to new and existing objects. (Existing objects are not automatically updated.)

Resizing

You resize objects by dragging the edge or corner of the object or providing specific dimensions. If you know the approximate size for an object, it's quicker—although perhaps less accurate—to resize the object manually and use the FrameMaker rulers as a guide. The ruler markings are updated as you resize the object, as are the dimensions displayed in the status bar. To change the specific height or width, you need to type new dimensions in the object properties or scaling properties.

FrameMaker has a handy feature called *gravity* that snaps objects together. Gravity is like "snap to object" and is called gravity because "objects attract." As you resize, reshape, or draw an object, the path or corner of the object meets the closest path or corner of the adjacent object. The centers of ovals and rectangles also have gravity. The following illustration shows how the paths of two objects meet as you resize the dark gray rectangle toward the hollow rectangle. To toggle gravity, select **Graphics > Gravity**.

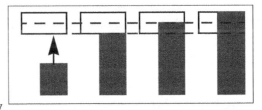

Manually resizing an object

To resize an object using the mouse, follow these steps:

1. Click one of the object selection handles. The cursor changes to an arrow.
2. Drag the handle to resize the object.

Typing a specific height or width

To type a specific height or width, do one of the following:

- Select the object, then select **Graphics > Scale** to display the Scale dialog. Type new dimensions in the Width or Height fields, then click the Scale button. The object is resized from the center.

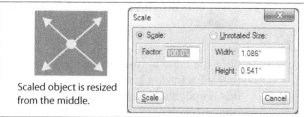

Scaled object is resized from the middle.

- Select the object, then select **Graphics > Object Properties** to display the Object Properties pod. Type new dimensions in the Width or Height fields, then click the Set button. The object is resized from the top-left corner.

Note When you rotate an object, the width and height in the Object Properties don't change to reflect the rotated width and height.

Resizing proportionately

To resize an object proportionately, do one of the following:

- Click a selection handle and press SHIFT while you drag the handle.
- Click the graphic, then select **Graphics > Scale** to display the Scale dialog. Type a percentage in the Factor field, then click the Scale button.

The object is resized proportionately.

Tip To resize several objects without grouping them, select them all and then use the Scale command.

Reshaping

You can alter the shape of many objects instead of redrawing them. You reshape objects using reshape handles and control points:

- Reshape handles look like selection handles, but they appear on corners and curves. By adding, moving, or deleting the reshape handles, you modify the direction, angle, and shape of freehand curves, arcs, polylines, and polygons.
- Control points let you change the direction and curve of freehand curves.

Display reshape handles and control points by clicking the object and selecting **Graphics > Reshape**. On the freehand curve, reshape handles and control points are displayed after you draw the object while it's still selected. Reshape handles and control points are no longer displayed when you deselect the object.

Modifying objects

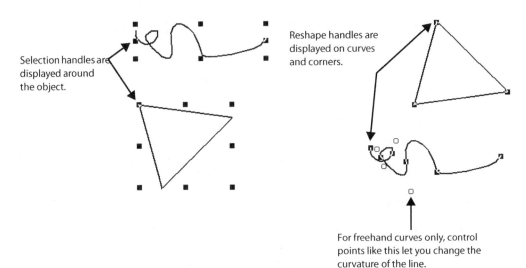

To alter the shape of a freehand curve, arc, polyline, or polygon, follow these steps:

1. Select the object, then select **Graphics > Reshape**. The reshape handles are displayed.
2. Do any of the following:
   - Drag a reshape handle or control point until a solid line is displayed, then change the length and direction of the line. The crosshair pointer marks the location of the reshape handle or control point as you move it. Lengthen the line and change the direction of the line until you like the shape. In the following figure, the freehand curve has both reshape handles and control points; the example shows a reshape handle being dragged.

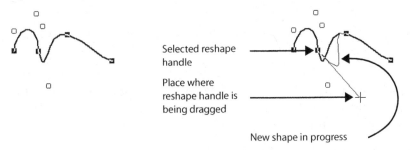

   - To add a reshape handle to a freehand curve, polyline, or polygon, CTRL+click the border where you want the handle. The new handle is displayed on the object.

305

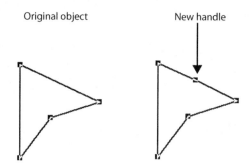

- To remove a reshape handle, CTRL+click an existing handle. The handle, which can change the shape of the object, is no longer displayed.

Smoothing corners

On rectangles, polylines, and polygons, you can convert the corners to curves. To smooth an object, select the object, then select **Graphics > Smooth**. The corners of the object are displayed as rounded edges, and the object type is updated in the object properties. To revert to the original shape, select **Graphics > Unsmooth**.

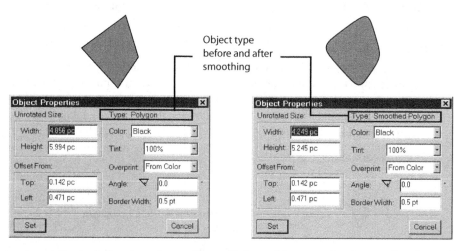

When you smooth a polyline or polygon, FrameMaker adds control points to the object. You change the curvature of the object by dragging a control point. For details on modifying a curve, see "Reshaping" on page 304.

Changing the number of sides

You can draw multisided objects, such as octagons and hexagons, using the Polyline or Polygon tool; however, it's quicker to draw an oval or rectangle and then change the number of sides. To configure the orientation of the object, you change the angle of the first side. For example, the following illustration shows an oval converted to an octagon with a 22-degree start angle.

Modifying objects

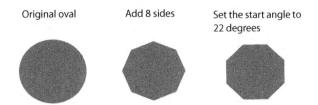

The object properties also show the angle of an object, but this angle is different from the start angle. The start angle is the degree at which the first side of the object is displayed; the angle in the object properties is the degree to which the object is rotated. If you rotate an object that has a start angle, the object is rotated from the start angle. The following illustration shows the octagon rotated 25 degrees.

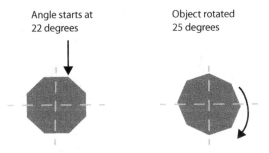

Tip You can change the number of sides on multiple objects by selecting them all at once, but a grouped or smoothed object's sides cannot be modified.

To change the number of sides on an object, follow these steps:

1. Select the oval or rectangle (not the rounded rectangle) you want to reshape, then select **Graphics > Set # Sides** to display the Set Number of Sides dialog.
2. Type a number in the Number of Sides field. For example, to convert a circle to a hexagon, type **6**.
3. To change the angle of the first side, type the number of degrees in the Start Angle field.
4. Click the Set button. The object is reshaped.

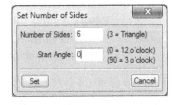

Joining lines

You can join two or more lines (straight lines, arcs, freehand curves, and polylines) to form a single object. Joining lets you create a shape from multiple lines.

When you join lines with different weights, FrameMaker assigns the width of the last selected line. The following illustration compares the result of selecting the thin or thick line last and then joining the lines.

307

Original lines   Thin line selected last   Thick line selected last

FrameMaker's gravity feature causes the ends of two lines attract to each other so you can join them. As you draw, resize, or reshape a line near another line, the closest paths or corners of the two lines meet. Gravity doesn't apply when you *move* a line near another line; it only works when you draw, resize, or reshape a line. If you need to move two lines together before joining them, align them and then move the lines until the edges meet. See "Aligning" on page 310 for details. To toggle gravity, select **Graphics > Gravity**.

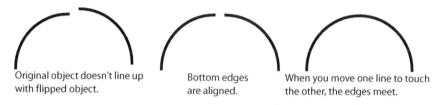

Original object doesn't line up with flipped object.   Bottom edges are aligned.   When you move one line to touch the other, the edges meet.

Caution   There is no "unjoin" command. You must undo the joined lines after you join them, or choose a change on the History list before the lines were joined. Select **Edit > History** or press CTRL-K to view the History list.

To join lines, follow these steps:

1. Drag one line toward the end of the line you want to join until the ends meet. The ends of the lines (not the paths) must touch for you to connect them.
2. Select both lines, then select **Graphics > Join**. The lines form one object. If the lines weren't touching, an error is displayed. Align the objects, then try to join them again.

To create a symmetrical shape, follow these steps

1. Draw a line that's half of the shape you need.
2. With the line selected, press SHIFT, then press CTRL and drag the line to copy it. (Pressing SHIFT while you copy a line keeps the alignment.)
3. Flip the copied line by selecting **Graphics > Flip Up/Down** or **Graphics > Left/Right**.
4. Move the flipped line toward the original line until the ends meet.
5. Select both lines, then select **Graphics > Join**. The lines form one object. If the lines weren't touching, an error is displayed. Align the objects, then try to join them again.

Modifying objects

Draw Copy Flip | Move one line to touch the other line and then join the lines. | Apply a fill, set the pen pattern (the border), then scale the object 50 percent.

After you create the symmetrical shape, you can format the object and, if necessary, scale the object. It's easier to create small items, such as the candle flame in the preceding example, by scaling a large joined object rather than trying to work with small lines.

Cropping and Masking Graphics

Graphic frames are handy for cropping or masking objects. In the following illustration, the graphic frame on the left is placed behind the sun. This crops the bottom-right edge of the sun to fit inside the windowpane. On the right, the graphic frame is placed over the clock to mask the object and create white space. The graphic frame has a white background by default, but you can apply a color and tint to the frame.

 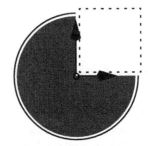

Cropped object   Masked object

To crop an object, follow these steps:

1. Click the Graphic Frame icon in the Tools panel and draw the frame.
2. Do one of the following:
   - Draw the object you're cropping inside the graphic frame and move the frame to crop the object.
   - Drag an existing object into the graphic frame and then adjust the frame to crop the object.

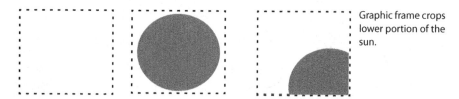

Graphic frame crops lower portion of the sun.

3. *(optional)* To apply a fill pattern and color to the graphic frame, select the frame and then click a fill pattern and color on the Tools panel.

To mask an object, follow these steps:

1. Click the Graphic Frame icon in the Tools panel.
2. Draw a graphic frame over the area of the object you want to mask. For a graphic consisting of several objects, you might need to force one of the objects to be displayed over the graphic frame. For example, placing the graphic frame over the following clock covers part of the clock hands. To display the hands on top of the frame, click the hands, then select **Graphics > Bring to Front**.

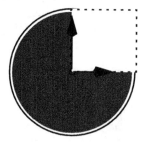

For more information on working with layered objects, see "Layering objects" on page 314.

Rearranging objects

Besides dragging and dropping objects to rearrange them, you can arrange objects by aligning, distributing, or moving them to the foreground or background.

Aligning

Alignment is based on the position of the *last* object you select. Suppose you want to align the tops of the line and triangle shown in the following illustration. If you select the line and then the triangle, the line moves down to align with the triangle. If you select the triangle first, the triangle moves up to align with the line.

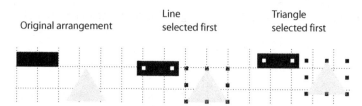

Rearranging objects

To align objects, follow these steps:

1. Select the objects you want to align.
2. Select **Graphics > Align** to display the Align dialog. (If you previously aligned objects, the last settings you used are displayed.)

3. Do any of the following:
   - To align graphics vertically, click the appropriate radio button in the Top/Bottom section.

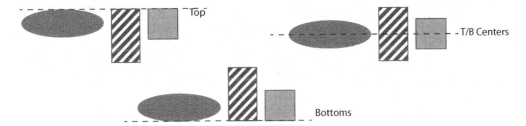

   - To align shapes horizontally, click the appropriate radio button in the Left/Right section.

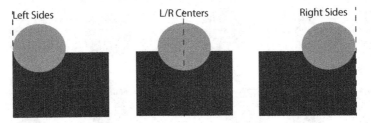

4. Click the Align button. The objects are aligned.

Distributing

You change the space among objects by distributing them. Distributing objects adds an equal amount of space vertically or horizontally. For example, if you specify a horizontal gap of 0.05 inches, FrameMaker adds 0.05 inches of space between each object horizontally.

Unlike aligning, which lines up objects with the last selected object, distributing objects doesn't move the left and right objects (in horizontal distribution) or top and bottom objects (in vertical distribution); only the objects in the middle move to distribute the space. The only exception occurs when you type in the amount of space between the objects. When aligned horizontally, the objects move from the right to the left. When aligned vertically, the objects move from the bottom up.

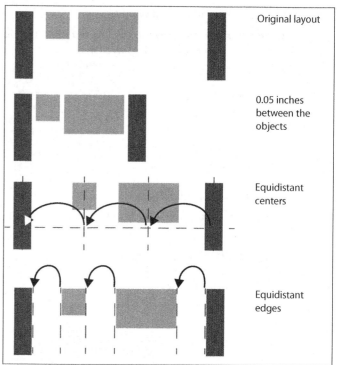

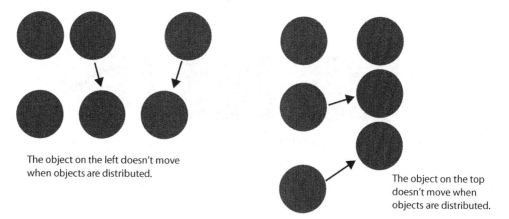

The object on the left doesn't move when objects are distributed.

The object on the top doesn't move when objects are distributed.

To distribute spacing among objects, follow these steps:

1. Select the objects you want to distribute.
2. Select **Graphics > Distribute** to display the Distribute dialog.
3. Do the following:
   - To align the objects horizontally, click the appropriate radio button in the Horizontal Spacing section.
   - To align the objects vertically, click the appropriate radio button in the Vertical Spacing section.
4. Click the Distribute button. The objects are distributed.

Tip To align objects so they are touching but not overlapping, set the edge gap to zero.

Grouping

Grouping is useful for creating one object from several shapes or for preventing objects from accidentally being moved on the page. You select the objects and then select **Graphics > Group** to group them. After objects are grouped, you can modify, copy, move, or delete the group as one unit. For example, you can format all grouped objects with a solid fill and 5-point border. You can also flip, scale, rotate, reshape, and smooth or unsmooth; you cannot, however, change the number of sides.

To ungroup a selected object, select **Graphics > Ungroup**.

Original grouped objects

Flipped, rotated, smoothed, scaled, and formatted with striped fill pattern

Note The Ungroup command is grayed out in the Graphics menu when you try to ungroup an object consisting of joined lines. See "Joining lines" on page 307 for details.

Running text around an object

You control how text is displayed around objects by configuring the object's runaround properties. The text either follows around the object's contour or the object's bounding box. (A bounding box is the smallest rectangle that contains the graphic.) You can also set the gap between the text and object. In the following illustration, the first object shows text running around the contour of the pyramid. The second object shows text that goes around the bounding box. Runaround properties aren't configured for the third object, so the text runs behind the object.

Anthropologists estimate that 2,000 years ago, the ancient peoples of Palacia lived in pyramids built of clay.

Text runs around the contour of the object.

Anthropologists estimate that 2,000 years ago, the ancient peoples of Palacia lived in pyramids built of clay.

Text runs around an imaginary box (represented by the dotted line).

Anthropologists estimate that 2,000 years ago, the ancient peoples of Palacia lived in pyramids built of clay.

Text does not run around the object.

Tip You can set runaround properties for several objects at once by selecting the objects and then changing the properties.

To configure runaround properties, follow these steps:

1. Select the object or objects, then select **Graphics > Runaround Properties** to display the Runaround Properties pod.

2. Do one of the following:
   - To flow text around the edges of the selected object, click the Run around Contour radio button.
   - To flow text around the bounding box, click the Run around Bounding Box radio button.
   - To avoid running text around the adjacent object, click the Don't Run Around radio button.

Tip Whenever possible, set the Don't Run Around option for your graphics and callout. If you add callouts to your graphics and the callouts won't consistently go where you want them, setting the Don't Run Around option for the graphic fixes the problem.

3. *(optional)* Type the number of points between the text and object in the Gap field. The default is 6 points. You cannot change the unit of measurement as you can with the default display unit.

4. Click the Set button. The text is displayed around the object as you specified.

Layering objects

To change the order in which layered objects are displayed, you send an object to the front or back. Some graphics programs let you send objects forward or backward one layer at a time, which is handy when you have several layers of objects. In FrameMaker, you have two options—to move the object to the front or the back.

Objects are layered in the order you draw them. For example, in a drawing with three layers, the object you draw first is displayed on the back layer, the second object is displayed in the middle layer, and the last object is displayed on the top layer. You can only move an object to the front or the back. As a result, if you want the top layer to be in the middle, select the middle layer and bring it to the front, as shown in the following illustration.

The sky is drawn first, the pyramid is drawn next, and the sun is drawn last.

To display the sun behind the pyramid, the pyramid is sent to the front.

To rearrange layers, do any of the following:

- To display an object on the back layer, click the object and select **Graphics > Send to Back**. The object is moved to the back.
- To display an object on the top layer, click the object and select **Graphics > Bring to Front**. The object is moved to the front.
- To display an object on a middle layer, do one of the following:
    - Click the object you want on top, then select **Graphics > Bring to Front**.
    - If you can't select the middle object, click the top layer, then select **Graphics > Send to Back**. For example, the rounded rectangle in the following illustration was previously on the back layer. When the top layer is sent to the back, the rounded rectangle is displayed in the middle layer.

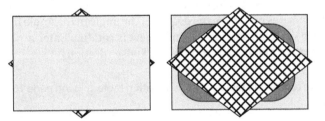

Changing the orientation

You change the orientation of objects by rotating or flipping them. Rotating an object modifies the angle of the object; flipping creates a mirror image of the object.

Rotating

There are two ways to rotate an object. You can use the mouse to drag a selection corner, or you can specify the degree of rotation.

To rotate with the mouse, follow these steps:

1. ALT+CLICK a corner of the object to display the rotation arrow.
2. Drag the corner to a new orientation. (If the rotation arrow isn't displayed, you'll reshape the object instead of rotating it. Try deselecting the object and then ALT-CLICK the corner again.)

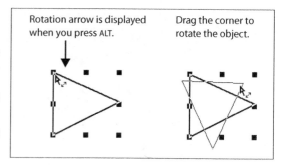

When you drag a corner to rotate the object with the graphics snap on, the degree to which you can rotate the object might be limited by the snap rotation in the View Options. For instance, a 25-degree snap rotation means that the object will snap to an imaginary grid every 25 degrees when you manually rotate the object. See "Working with grids" on page 297 for details.

To rotate an object by specifying the degree of rotation, follow these steps:

1. Select the object you want to rotate.
2. To rotate the object do one of the following:
   - Select **Graphics > Rotate** to display the Rotate Selected Objects dialog. Type the degree in the Rotate By dialog box, click the Clockwise or Counterclockwise radio button, then click the Rotate button. The object is rotated.
   - Select **Graphics > Object Properties** to display the Object Properties pod. Type the degree of the angle in the Angle field, then click the Set button. The object is rotated. Enter a negative number for counterclockwise rotation.

Note To rotate text within table cells, see "Rotating table cells" on page 167.

Flipping horizontally or vertically

When you flip an object, FrameMaker changes the object into a mirror image of itself. The object is flipped left-to-right (horizontally) or up-and-down (vertically).

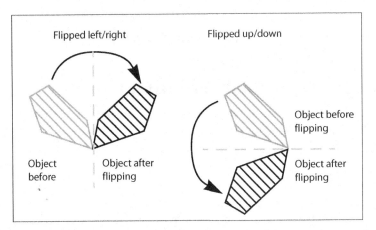

Text lines inside the object are not flipped; only the alignment of the text changes. In the following illustration, each text line is left aligned. After flipping the object horizontally, the grouped text lines are right aligned. If you ungroup the text lines and view the object properties, you'll see that the alignment has changed from left to right.

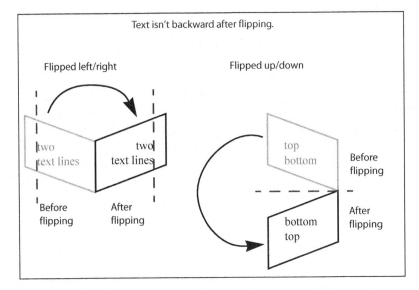

Note   Text frames are treated differently from text lines. Flipping an object with a text frame does not change the text alignment because the paragraph tag applied inside the frame controls the alignment of the text.

To flip an object, follow these steps:

1. Select the object or objects you want to flip.
2. Do one of the following:
   - To flip the object horizontally, select **Graphics > Flip Left/Right**.
   - To flip the object vertically, select **Graphics > Flip Up/Down**.

   The object is flipped as you specified.

Transparency in FrameMaker graphics

Although FrameMaker respects transparency in placed objects, graphics created within FrameMaker have no option for creating transparency. With the exception of creating overprint, FrameMaker graphic objects are opaque.

Image hotspots

Image hotspots were introduced with FrameMaker 11 and can be created using existing objects or by activating the Hotspot Mode ( ) in the Graphics Tools pod. With the Hotspot Mode on, shapes created with the graphics tools will automatically prompt you for hotspot properties.

Hotspots allow you to create one or more hyperlinks on top of images and graphics. For more information on setting up hotspots, see "Creating hotspots" on page 490.

You can ungroup vector formats like CGM so that individual pieces of the graphic can be assigned hotspot properties. To do this, right+click on the ungrouped part and choose "Hotspot Properties."

Hotspots work in PDF and other electronic outputs.

Part IV

Building Books

Chapter 17: Setting up book files

FrameMaker does well with short documents, but it really shines with long, complex documents. One example of this is FrameMaker's book feature, which lets you group related documents. After you put a collection of documents in a book, you can control pagination and create tables of contents (and other "generated" files), for the entire book. You can also perform many actions on the book instead of on individual files. For example, you can use the book to change conditional text settings across all the files in the book in a single step.

Creating a book file does not by itself affect the component files; the book file is little more than a file list. Opening a file by double-clicking its icon in the book is exactly the same as using the **File > Open** command. The book file contains pointers to the files—FrameMaker does not embed copies of the files in the book file.

Here's a look at what's in this chapter:

Advantages of using book files	.321
Creating a book file	.322
What the book window tells you	.322
Managing files in a book	.323
Adding files	.323
Adding special book structures	.325
Opening, closing, and saving all files	.327
Removing files	.327
Rearranging files	.327
Renaming files	.328
Updating a book	.329
Troubleshooting book updates	.330
Managing numbering	.332
Printing a book	.341
Modifying files from the book	.342
Spell-checking and finding/changing items in book files	.342
Book-level features	.343
Choosing files for other book-level operations	.343
Book features available inside files	.344
Paging through files in a book	.344

Advantages of using book files

Book files work reliably for small and large books; books with thousands of pages and hundreds of component files are not unusual. During updates, FrameMaker scans each file in the book, so updating a very large book can take some time. If you have a computer with plenty of system resources, you can speed things up by opening all the files before updating.

To adjust pagination, you must update the book file. The book file does not update pagination automatically when you add a new file to the book or when you modify a component file; you must explicitly tell FrameMaker to update the files.

Creating a book file

To take advantage of book features, you need to work with a book file!
To create a book file, follow these steps:

1. Select **File > New > Book**. If you had a file open, FrameMaker asks whether you want to add that file to the book. Click the Yes button to include the file or the No button to create the book file without it. FrameMaker creates the book and displays it.

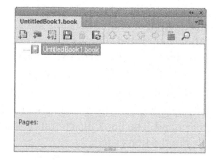

2. Save the book file. When you save your book file, notice that the book file name is listed with a complete path. Component files (chapters) will be listed with the path relative to the book file.

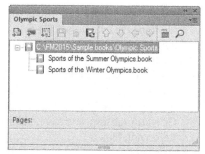

What the book window tells you

FrameMaker packs a lot of information about your files into the book window.

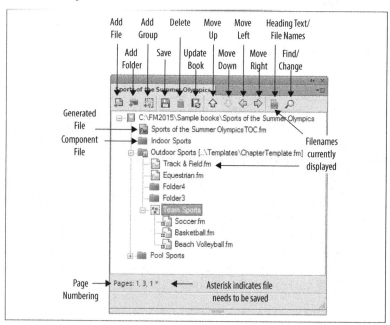

Managing files in a book

After you create a book file, you can add files to it, delete files that are listed in the book, change the order of files, change numbering and other properties, and rename files. Except for renaming, these features do not affect the actual files; for example, deleting a file from the book *does not* delete it from your computer. Deleting merely removes that component from the book. (Renaming a file does rename it on the system as well as in the book file.)

Adding files

FrameMaker makes a distinction between adding "regular" content or component files and adding generated files. Generated files are derived from information in the book—tables of contents, indexes, and lists of figures are all examples of generated files.

Regular files, often chapters or appendixes, are added to a book to help you control file order, pagination, and the like. Generated files do not exist until you create them through the book file.

Note Chapter 18, "Creating tables of contents," Chapter 19, "Creating indexes," and Chapter 21, "Creating Other Generated Files," describe how to add generated files to your book. This chapter focuses exclusively on managing regular files.

To add a content file (such as a chapter or appendix file), you must tell FrameMaker where the file is located and then position it in the book file, as described in the following steps:

1. Make sure the book window is the active window.
2. *(optional)* To add the file at a particular location, select the file immediately before where you want to place the new file.

Note By default, files are added after the selected file. It's easy to rearrange files, though, so you may want to add your files and worry about their order later.

3. Select **Add > Files** (or click the Add Files icon ( ) at the bottom of the window) to display the Add Files dialog.

4. Locate the file you want to add and select it. To add several files, CTRL+CLICK them.

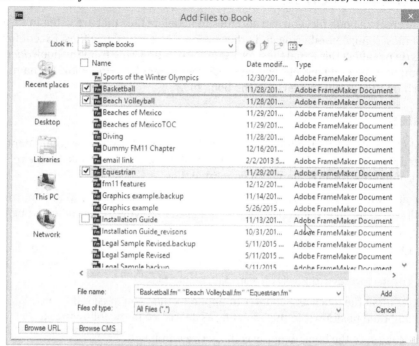

5. Click the Add button to add the files to the book.

If the files are not in the right order, you can move them around. See "Rearranging files" on page 327 for details.

Note   You can drag and drop files from the desktop or Explorer onto the book window to add them. To copy a file from one FrameMaker book file to another, open both books, select one or more files in the first book, then press CTRL and drag to the other book. (If you don't press CTRL, the file references are moved instead of copied.)

Adding non-FrameMaker files

You can include non-FrameMaker files in your book file. Only FrameMaker files are counted in the book's pagination or numbering; however, adding a non-FrameMaker file does provide you with a few useful features, including the following:

- **Keeping track of related files.** While working on a book, you'll have many ancillary files that you need to keep track of. Adding them to the book helps to keep those files organized. Files you might want to add to the book file include outlines, style guidelines, and other "extras" that might not be in FrameMaker format.

- **Opening from the book file.** You can open the non-FrameMaker files from the book file. FrameMaker prompts you to specify whether you want to open them in FrameMaker (which probably means a conversion) or in the native application (such as Adobe Reader for a PDF file).

When you print the book, only the FrameMaker files are printed.

Adding special book structures

Starting with FrameMaker 9, you have several options for organizing your book files. Read the sections below for more information.

Add a child book

You can add books within books. A child book is treated as a placeholder within the parent book. You cannot edit a child book from within the parent book view. All maintenance tasks must be performed in the child book separately. For example, any book-wide operation on the parent book, such as spell check or find/replace works only on parent book components and not on the files inside the child book. You should search and update child books separately.

Add a folder

You can add folders to your book and organize related documents in it. FrameMaker treats a folder as a logical container; it does not create a folder on the file system. A folder can have one or more folders, groups, or files within it.

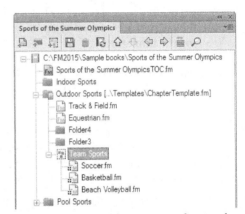

Folders change the hierarchy of your files. Depending on the level at which you add a folder, it can act like a chapter, section, or sub-section. You can set special numbering styles for a folder and all its contents. You can also choose to exclude files from being published by selecting Exclude from the context menu.

To add a folder to a book, follow these steps:

1. *(optional)* Select the file closest to where you would like to place the folder. You can change the hierarchy and ordering later if you like.
2. Click the Add Folder icon ( ) in the book file.
3. Right-click the folder and click the Rename option.
4. Position the folder and child elements as needed by dragging up or down, or by using the four arrow buttons ( ) at the top of the book window.

Associate a template with a folder. You can automate the process of creating and updating section cover pages by associating a template file with folders. The folder stores its associated title in the <$chaptertitlename> variable. By setting up a template file that uses the <$chaptertitlename> variable, you can control the formatting of all section cover pages from a single template file. This is useful for complex projects.

At time of printing, a video of this process is available at

blogs.adobe.com/techcomm/2009/09/fm9_hierarchical_books_new_variable.html

To associate a template with a folder, follow these steps:

1. Right-click the folder and select Properties to display the Container Properties dialog.
2. Select the Template Path checkbox.
3. Click the ellipsis button ([...]), specify the template file location, and click Open.
4. Click Set. The folder icon changes to indicate that a template is associated with the folder.

Note   The template associated with a folder creates pages in your content like any other file in a book. It is different from the collection of styles and resources that we have described elsewhere in this book.

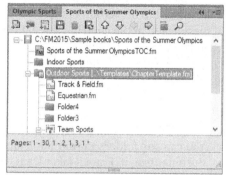

Open a folder template. You can open the template associated with a folder either by the traditional **File > Open** method, or by opening all files in the folder via the book file.

Open all files in a folder. To open a template associated with a folder, right-click on the folder and select Open. This opens all the files contained in the folder, including the template file if specified.

Add file information for a folder template. You can specify file information, such as author name, file title, keywords for a template associated with a folder. To set file information, follow these steps:

1. Right-click the folder that has a template associated with it and select File Info to display the File Info for Selected Files dialog.
2. Specify the file information and click Set.

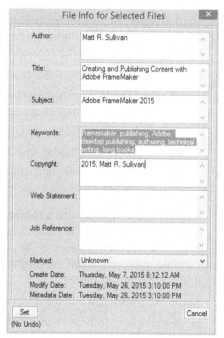

Adding a group

Unlike a folder, a group does not change the hierarchy of the files within it. Groups only provide an organizational tool within your book file. Groups do not impact numbering and the value of the <$section> and <$subsection> variables, nor do they change how numbers are handled between chapters.

To add a folder to a book, follow these steps:

1. *(optional)* Select the file closest to where you would like to place the folder. You can change the hierarchy and ordering later if you like.
2. Click the Add Folder icon ([icon]) in the book file.
3. Right-click the folder and click the Rename option.

Managing files in a book

4. Position the folder and child elements as needed by dragging up or down, or by using the four arrow buttons ( ) at the top of the book window.

Opening, closing, and saving all files

Manually opening each file in the book is a tedious process. Fortunately, FrameMaker provides a convenient set of shortcuts to open, close, and save all files at once. Make the book file active, hold down the SHIFT key, then select the File menu to see the global commands:

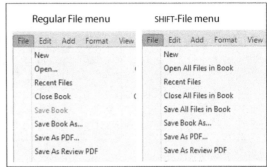

- Open All Files in Book
- Close All Files in Book
- Save All Files in Book

Removing files

Removing a file from the book file does not delete the file from the computer; it only removes the reference to the file from the book file.

To remove a file from the book, follow these steps:

1. In the book file, select the file or files you want to remove.
2. Select **Edit > Delete**, the Delete key, or click the Delete icon ( ) in the buttons at the top of the book window. The file disappears from the book window. There is no confirmation prompt, and the file is not deleted from your files system.

Rearranging files

When you add files to the book, they may not appear in the order that you want them. To correct this, you can drag and drop in the book window to rearrange the files. To rearrange files, follow these steps:

1. Select the file you need to move.
2. Click and drag to reposition the file. A bar shows you where the file will go.
3. Release the mouse button to drop the file in its new location.

Click the file...

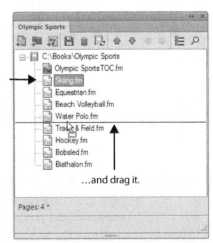

...and drag it.

327

Renaming files

You can rename files from within the book file. When you rename a file, FrameMaker updates references to that file throughout the book. FrameMaker also creates a backup of your original file.

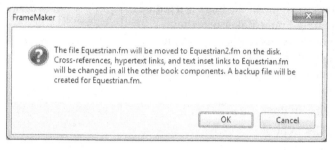

Caution FrameMaker updates only references within the current book file when you rename. If you rename a file that is referenced *by files outside the current book*, those references are broken by the renaming process. As illustrated in the following figure, assume that you have a file called apple.fm. There are a number of cross-references from pie.fm and from horse_favorites.fm that point to apple.fm. The apple.fm and pie.fm files are in the book called baking.book. The horse_favorites.fm file is not part of this book. If you use the book-level renaming feature from baking.book, references between apple.fm and pie.fm are updated correctly. However, the links from horse_favorites.fm are broken.

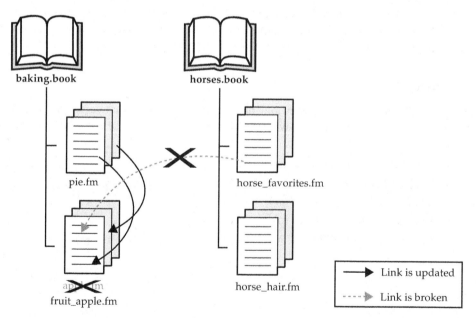

To rename a file, follow these steps:

1. *(optional, but recommended)* Open all files in the book (SHIFT-**File > Open**). Opening all files speeds up the process and also ensures that any missing font problems are avoided.

Updating a book

Caution If you do not open all files, missing fonts or other opening errors (such as files in an earlier version of FrameMaker) may cause the renaming operation to fail.

2. Make sure that file names are displayed in the book window. If they are not, click the Display Filenames button ( ) to see them.

3. Select the file you want to rename. (Don't double-click; that opens the file.)
4. Click the file name to make it editable.
5. Type the new file name and press ENTER. FrameMaker prompts you to confirm that you want to rename ("move") the file and Update All references to that file with the new name.

Caution Make sure you press ENTER before you click anywhere else in the book, or FrameMaker will reorder the files.

6. Click the OK button. FrameMaker scans all of the files in the book for cross-references, hypertext links, and text insets that refer to the old file name and updates them.

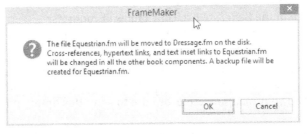

After completing the updating process, FrameMaker notifies you that it has updated the files.

Updating a book

Periodically, you will need to update the book. During an update, FrameMaker can modify the following items:

- **Numbering.** Chapter numbers, volume numbers, section numbers, and paragraph-based numbering (autonumbers) are updated.
- **Pagination.** Page numbers updated based on the settings in the book.
- **Cross-references.** Referenced page numbers and text are updated to reflect any changes.
- **Text insets.** If imported by reference, FrameMaker files have been updated, the insets are refreshed (see Chapter 29, "Content reuse with text insets").
- **OLE links.** Any OLE links are refreshed.
- **Generated files.** The entire book is scanned, and generated files are replaced with new versions.
- **Master pages.** Any master pages applied based on paragraph tags are updated.

To update your book, follow these steps:

1. Display the book file.
2. *(optional)* To speed up the generation process, and to ensure that it works even if you have missing fonts in some of the files, open all of the files in the book (SHIFT-**File > Open All Files in Book**).
3. Select **Edit > Update Book**, or click the Update icon ( ) at the top of the book window to display the Update Book dialog.
4. Check each of the items that you want to update. Your choices are as follows:

   - **Update Numbering:** If checked, all page, chapter, volume, and paragraph numbering is updated.
   - Update All **Cross-References:** If checked, all cross-references are updated.
   - Update All **Text Insets:** If checked, all text insets are updated.
   - **Update OLE Links**: If checked, all files that are imported with OLE links are updated.
   - **Generate Table of Contents, Lists, and Indexes:** If checked, FrameMaker generates the files that are listed in the Generate list. Only tables of contents, indexes, and similar generated files are available to generate or not generate.
   - **Apply Master Pages:** If checked, FrameMaker updates master page assignments. For details, see "Mapping paragraph tags to master pages" on page 211. If you haven't set up your mapping table, be sure to leave this option unchecked.

Note If your book is very large or you work across a slow network, you can save some time by not generating tables of contents and the like. Be sure, though, to update them before you finalize the book to ensure that the most current information is displayed in generated files.

Troubleshooting book updates

When updating a book, FrameMaker reports any problems in the book error log. Any issues requiring user intervention (such as dialog boxes displayed due to missing resources) will cause problems. To reduce confusion later while editing, you may find it helpful to close these logs after correcting the errors.

Here are some of the most common errors that can cause the update process to fail:

- Missing fonts or graphics
- FrameMaker files from an earlier version of FrameMaker
- FrameMaker files in use by someone else
- Read/write permissions on the files

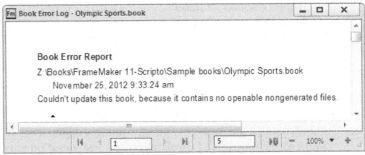

Other generation problems do not cause the process to fail but are reported in the book error log. They include the following:

- Unresolved cross-references
- Inconsistent conditional text settings
- Inconsistent color settings
- Inconsistent numbering properties

The following sections explain how to correct these problems.

Missing fonts. A missing fonts message is displayed if FrameMaker attempts to open a file, but the file calls for fonts that are not currently available on your computer. The missing font message stops the file-opening process, which in turn causes the entire book update to fail. To correct the missing fonts problem, you have three choices:

- **Install the missing fonts on your system.** This is an operating system task. Refer to your system documentation or IT department for more information.
- **Remove the missing fonts from the document.** You can change the FrameMaker preferences so that missing fonts are permanently replaced. If the fonts are not assigned to body content or specified in paragraph or catalog entries, you may need to refer to "Handling missing fonts" on page 25 for instructions.
- **Open all files before updating the book.** If you open the files and dismiss the missing font dialogs, FrameMaker doesn't have to open them during the update, and you avoid this error. Keep in mind that the fonts used and the pagination of your content may be affected by a font substitution.

FrameMaker files from an earlier version of FrameMaker. Before updating the book, open all of the files and save them in the latest version of FrameMaker. That will eliminate this message. FrameMaker will not update a book that contains files from earlier versions of FrameMaker.

FrameMaker files in use. This message indicates that one of the files has been opened by another user. If you attempt to open the file directly, FrameMaker will tell the machine name or user name that has locked the file. Have the other user close the file so that you can update the book. If the message is a result of a previous FrameMaker or system crash, you'll need to open the file and reset the lock prior to updating the book.

Read/write permissions on the files. To open the files, FrameMaker needs read permissions. To save them, FrameMaker needs write permissions. Make sure that you have the appropriate permissions on the files. Files that are set to be read-only will not update properly. (This often happened on older Windows systems with files copied from a CD.) You can also create view-only files inside FrameMaker; these will not update, either. To toggle the view-only setting, open the file and press ESC SHIFT-F L K to make it editable.

Unresolved cross-references. This message indicates that FrameMaker was unable to resolve cross-references in the book. This message can occur because of problems opening the file due to missing resources in the target file, missing or deleted cross-reference markers, or missing files. Before replacing or relinking the reference, open the referenced file. If the file requires user input, resolve the problem and try updating again. If the cross-reference is still unresolved, see "Correcting unresolved (broken) cross-references" on page 187.

Inconsistent conditional text settings. This message indicates that a particular conditional text tag is set to be shown in one file and hidden in another. The message is informational; if you need to have your files set up this way, you can ignore it (for one or all files). If the conditional text settings should be consistent across the entire book, select all the files in the book, select **View > Show/Hide Conditional Text** and choose the settings you want for all the files. For more information, see "Showing and hiding conditional text" on page 454.

Inconsistent color settings. This message indicates that color definitions or color print settings are inconsistent. To correct the problem, apply consistent settings across the entire book (see Chapter 24, "Color output"). The message is irritating, but you can safely ignore this message if your files do not actually use the inconsistent color.

Inconsistent numbering properties. Indicates that the numbering set in the file conflicts with the numbering in the book. Since FrameMaker will use the book's numbering properties, after updating a "Save All Files in Book" will often eliminate this message. If you still need to correct numbering, see the next section.

Managing numbering

FrameMaker uses the order of files in the book and information you provide about numbering to control the numbering sequence for your book.

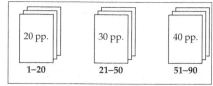

You can, for example, specify that the files in the book should be numbered sequentially as shown in the first example. When you make changes to the files, FrameMaker manages the pagination to keep everything in sequence, as shown in the second example.

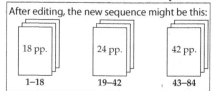

Managing numbering

Another numbering setup might incorporate both the chapter number and page number, requiring each chapter's page numbering to restart at 1. The page numbering would look as shown in the third example.

Numbering is controlled in the Numbering Properties dialog. For files, you access this by navigating to **Format > Document > Numbering**, while at the book level, you can also select the file(s) to number and have numbering available in the context menu when right-clicking. There are sheets for controlling the various types of numbering.

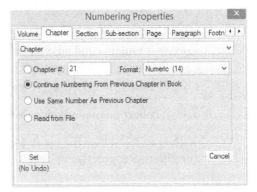

Tip A few sheets include the Continue Numbering option. Selecting this option causes the numbering to increase by one from the previous file. To keep the same number as the previous file, select the Use Same Number option.

Because FrameMaker's numbering is calculated based on the file sequence, a missing file will throw off the numbering. You can eliminate this problem by assigning fixed values to a chapter (instead of letting FrameMaker calculate the value that should be there) but this can make life difficult later when you need to move chapters around. I recommend that you stick with calculated numbering as much as possible. If this is unacceptable to you, consider creating dummy (empty) content as needed to account for missing content.

The following sections explain how to set up numbering for each of the numbering types available in FrameMaker:

- **Pages.** Controls the value of the <$pagenum> system variable inside the files. Page numbers increase from page to page within a file.
- **Sectioning variables.** A series of variables allow more explicit control of book numbering. The variables are available to control the following items:
    - **Chapters.** The <$chapnum> system variable is independent for each file. This variable is constant inside a single file. (In other words, you cannot embed both chapter 1 and chapter 2 in a single file if you want to use <$chapnum> to control the chapter numbering.)
    - **Volumes.** The <$volnum> system variable allows you to effectively nest chapters into groups. A volume usually contains several files. Like <$chapnum>, <$volnum> can have only one value inside a file. The most common use of the <$volnum> building block is for books with multiple parts, such as this book.

> ## Format choices for numbers
>
> The Format drop-down list, available in several of the numbering tabs, lets you specify the type of numbering you want to use. Your choices are:
>
> - Numeric (14). Creates a numeric sequence: 1, 2, 3, …
> - roman (xiv). Creates lowercase Roman numerals: i, ii, iii, …
> - ROMAN (XIV). Creates uppercase Roman numerals: I, II, III, …
> - alphabetic (n). Creates lowercase letters: a, b, c, …
> - ALPHABETIC (N). Creates uppercase letters: A, B, C, …
> - Text (volume and chapter only). Creates a custom label. If you select text, you must specify the label you want for each item. You could, for example, have a volume number of Read Me.
> - Custom (footnote only). Creates a custom footnote sequence.
> - A collection of more complex double-byte (Asian language) numbering options
>
> Each example shows the result for the number 14 in parentheses, presumably because it shows the differences clearly.

- **Sections and Subsections.** The <$sectionnum> and <$subsectionnum> variables are used to control numbering within nested (also referred to as hierarchical child books).
- **Paragraphs.** Controls when the numbering properties for paragraph tags are reset.
- **Footnotes and Table Footnotes.** Controls the numbering behavior of these objects.

Page numbers

FrameMaker provides three options for page numbering. The numbering can begin on a specified page (with a specified format), continue whatever numbering was specified in the previous file, or start on the page specified in the file-level numbering properties (Read from File). To set up numbering that works properly, you usually want the page numbers to continue from the previous document. Exceptions include folio-by-chapter numbering (for example, 1-12) and numbering in chapters that include multiple files per chapter.

Consider the numbering example shown in the following table. The first few files in the book (table of contents, preface, and other front matter) use Roman numerals. After that, the first chapter starts with a numeric 1. All other documents are numbered straight through.

Document	Page Numbers	Setting Needed
Title page	i–ii	Start at: 1 roman (xiv)
Table of contents	iii–vi	Continue
Preface	vii–x	Continue

Managing numbering

Document	Page Numbers	Setting Needed
Chapter 1	1–16	Start at: 1 Numeric (14)
Chapter 2	17–28	Continue
Chapter 3	29–34	Continue
Chapter 4	35–46	Continue

To match this numbering, you need to set most of the files to "continue numbering." The files will pick up the last number in the previous file and increment by 1 to continue. Set the first file in the book to start at 1 and use Roman numerals. The first chapter file needs to restart at 1 and use numerics. To set up this numbering, follow these steps:

1. Select all of the files in the book window (click in the book window and press CTRL+A or ESC E A to select all the files).
2. Select **Format > Document > Numbering** to display the numbering properties.
3. Click the Page tab to display the page numbering properties.
4. Click the Continue Numbering from Previous Page in Book radio button and click the Set button. FrameMaker applies the specified numbering scheme to every file in the book.

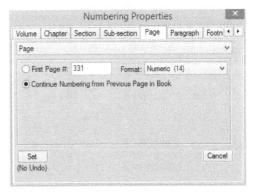

5. Now, set the numbering exceptions. Select only the table of contents file and display the Page tab in the Numbering dialog box again. In the First Page # field, type **1**. In the Format drop-down list, click roman (xiv). Click the Set button.
6. Select the first chapter file. Display the Page tab. In the First Page # field, type **1**. In the Format drop-down list, click Numeric (14). Click the Set button.
7. Update the book. Your new settings are applied throughout the book.

Tip To set up page numbers that restart with every chapter, select all the files and set the first page number as 1 for each file. Separately, you'll need to set the chapter numbering to continue. The result will be the correct sequence:

   1-1, 1-2, 1-3, …, 2-1, 2-2, 2-3, …, 3-1, 3-2, 3-3, …

You'll also need to set up the generated files and the master pages to use the chapter numbers. See those sections for details.

Chapter numbers

For chapter numbers, FrameMaker provides the same "restart" and "continue" settings as for page numbers. In addition, you also have the option of retaining the same chapter number as for the previous file. This allows you to split a large chapter into two files and assign the same chapter number to both files.

If a file already has a chapter number assigned at the file level when the file is added to a book, FrameMaker will retain it. Be sure to verify appropriate chapter number settings in the book and the files to avoid inconsistency messages during book updates.

You will probably need to customize the numbering to ensure that you get the correct numbers in your book. Consider the book shown in the following table. Because each file initially is assigned a chapter number, you end up with the table of contents, preface, and other front matter incrementing the chapter number. In this example, the first three files do not display a chapter number, so their settings are irrelevant.

Document	Default Chapter Number	Correct Chapter Number	Setting Needed
Title page	1	N/A	Doesn't matter
Table of contents	2	N/A	Doesn't matter
Preface	3	N/A	Doesn't matter
Chapter 1	4	1	Start at: 1 Numeric (14)
Chapter 2	5	2	Continue
Chapter 3	6	3	Continue
Chapter 4	7	4	Continue

To correct this, you need to set the chapter numbering properties so that the first chapter file starts at 1. Follow these steps:

1. Select all of the files in the book window.
2. Select **Format > Document > Numbering** to display the numbering properties.
3. Click the Chapter tab to display the chapter numbering properties.
4. Select the Continue Numbering from Previous File in Book radio button and click the Set button. FrameMaker now assigns an incrementing chapter number to each file in the book.

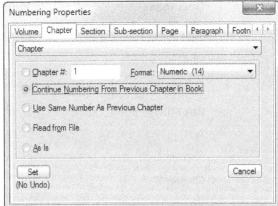

   Note This is the default setting for chapter numbers, but it's a good idea to reset them to ensure that any custom numbering is stripped.

5. Select only the first chapter file. Display the Chapter tab again. In the Chapter # field, type **1**, and in the Format drop-down list, select Numeric (14). Click the Set button.
6. Update the book. Your new settings are applied throughout the book. Inside the files, every reference to <$chapnum> is also updated.

The chapter number in the front matter is still incorrect, but because the front matter doesn't normally use the chapter number, you can ignore this issue.

Managing numbering for multifile chapters

You may need to split some chapters across two or more files. For example, a foldout page with a large diagram would reqire a different file because FrameMaker cannot mix page sizes in a single file. If this occurs, make sure that the second and subsequent pieces of a chapter are set to use the same chapter number as the previous file.

Volume numbers

For volume numbers, FrameMaker provides the same options as for chapter numbers. FrameMaker assumes that your volumes would typically contain multiple chapters (files). To accommodate this, each file in the book defaults to using the same volume number as the preceding file (the chapter numbers increment by default), unless the volume number is set in the file. The <$volnum> building block reference the current volume number in a document.

If your book uses volume (or part) numbers, you need to customize the numbering to set the volume numbers. Consider the book shown in the following table. The book defaults to a single volume number for each file, but you need the volume number to increment with each part of the book.

Document	Volume Number	Setting Needed
Title page		Doesn't matter
Table of contents		Doesn't matter
Preface		Doesn't matter
Part I title page	I	Start at: 1 ROMAN (XIV)
Chapter 1	I	Same As Previous File
Chapter 2	I	Same As Previous File
Part II title page	II	Continue
Chapter 3	II	Same As Previous File
Chapter 4	II	Same As Previous File

To set the volume numbering properties, follow these steps:

1. Select all of the files in the book window.
2. Select **Format > Document > Numbering** to display the numbering properties.
3. Click the Volume tab.
4. Select the Use Same Number As Previous File radio button, and click the Set button. FrameMaker now assigns the same volume number (1) to each file in the book.

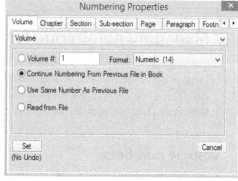

Note This is the default setting for volume numbers, but it's a good idea to reset them to ensure that any custom numbering is stripped.

5. Select only the first file in the first part. Display the Volume tab again.
In the Volume # field, type **1**, and in the Format drop-down list, select ROMAN (XIV). Click the Set button.
6. Select the first file in the second part. Display the Volume tab. Select the Continue Numbering from Previous File in Book radio button, and click the Set button.
7. Repeat step 6 for each additional part.

Managing numbering

8. Update the book. Your new settings are applied throughout the book. Inside the files, every reference to <$volnum> is also updated.

Section and subsection numbers

The <$sectionnum> and <$subsectionnum> variables are nearly identical in function to the <$chapnum> variable. They are used in place of the chapter number variable for files that are nested within folders.

You create Folders and Groups using the buttons circled below
Outdoor Sports is a folder with an attached template
Team Sports is a Group

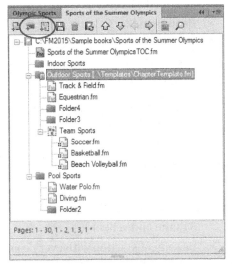

When you nest a file within one folder, you have the option of setting numbering for the section variable, but the content (not the tabs) for the chapter and subsection tabs will be grayed out. Similarly, when you nest a file within two folders, your subsection content will be active, but the content in the chapter and section tabs will be grayed out.

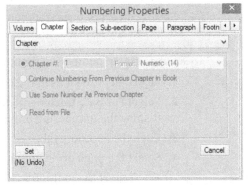

Paragraph numbers

The paragraph numbering properties determine whether autonumbering continues or restarts from one file to the next. The setting you choose here matters only if you have a numbering stream that carries over from one chapter to the next. If, for example, you have figure or table numbers that carry through the entire book (1, 2, 3, ..., 57, 58, 59, ...), make sure that paragraph numbering is set to Continue. The paragraph numbering properties control the values of the autonumbering counters.

339

For paragraph numbering, you have the following choices:

- **Restart paragraph numbering:** All counters are reset to zero at the beginning of the selected file.
- **Continue numbering from previous paragraph in book:** Counter values are carried over from the previous file.
- **Read from file:** Counter values are picked up from the current file.

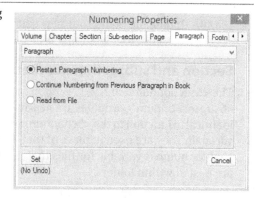

Note If you have set up your files to track chapter and section/subsection numbering with paragraph-based numbering instead of using the supplied variables, changing the paragraph number settings might affect those items in your book. For best results, use the variables when possible. If your files were created in a version of FrameMaker before version 9 (when they were introduced), you may need to rework your numbering scheme before you use these variables.

Footnote numbers

FrameMaker's footnote numbering gives you three alternatives. Footnotes can:

- Increment through the entire book (which means there would be only one footnote number 3 in the entire book).
- Increment starting with every chapter (footnotes would start with 1, 2, 3 in each chapter, meaning you could have several footnotes labeled 3—one per chapter).
- Restart on every page (footnotes would start with 1, 2, 3 on each page, so you could theoretically have a footnote number 3 on each page in the file).

To set up footnote numbering properties for all of the files in your book, follow these steps:

1. Select all of the files in the book window.
2. Select **Format > Document > Numbering** to display the numbering properties.
3. Click the Footnote tab.
4. Select the pagination option you want:

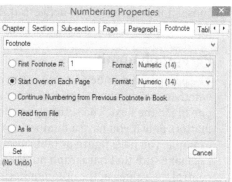

- To start the footnotes with a particular number (usually 1) for each file, select First Footnote, type in the number, and select a format from the drop-down list.
- To restart footnotes on every page, select Start Over on Each Page and select a format from the drop-down list.
- To number footnotes consecutively through the book, select Continue Numbering from Previous Footnote in Book. (If you do this, you should also set the first file to start footnotes at 1. The file *should* default to this setting, but it's safer to apply it explicitly.)

5. Click the Set button.
6. Update the book. Your new settings are applied throughout the book.

Table footnote numbers

Table footnotes are footnotes that are created for text inside a table. You cannot change the numbering for table footnotes; they always restart for each table. You can, however, set the format of the number at the book level.

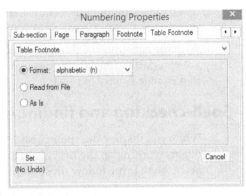

To set the formatting for table footnotes, follow these steps:

1. Select all of the files in the book window.
2. Select **Format > Document > Numbering** to display the numbering properties.
3. Click the Table Footnote tab to display the table footnote numbering properties.
4. In the Format drop-down list, select the option you want. Click the Set button.
5. Update the book. Your new settings are applied throughout the book.

Printing a book

When the book file is active, selecting **File > Print Book** results in printing all the files in the book. But you can also choose to print just a portion of the book. To print specific files in a book, follow these steps:

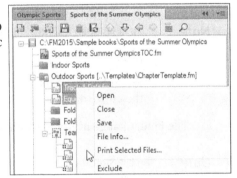

1. In the book window, select the files you want to print by CTRL+CLICKING each file.
2. Select **File > Print Selected Files**.
3. Proceed as usual in selecting your print options.
4. Click the Print button to print the selected files.

Modifying files from the book

From the book window, FrameMaker provides a subset of the regular FrameMaker commands on the menus. These include spell-checking, find/change, conditional text settings, and others. Using these commands is equivalent to opening each file in the book and issuing the command there, so the book file can save you lots of time.

With the exception of the spell-checking and the find/change utility, the book-level commands are reversible. You can, for example, change your view settings for all the files in the book, but that change is easy to reset.

Spell-checking and finding/changing items in book files

When you display the spell-checker or the Find/Change window from the book file, the utility begins working on the first file in the book and works through every file in order. To spell-check at the book level, follow these steps:

1. Select **Edit > Spelling Checker** to display the Spelling Checker pod.
2. Verify that the Book radio button is selected on the left.
3. Click the Start Checking button to begin checking spelling.

    FrameMaker attempts to open each document in the book and spell-check its contents. If FrameMaker cannot open a file, an entry is made in the book error log, and the spell-checker continues with the next file. After completing spell-checking in a file, that file is saved and closed, and FrameMaker moves on to the next file.

Note If a file is open during a spell check, FrameMaker will not save or close the file after spell checking it.

4. Use the spell-checking options as normal. See page 43 for details.
5. When FrameMaker reaches the end of the book, (or when FrameMaker loops through the entire document) the spell-check is complete.

The Find/Change utility works the same way as the spell-checker. See page 38 for details.

Note You can start the spell-checker or find/change process at a specific location by opening a file, displaying the spell-checker or Find/Change pod, and then selecting the Book radio button to specifiy that the operation should be applied to all the files in the book. This works only if the book file is open when you open a component file; otherwise, the Book option is grayed out.

Book-level features

There are a number of functions you can apply at the book level to any or all the chapters in a book. Among them are:

- Open, save, and close files
- Import formats (template items)
- Set **View > Options** items
- Set color views
- Set numbering and pagination options
- Control conditional text options
- Exclude items from processing/printing
- Set file info
- Rename and delete from book

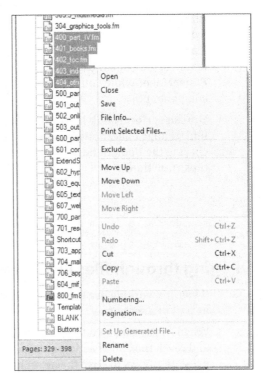

Choosing files for other book-level operations

For most book-level features, you can select which files the operation is performed on. If you want to make a change to all of the files in the book, you must select all of the files in the book before applying the command.

To apply a book-level command, follow these steps:

1. In the book window, select the files that you want to make changes to.
2. Select the command you want from the menus. For example, to toggle the display of document borders, select **View > Show Borders** (or **Hide Borders**). FrameMaker opens each file in turn and makes the change in the file.

Note If you open all your files before making changes, remember to save them (SHIFT-**File > Save All Open Files**).

Book features available inside files

Although most book features are available only from the book window, a few are available from inside component files. To use book-level features, the book file must be open in FrameMaker while you're working inside the component file.

You can perform book-level spell-checking and find/change operations from inside the file by clicking the Book radio button in the pods for those functions.

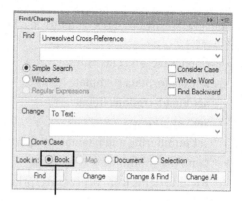

Select the Book option to start a global find/change operation.

Paging through files in a book

If you are working in a file that belongs to an open book, FrameMaker tracks that the current file as part of a sequence.

When you scroll down through a file and reach the end of a file, you can press PAGE DOWN to go to the next file in the book.

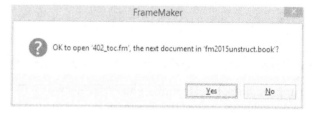

If the file is already open, FrameMaker displays the next file automatically (which can be quite disconcerting). If the file is not open, FrameMaker prompts you to open it.

Press PAGE UP at the beginning of a file to open the preceding file.

Chapter 18: Creating tables of contents

FrameMaker makes it easy to create and automatically maintain generated tables of contents. When setting up a table of contents, you specify which paragraphs will be included. FrameMaker takes care of the rest, scanning the documents to locate every instance of those paragraphs, and then listing them in the table of contents as you've specified, with information like the corresponding page number.

Starting with FrameMaker 2015, you can also insert Mini TOC into your documents. In fact, I've used the Mini TOC at the beginning of each page of the print version of this book. It acts just like a regular TOC, and looks something like this:

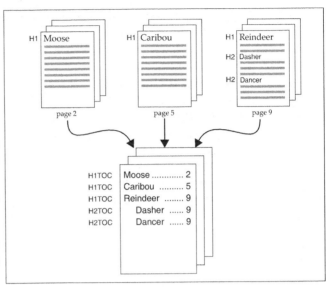

Here's a look at what's in this chapter:

```
Setting up table of contents file  . . . . . . . . . . . . . . . . . . . . . . . . . . . . . . . . .345
    The initial table of contents file  . . . . . . . . . . . . . . . . . . . . . . . . . . . . . .347
Customizing the table of contents. . . . . . . . . . . . . . . . . . . . . . . . . . . . . . . .348
    Locating the TOC reference flow . . . . . . . . . . . . . . . . . . . . . . . . . . . . . .349
    Understanding the TOC flow entries . . . . . . . . . . . . . . . . . . . . . . . . . .350
    Formatting the table of contents . . . . . . . . . . . . . . . . . . . . . . . . . . . . .352
    Using character-level formatting . . . . . . . . . . . . . . . . . . . . . . . . . . . .352
    Formatting examples. . . . . . . . . . . . . . . . . . . . . . . . . . . . . . . . . . . . . .352
Mini TOC . . . . . . . . . . . . . . . . . . . . . . . . . . . . . . . . . . . . . . . . . . . . . . . . . .356
```

Setting up table of contents file

When you create a table of contents, you specify which paragraph tags to include in the document. This is an all-or-nothing operation—if you include the Heading1 paragraph tag, every Heading1 in the book will be included in the generated table of contents.

Tip If you want to eliminate a specific heading from the table of contents, you can create two paragraph tags (with identical settings): for example, Heading1 and Heading1NoTOCEntry. Tag the paragraphs appropriately, and include only the Heading1 in your generated list.
For example, if you want to exclude the "Caribou" entry in the preceding example, you would tag Moose and Reindeer as Heading1 and tag Caribou as Heading1NoTOCEntry. This lets you exclude Caribou, even though it looks just like the other first-level headings.

To set up a table of contents, follow these steps:

1. Open the book file.
2. *(optional)* To add the table of contents at a particular location, select the file immediately before or after where you want the table of contents. You can move the table of contents after creating it, so this is not a critical step.
3. Select **Add > Create Stanalone TOC** to display the Set Up Table of Contents dialog.
4. The default file name suffix for the table of contents is TOC. You can change it if you want to. If you create several tables of contents, the suffix increments (TOC, TOC2, TOC3, etc.).
5. *(optional)* In the Add File drop-down list, specify whether you want the table of contents before or after the file you selected in step 2.

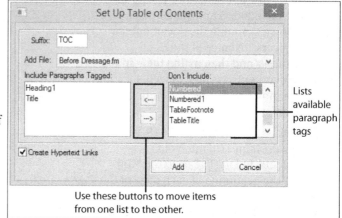

6. Move the items that you want to include to the Include list on the left.

   You can move items by doing any of the following:

   - Select an item, then click the left or right arrow button as needed.
   - Select an item, then double-click it to move it to the other list.
   - To move all the items from one list to another, press and hold SHIFT, then click the appropriate arrow button.

7. Make sure that the Create Hypertext Links checkbox is checked. This creates hyperlinks for each entry in the table of contents. It also enables hyperlinks when converting to other electronic formats (like PDF, help, and apps).
8. Click the Add button, which displays the Update Book dialog.

Note The file name of the TOC file is based on the book file name. If your book file is foo.book, the TOC file will be fooTOC.fm.

9. Click the Update button to create the table of contents. (See page 329 for a detailed discussion of updates.) The table of contents file now is displayed in the book; you can move it just like any regular file; see "Rearranging files" on page 327.

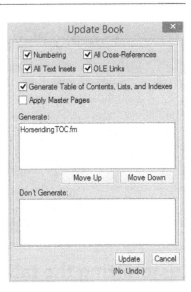

To change the items included in the table of contents, select the table of contents file, and select **Edit > Set Up Table of Contents** to display the setup dialog box again.

The initial table of contents file

After creating the table of contents, double-click it in the book file to open. By default, your table of contents will look something like this accompanying illustration.

Unless you deselected the Create Hypertext Links checkbox, your generated table of contents contains hyperlinks on each of the generated paragraphs. You can navigate to a topic from the table of contents by pressing CTRL-ALT and clicking the entry.

FrameMaker uses reference pages and paragraph tags to determine the appearance of the table of contents. When you create the table of contents, FrameMaker will take the resources available in the first file in the book and use them to format your table of contents file. If that file is set up with all the needed table of contents settings (that is, if the reference pages and paragraph tags contain the needed definitions), your work may be done. In most cases, however, you may find that maintaining a separate table of contents template file will help you manage the entire process more efficiently.

If you have a table of contents template, you can import formats from that template into the table of contents file and update the book again. At that point you should have a table of contents that's formatted correctly.

The rest of this chapter describes how to set up the formatting in the table of contents.

Tip You can insert a title for the table of contents (such as "Contents") and apply a separate paragraph tag (such as ContentsTitle) to it. Make sure that the paragraph tag applied does not use the file suffix (such as TOC or TOC1) associated with the file, as paragraphs formatted with tags containing the suffix in their label are deleted upon generating the file.

> **A Sneaky Way of Avoiding Formatting Work**
>
> If you have a template file that contains the table of contents settings you need, you can save yourself some time. Instead of generating the table of contents and then importing from the template file, try this:
>
> 1. Copy the template file to the book directory.
> 2. Rename the template file to match the name that FrameMaker assigns to the table of contents. For example, if your book is named long.book, the table of contents name will be longTOC.fm.
> 3. Open the book file and add the table of contents file as usual.
> 4. Update the book. FrameMaker automatically uses the file that you've snuck into the book's directory and picks up all the formatting from that file.

Customizing the table of contents

After creating the initial file, you have several ways to customize the table of contents. You can do any of the following:

- Include or exclude paragraphs when you set up the table of contents.
- Specify on the reference page the information to include for each item (for example, paragraph text, paragraph autonumber, page number, or chapter number).
- Format the included information.

To accomplish the last two items, you must modify the reference pages and the paragraph tag definitions stored in the table of contents file.

The content on the reference pages is generated based on the paragraph tag settings, so often you have to go back and forth to get everything exactly right. You must update the book after you make changes on the reference pages to see the changes in the table of contents.

Remember Because the table of contents generation process is based on paragraph tag names, it's important (once again) to tag your document consistently. Using blank paragraphs to provide vertical white space in a document is a bad idea, and can cause problems in the table of contents. If the blank paragraphs use a tag that's included in the table of contents, you will see empty entries in the table of contents.
To remove empty TOC lines, go to the source file (CTRL+ALT+CLICK on the blank TOC entry) and delete the empty paragraphs. Use the space above and space below settings in the Paragraph Designer instead (see "Avoiding formatting overrides" on page 112 for details).

Locating the TOC reference flow

The information that's displayed for each paragraph is controlled by the reference pages of the generated file. For the table of contents, you will find a page named TOC in the reference pages. The TOC page contains a text frame with a TOC flow.

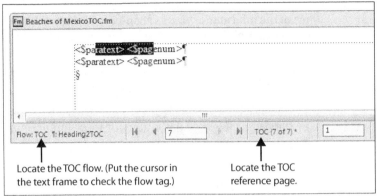

Locate the TOC flow. (Put the cursor in the text frame to check the flow tag.)

Locate the TOC reference page.

Each paragraph that you include in the table of contents has its own entry in the reference pages. This entry determines exactly what is displayed for each item.

To locate the TOC flow on the reference pages, follow these steps:

1. Display the reference pages by selecting **View > Reference Pages**.
2. Find the TOC reference page. Notice that the status bar tells you the name of each page along with the number of pages as you move from page to page (as shown in the graphic earlier). To quickly find a reference page, use the CTRL+G shortcut and select the page name from the drop-down menu. Click Go to jump to that page.

On the TOC reference page, locate the TOC flow. Click inside the flow, and check the status bar to verify that you are in the TOC flow.

Caution If you have *two* TOC reference page entries in the file, FrameMaker uses the one it finds first. If you update the book but do not see an updated TOC, check the reference pages for an extra TOC text frame. FrameMaker might be using that frame instead of the one you want.
Sort order is not necessarily the same as the page order of the reference pages, so a frame that occurs later in the reference pages could be considered the "first" TOC frame. Check every text frame's flow to confirm that only one text flow in the reference pages is labeled TOC. Keep in mind that the renegade extra flow might be in a very small text frame. If you do find an extra TOC flow, we recommend that you copy the version you want, delete *both* flows and regenerate. FrameMaker will recreate the TOC text frame (and flow), and you can paste the good information into this new text frame.

> ### Book vs. stand-alone table of contents
>
> Most of the information in this chapter focuses on tables of contents stored in a book file. You can, however, also create a table of contents without a book file. Without a book file, you obviously can't update from the book.
>
> To create a TOC for a single FrameMaker file, open the file and select **Special > Table of Contents**. When needed, choose **Special > Table of Contents** to update the file.
>
> Unless you change the name of the source file, any changes you made to the list of included tags, the paragraph style definitions, or the reference pages are preserved.

Remember   After changing the reference pages, you must update the book before your changes are shown in the table of contents entries.

Understanding the TOC flow entries

In the TOC flow, you will find a paragraph for each of the tags you've included in the table of contents. You can determine which paragraph belongs to an included item based on the paragraph tag. For example, if you included ChapterTitle in your table of contents and have updated your book, you will find a ChapterTitleTOC paragraph in the TOC flow.

Note   FrameMaker does not delete paragraphs from the TOC flow. If you added a paragraph tag and later removed it from the setup dialog box, you will still see an entry on the TOC reference page. You can safely remove that entry if you want to.

Each paragraph in the TOC flow determines what information is included for a specific paragraph. For instance, you might have a paragraph such as the following with a paragraph tag of Heading1TOC:

    <$paratext> <$pagenum>

For each Heading1, FrameMaker will retrieve the contents of the paragraph (<$paratext>), followed by a space, followed by the page number of the paragraph (<$pagenum>).

Most TOC flow definitions consist of a mixture of text and building blocks. For instance, you could use the following definition for a chapter title:

    Chapter <$chapnum>: <$paratext>\t<$pagenum>

In the chapter title tag, the word "Chapter," the colon, and the various spaces are regular, literal text. The <$chapnum>, <$paratext>, and <$pagenum> building blocks are processed for each TOC entry to determine the current values for those items. The \t is a tab (you can press the TAB key to insert a tab instead of typing \t, if you choose).

Several FrameMaker features—such as cross-references and variables—use building blocks. The building blocks for the table of contents are special in a few ways:

- FrameMaker does not provide you with a way to insert building blocks by choosing them from a list. Instead, you must type the building block.
- In table of contents definitions, FrameMaker can retrieve information only from the paragraph that's currently being processed. You cannot use building blocks that call other paragraph tags, such as:

    <$paratext[*some_other_para*]>

- Because you are typing in a text frame and not a dialog box, you can use the normal keyboard shortcuts for special characters. For more information about typing special characters, see "Inserting special characters" on page 35.

Table 18-1 lists the available building blocks that you can use to customize table of contents definitions.

Table 18-1. Building Blocks for Table of Contents Definitions

Building Block	Description
<$paratext>	Inserts the text of the source paragraph (for example, "Reindeer" or "Caribou" from the example).
<$paranum>	Inserts the entire autonumber (if any) of the source paragraph.
<$paranumonly>	Inserts the numeric portion of the autonumber from the source paragraph.
<$pagenum>	Inserts the page number on which the source paragraph occurs.
<$chapnum> <$volnum> <$sectionnum> <$subsectionnum>	Inserts the requested variable from the source paragraph's file.
<*char_tag*>	Applies the specified character tag to the text that follows in the paragraph.
<Default Para Font> or </>	Removes any character tagging and returns formatting to the paragraph's catalog definition.
\t	Inserts a tab. You can get the same result by typing a tab on the keyboard.

Formatting the table of contents

For each tag you include in the table of contents, FrameMaker automatically creates a paragraph tag. The new tag name is the name of the source tag with "TOC" appended. If, for example, you include Heading1 in your table of contents, all of the Heading1 entries will use the Heading1TOC paragraph tag. To format the table of contents, you modify these tag definitions in the table of contents file. Refer to Chapter 6, "Formatting text with paragraph tags," for details on how to format the paragraph tags.

Tip Because of FrameMaker's convention of appending "TOC" to paragraph tag names, I recommend against creating any tag names for regular content that end in TOC. This recommendation applies to all generated files. The default file name suffix of the generated file is used for the paragraph tags that control the formatting in the generated file. (For example, by default in the index, the tag names end with IX; in lists of references, they end with LOR.) Different letters are appended to generated files if you are using a localized version of FrameMaker. (For example, in German, the suffix IVZ for Inhaltsverzeichnis is used for the table of contents.)

Using character-level formatting

You can use character tags as building blocks to format a portion of a TOC entry differently from the rest of the paragraph. For example, to allow all tab leaders and page numbers to have consistent formatting, you would format each entry to look something like this:

    <$paratext> <tocNumber><$pagenum></>

Note The character tag used must be defined in the Character Catalog of the table of contents file.

Caution Although this works quite well for print, FrameMaker recognizes character tags as limiting the active area of hyperlinks. As a result, the hyperlinks in your TOC may not be active for the entire TOC entry paragraph. Make sure this will be acceptable in PDF and other electronic output.

Formatting examples

You may need to make changes in two different places—the reference pages and the paragraph tags—to get your TOC formatted to your liking. Reference-page changes don't take effect until you update the book, but paragraph tag changes are immediate. In this section, you'll find explanation of two common formatting issues:

- Dot leaders
- Putting two paragraphs on a single line

Creating entries with dot leaders

Setting up table of contents entries with dot leaders is a common requirement. FrameMaker can handle these automatically, but setting them up requires a little work.

```
Baking a carrot cake}..................................3
    Kitchen appliances}............................3
    Baking ingredients}............................4
    Preparation}...................................5
    Making the cake}...............................6
    Making the icing}..............................7
    Frosting the cake}.............................7
    Technical support for your cake}...............7
```

To create a table of contents entry with dot leaders, follow these steps:

1. Go to the reference pages (**View > Reference Pages**) and locate the TOC flow.

2. Modify the TOC definition as follows for each paragraph that should have a dot leader:

   <$paratext>\t<$pagenum>

   In other words, place a tab between the <$paratext> and <$pagenum> building blocks. If you prefer, you can insert a tab character (press the TAB key) between the building blocks instead of the \t. If your page numbers reference chapter numbers, your definitions might look like this:

   <$paratext>\t<$chapnum>-<$pagenum>

3. Update the book file, and then view the body pages for the table of contents. The tabs now appear in your table of contents entries, but they probably aren't yet moving the page numbers.

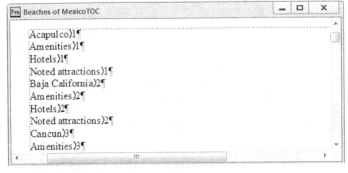

4. Modify the paragraph tags in the table of contents file to include a right-aligned tab stop with a dot leader. The following illustration shows a sample paragraph tag with the appropriate settings. You need to make this change in every TOC paragraph tag that includes a tab.

Your table of contents entries should now line up correctly.

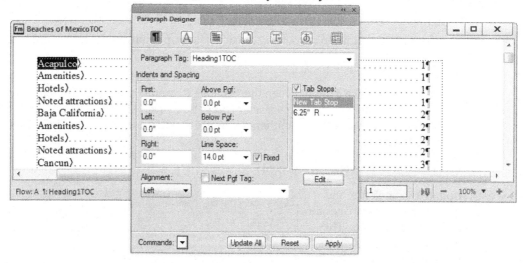

After you set up the dot leaders for each paragraph, FrameMaker generates them every time you update the table of contents.

Note Click the Update All button to save your changes for each paragraph tag you change.

Autonumbering in the table of contents

Autonumbering is available for the paragraph tags in the table of contents, just like in any regular document. Therefore, you can have the same result from two different table of contents paragraph definitions.

For example, let's say you want the ChapterTitle paragraph text to look like the following sample in the table of contents:

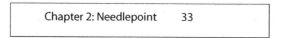

I prefer to have FrameMaker retrieve the numbering for the table of contents from the chapter, and not from the position of a paragraph from within the TOC.

Customizing the table of contents

I like to do this in the ChapterTitleTOC definition on the reference page:

> Chapter <$chapnum>: <$paratext>\t<$pagenum>

Another way to do this would be to set the autonumbering in the ChapterTitleTOC paragraph tag itself.

For this, first create a ChapterTitleTOC definition on the reference page as follows:

> <$paratext>\t<$pagenum>

Since the ChapterTitleTOC definition contains no building blocks to pick up the autonumbering from the chapters, you then need to modify the ChapterTitleTOC autonumbering properties.

In the paragraph designer, modify the Numbering sheet ( ) in the ChapterTitleTOC paragraph tag as follows:

Note that the ChapterTitleTOC definition can't use <$chapnum> because FrameMaker would pick up the chapter number of the *current* file (that is, the <$chapnum> of the TOC itself).

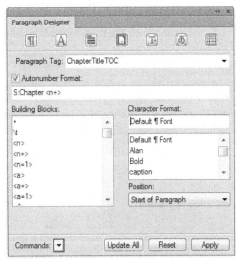

In general, I don't like this second technique because the autonumbering stream is completely unrelated to the actual chapter number. You might mask problems with the chapter numbering because you do not see a true chapter number; you see only the number of TOC paragraphs in the TOC file.

You might need to use this technique, though, if your chapter numbering and chapter titles are in separate paragraphs in your files and those paragraphs do not appear in the order you need for the table of contents. In that case, using autonumbering in the table of contents paragraph definitions might let you create the output you need.

Putting Two or More Paragraphs on a Single Line

A common technique used in cross-references and variables is not available inside the table of contents. In cross-references and variables, you can use building blocks that call a particular paragraph tag, such as this:

<$paratext[ChapTitle]>

These options are not available inside the table of contents definition. While generating the table of contents, FrameMaker can pick up information only from the current paragraph.

This presents a problem when, for example, your chapter number and chapter title are managed in two separate paragraphs, on separate lines. In the table of contents, you want something like this:

Chapter 7: Important Information

But in the document, you have two paragraphs:

355

- chap_number: Chapter 7
- chap_title: Important Information

If the paragraphs are in sequence, you can use a run-in heading format in the table of contents. If the chapter_numberTOC paragraph uses a run-in heading, the chapter number and title are displayed on the same line.

> **Managing Line Breaks in the Table of Contents**
>
> After generating the table of contents, you'll probably have a few lines that break in interesting and unfortunate ways. It's tempting to correct them in the TOC by inserting forced line breaks (SHIFT-ENTER), but those changes are lost when you regenerate. A better strategy is to use nonbreaking spaces (CONTROL-SPACE) in the source paragraph to "glue together" the relevant pieces so that they break properly in the generated table of contents.
>
> ```
> Baking a carrot cake............................................3
>     Kitchen appliances..........................................3
>     Baking ingredients..........................................4
>     Preparation.................................................5
>     Making the cake.............................................6
>         Preparing the ingredients...............................6
>         Putting it all together, along with some tips on how to prevent common problems with
>             carrot cakes........................................6
>     Making the icing............................................7
>
> Baking a carrot cake............................................3
>     Kitchen appliances..........................................3
>     Baking ingredients..........................................4
>     Preparation.................................................5
>     Making the cake.............................................6
>         Preparing the ingredients...............................6
>         Putting it all together, along with some tips on how
>             to prevent common problems with carrot cakes........6
>     Making the icing............................................7
> ```
>
> Another alternative is to insert forced line breaks in the content itself, which will force line breaks in both the chapter and in the TOC.

Tip Chris Despopolous also has the freeware TocBreaker plug-in available at: www.cudspan.net/plugins/

Mini TOC

Until FrameMaker 2015, if you wanted a document-level table of contents, you had to either purchase a plugin, or do labor intensive TOC generation for each document in your project. Now, you just need to use the **Special > Table of contents > Create Mini TOC** command. The interface is similar to the standard TOC interface, and will create its own reference page for editing. See "Locating the TOC reference flow" on page 349 and "Understanding the TOC flow entries" on page 350 for details on using the Mini TOC.

Chapter 19: Creating indexes

Many readers of technical books consider the index to be the most important part of the book. A complete index lets readers locate information quickly. You can create indexes for your FrameMaker books (or files). An index, like a table of contents, is a generated file. Creating indexes, however, takes more work than creating a table of contents. Before you can generate the index, you must insert index markers (hidden text with index information) throughout the document to identify terms that should appear in the index.

Here's a look at what's in this chapter:

The mechanics of a generated index 357
Creating the index file 358
Creating index entries 359
    Basic entries 361
    Inserting Unicode in index entries 362
    Editing and deleting index entries 362
    Creating subentries 363
    Stacking multiple entries in a single index marker 363
    Creating ranges 363
    Creating references to synonyms ("See") .364
    Changing sorting order for a single entry 365
Formatting the index 365
    Ignoring characters while sorting 367
    Modifying page separators 367
    Changing the sort order 368
    Changing the group titles 369
    Formatting the page number 370
    Creating ranges automatically 370
    Eliminating unwanted chapter numbers .371

The mechanics of a generated index

FrameMaker doesn't automatically create index entries for you, but it does provide automatic sorting and concatenation of index entries. Specifically, FrameMaker takes your index markers and does the following:

- Alphabetizes entries
- Inserts group titles (A, B, C, D, and so on)
- Groups identical entries into a single entry with multiple page numbers
- Groups secondary entries under their parent entry
- Inserts the correct page number for each entry (based on the marker location)

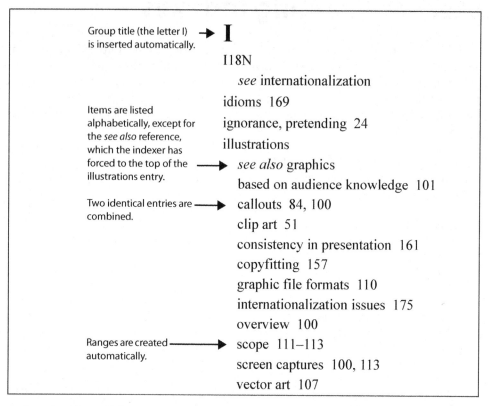

Creating the index file

To create an index, you must insert index markers into the document files to create content for the index. FrameMaker gathers text from the index markers and places the text in the index file.

To add an index file to a book, follow these steps:

1. Open the book file.
2. *(optional)* Click the file that will be immediately before or after the index.
3. Select **Add > Standard Index** to display the Set Up Standard Index dialog.
4. *(optional)* In the Add File drop-down list, select whether you want the index before or after the file you selected in step 2.
5. Verify that the Index marker is in the Include Markers of Type list.
6. Make sure that the Create Hypertext Links checkbox is checked. This feature creates the index entries as

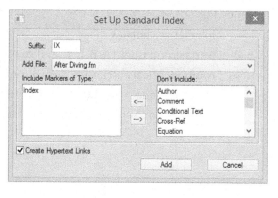

hyperlinks. It also ensures that if you convert to other formats (PDF, HTML, etc.), the links in the index are preserved.

Caution If you are generating indexes with a very large number of entries (more than 50,000 or so), and index generation fails or takes a very long time, uncheck the Create Hypertext Links checkbox.

7. Click the Add button. The Update Book dialog box is displayed (see page 330). The index you're creating is shown in the generated files list.

8. Make sure that the index is in the Generate list, then click the Update button. The file name of the index file is based on the book file name. If your book file is foo.book, the index file will be fooIX.fm.

The index file now is displayed in the book; you can move it just like any regular file (see "Rearranging files" on page 327).

Creating index entries

When you insert an index marker, FrameMaker creates a corresponding index entry. The page number that is displayed in the index matches the page on which the index marker occurs. In addition to basic index markers, FrameMaker provides a variety of building blocks and commands that let you create secondary entries, entries with character formatting, and the like. Table 19-1 provides a quick reference for indexing commands; the sections that follow explain the commands in more detail.

Table 19-1. Indexing Command Examples

Command	Marker Text Example	Result
: (colon) Creates a second-level entry	apple:pie	A apple pie 7

	apple:pie:a la mode	A
		apple
		  pie
		    a la mode 7

; (semicolon) Creates two entries in a single marker	apple;banana	A apple 7 B banana 7

359

Table 19-1. Indexing Command Examples (Continued)

Command	Marker Text Example	Result

	pie:apple;cream pie:banana	C
		cream pie
		  banana 7
		
		P
		pie
		  apple 7

<$nopage> Suppresses display of the page number	torte, see cake<$nopage>	T torte, see pie
<character tag> Assigns the specified character tag to the marker text that follows	torte, <Emphasis>see</> pie<$nopage>	T torte, see pie

[sort order]
Sorts the entry as if it used the text shown in brackets
	pie:<Emphasis>see also</>
	cream pie<$nopage>[pie:aa]
		P
		pie
		  see also cream pie
		  apple 7

<$startrange> <$endrange> Lets you create entries with ranges	On page 8: pear<$startrange> On page 12: pear<$endrange>	P pear 8–12

File-level (stand-alone) indexes

Although you'll mostly create indexes for book files, you can also create an index for a single file. FrameMaker calls this a *stand-alone* index because the index stands on its own and is not part of a book.

If you're working on a very large index, consider creating a stand-alone index for each chapter as you're inserting the index markers. This lets you check the index entries chapter by chapter (instead of having to plow through the entire index).

To create a stand-alone index, open a document, then select **Special > Standard Index** and follow the prompts.

Basic entries

To create a basic index entry, follow these steps:

1. Position the cursor where the relevant information occurs in the document.
2. *(optional)* Select the term you want to index. FrameMaker automatically puts the selected text in the marker text field.

Tip Place the index marker at the beginning or end of words. If you place the marker in the middle of a word, the text will look fine in your PDF files, but PDF searches may not find the word and conversion to other electronic formats (Help, EPUB) may not work as expected. If localizing content, consider putting markers at the beginning or end of paragraphs.

3. Select **Special > Marker** to display the Marker pod.
4. In the Marker Type drop-down list, make sure that Index is selected.

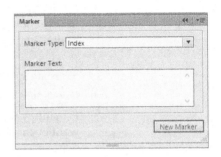

Remember The marker type defaults to whatever type was last used, so if you just inserted a hypertext marker, you'll need to change the marker type from Hypertext back to Index.

5. In the Marker Text field, type the text you want in the index. For example, to create an entry under A for apple on page 7, go to page 7, insert a marker in the apple-related content, and type in:

   apple

6. Click the New Marker button to insert the marker. A marker symbol is displayed in your text. It looks like a boldface T, but it does not print.

7. To see the result in your index, update the book file. The new index entry is shown in the generated index. The letter A and the page number are inserted automatically.

A
apple 7

Note If the entry you want is displayed in the text, you can insert a blank index marker immediately to the left of the relevant word. If the index marker is blank, it picks up the word to the right of the index marker during generation. However, we do not recommend this method because it makes index maintenance and localization more difficult. You cannot search for index text; instead, you must locate the blank index markers and then check the words next to them.

Tip To insert a command character (such as colon, semicolon, or brackets) as a literal, precede it with a backslash. For example:
   \<HTML\>
For information about inserting other special characters, see "Entering Special Characters in dialog boxes" on page 100.

> ### A few words about good indexing
>
> The index provides readers with *information access points*—a way to locate the content that's of interest. Creating a detailed, useful index is critical to the success of your document. The longer the document, the more important the index becomes to the reader. A complete discussion of indexing is beyond the scope of this book, but here are a few pointers:
>
> - Index terms where they are defined and when important information about the term is provided. Do not index every occurrence of a term (unless you're creating a specialized concordance index).
> - Index actions by the action (gerund) and the item (noun). For example, for information about steeping tea, put an entry under "steeping tea" and another under "tea, steeping."
> - If you have more than two or three entries for a particular item, modify the entries to create secondary entries under the primary entry.
> - Use ranges to cue readers that a detailed discussion of a topic is available.
> - Provide synonyms ("see" references) to help readers locate the information they want even if they don't know your terminology.
>
> Refer to Appendix A, "Resources," for tools to help you index FrameMaker files. If you do not want to create the index yourself, consider hiring a professional indexer. The American Society of Indexers (www.asi.org) is a good place to start.

Inserting Unicode in index entries

For characters outside the usual range, you can insert content using a number of FrameMaker utilities and copy/paste the content into dialog boxes. For instance, to insert the marker symbol discussed in the previous section, Unicode-32 character 02D5, use the File > Utility > Hex Input dialog to insert the character (**T**) into your text, then cut/paste the character into your index marker dialog.

Editing and deleting index entries

To edit an existing marker, select the marker (or select a block of text that includes the marker). Then, select **Special > Marker** to display the Marker pod, make your changes, and click the Edit Marker button to save them.

Tip If you are planning to do a lot of indexing, consider the IXgen third-party plug-in, which helps automate index maintenance. For details, see Appendix A, "Resources."

To delete an index entry, delete the marker. Locating and selecting index markers can be tedious, so here are some tips:

- Use the Find command to search marker text or to search for a marker of type Index.
- Take advantage of the hypertext links in the index. When you CTRL+ALT+CLICK on a page entry, FrameMaker opens the associated file, and selects the index marker.

- Leave the Marker pod open. If open when a marker is selected, the marker text displays in the Marker pod automatically.

Creating subentries

To create a subentry, separate the primary and secondary entry with a colon. For example, type the following:

soup:chicken noodle

The entry will look like this:

> S
> soup
>   chicken noodle 27

Additional colons in the same entry will result in additional nesting of index levels. Be careful when using colons and semicolons in index markers, as they perform different tasks and can be easily mixed up in the Marker pod.

Stacking multiple entries in a single index marker

Use a semicolon to separate multiple, independent entries in a single index marker. For example:

food;drink

Results in an entry like this:

> D
> drink 36
> F
> food 36

This is equivalent to creating two separate markers, each of which contains one entry. Be careful when using colons and semicolons in index markers, as they perform different tasks and can be easily mixed up in the Marker pod.

Creating ranges

Page ranges in indexes indicate a significant amount of information available about a particular topic. You indicate the beginning and the end of the range with separate, matching markers. A range looks like this:

> G
> gluttony 121–135

There are two ways to create ranges—you can use index markers that define ranges, or you can use the <$autorange> command. This section discusses how to create ranges with index markers; refer to page 370 for details on the <$autorange> command, which is inserted in the IX reference page flow.

To create an index entry with a range of pages, follow these steps:

1. On the first page of the range, insert an index marker with the relevant text and the <$startrange> command:

   gluttony<$startrange>

2. On the last page of the range, insert a matching index marker with the <$endrange> command:

   gluttony<$endrange>

FrameMaker matches the <$startrange> and <$endrange> commands and puts a single index entry in the index.

Caution The marker text (other than the range command) must match exactly, otherwise FrameMaker can't match the two ends of the range. Unmatched markers will cause double question marks in the index.

G
gluttony ??–135

To correct this problem, make sure that every <$startrange> marker has a corresponding <$endrange> marker and that the marker text matches in those entries. A common mistake is to create one singular and one plural entry:

    apple<$startrange>
    apples<$endrange>

The result in the index is two entries, each with half a range:

A
apple 121–??
apples ??–125

When you insert a range command, the command is applied to the current index entry in the marker and all that follow to the right in that marker. For example, assume you have the following two markers:

    land<$startrange>
    sea;land<$endrange>

In this case, the "land" range resolves correctly, and you see a single page entry under "sea." However, the following will cause problems:

    land<$startrange>
    land<$endrange>;sea

The <$endrange> command is applied to the "land" entry and also to the "sea" entry, so the result is a broken range because you do not have a "sea<$startrange>" entry anywhere.

L
land 2–10

S
sea ??–10

To prevent this problem, put all range commands in individual markers instead of stacking them with other entries in a single marker.

Creating references to synonyms ("See")

In an index, *see* references or synonyms are very helpful because they point the reader in the right direction. For example, a reader might find something like this:

These references are inserted just like regular index markers, except that you need to suppress the page number. For this, you need the <$nopage> command:

G
Granny Smith, see apple

    Granny Smith, see apple<$nopage>

If you want to italicize the word *see*, use a character tag inside the index marker and the <Default Para Font> tag or its abbreviation </> to turn off the character tagging:

    Granny Smith, <Emphasis>see </>apple<$nopage>

Remember The character tag you specify must be available in the index file's character catalog. It does *not* have to be available in the file where you insert the marker.

Like the range commands, <$nopage> applies to all the text that follows in the marker. You can restart pagination with the <$singlepage> building block, but it's usually easier to separate the <$nopage> entries into individual markers.

Changing sorting order for a single entry

The *see also* reference is a variation on *see* references. You generally use *see* references when you need to point readers to a different entry in an index. By contrast, *see also* references are used when the current entry has some information, but there is also another relevant possibility. In the index example on page 358, the index includes a *see also* reference (illustrations, see also graphics) and a *see* reference (I18N, see internationalization).

The problem with *see also* references is that, by default, FrameMaker sorts them under S in the subentry. Instead, you probably want them to appear at the top of the list. You accomplish this by modifying the alphabetization of that entry. Instead of using the actual text of the entry for sorting, you provide alternate text that's used when sorting the entry.

The *see also* entry uses the following syntax:

    illustrations:<Emphasis>see also</> graphics<$nopage>[illustrations:aa]

The text in brackets at the end of the entry controls where the information is sorted in the index. In the preceding example, the aa text ensures that the entry is sorted to the top of the illustrations subentries.

The text in brackets is only for the current entry in the marker. For example, consider the following:

    Korean War;'50s[Fifties]

The Korean War entry is displayed under K, but the '50s entry shows up under F.

Formatting the index

To format the index, you modify the index paragraph tags and the reference pages. The paragraph tags are as follows:

- **GroupTitlesIX** is applied to the lettered headings (A, B, C, and so on) that separate the alphabetical groupings in the index.
- **Level1IX** is applied to the first-level index entries.
- **Level2IX** is applied to the second-level index entries.
- If you have deeper index levels, FrameMaker creates the appropriate **Level3IX**, **Level4IX**, and so on.

You can change these paragraph tags inside the index file to create the look you want.

Tip When you generate the index, FrameMaker replaces the entire generated flow in the index. If you want to insert text preceding the index (such as the title "Index"), type this text into the index file before the generated text. Be sure that the first marker in the flow is after the title. FrameMaker will not delete any text that precedes the generated flow.

In addition to the paragraph tags, some formatting information is controlled by the IX flow on the reference pages (**View > Reference Pages**). The following figure shows a sample IX flow.

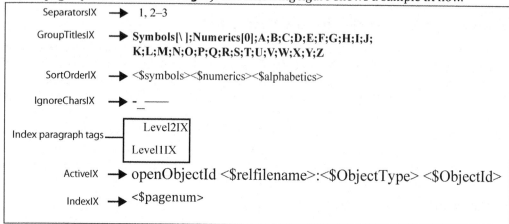

Table 19-2 describes the function of each paragraph in the flow.

Most likely, you'll make fewer changes in the IX reference flow than you do in the TOC reference flow. A common change is to control the formatting of the index page number (see page 370).

Table 19-2. IX Reference Flow Paragraphs

Paragraph Tag on Reference Page	Text in Flow	Function
IgnoreCharsIX	- _ – — (hyphen, underscore, en dash, em dash)	Lists the characters that are ignored when FrameMaker sorts index entries
ActiveIX	open ObjectId <$relfilename>: <$ObjectType> <$ObjectId>	Inserts the hypertext markers needed to make the index markers into links
Level3IX, Level2IX, Level1IX	Level3IX Level2IX Level1IX	Lists the levels of index
SeparatorsIX	1, 2–3	Lists the characters that separate index entries and page references.
SortOrderIX	<$symbols><$numerics><$alphabetics>	Determines the order in which symbol entries, numeric entries, and alphabetic entries are presented

Table 19-2. IX Reference Flow Paragraphs (Continued)

Paragraph Tag on Reference Page	Text in Flow	Function
GroupTitlesIX	Symbols[\ ];Numerics [0];A;B;C;D;E;F;G;H;I;J;K;L;M;N;O;P; Q;R;S;T;U;V;W;X;Y;Z	Lists the group titles (usually A, B, C, and so on)
IndexIX	<$pagenum>	Controls the formatting of the page numbers in the index

Ignoring characters while sorting

The IgnoreCharsIX paragraph in the IX flow controls which characters are ignored for alphabetization purposes when the index is sorted. By default, the IgnoreCharsIX contains these characters:

-_–—

The characters are hyphen, underscore, en dash, and em dash.

To change which characters are considered (or not) when sorting the index, modify the items in the list. For example, to eliminate the forward slash (/) from consideration during sorting, add it to the IgnoreCharsIX line, like this:

-_–—/

When you regenerate the index, FrameMaker ignores any forward slashes while sorting index entries. The slashes and other characters listed here are still printed as part of the index entry.

By default, FrameMaker sorts index entries word-by-word. That is, the spaces between words are taken into account when sorting. To change your sort to letter-by-letter, where spaces are ignored, add a space to the IgnoreCharIX paragraph. To change how a specific entry is sorted, see "Changing the sort order" on page 368.

Modifying page separators

An index entry has several different components that control the overall format. The SeparatorsIX entry in the IX reference flow sets the three following parameters:

- The separator between the index entry text and the first page number
- The separator between multiple page numbers
- The separator in a range

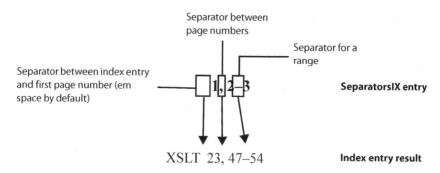

Index entry result

The default SeparatorsIX entry looks like this:

    1, 2–3

Its settings are as follows:

- An em space between the index entry and first page number
- A comma followed by a space between separate page entries
- An en dash in ranges

To change the separators, you change the items in this line.

Note   If you need to change the formatting of the page number, you need to modify the IndexIX entry. See "Formatting the page number" on page 370.

Changing the sort order

The SortOrderIX entry in the reference flow determines how entries are sorted. By default, it contains the following building blocks:

    <$symbols><$numerics><$alphabetics>

The building blocks are placeholders for the following characters:

<$symbols>	All characters not listed under <$numerics> or <$alphabetics>, sorted by ASCII value
<$numerics>	0 1 2 3 4 5 6 7 8 9
<$alphabetics>	AÁÀÂÄÅaáàâäåªBbCÇcçDdEÉÈÊËeéèêëFfƒGgHhIÍÎÏiíîïJjKkLlMmNÑnñ OÓÒÔÖÕØoóòôöõø°PpQqRrSsTtUÚÙÛÜuúùûüVvWwXxYŸyÿZz
	This is the sort order for the U.S. English interface; other languages are slightly different. You can find the specifics for other languages in the localized help.

To put the symbol characters last in the index, change the building block order:

    <$numerics><$alphabetics><$symbols>

If you need finer control over the sort order, you can replace the building blocks with an expanded list. Assume, for instance, that you have a number of index entries that start with pound signs (#98). Normally, those would be sorted in the <$symbols> sections, but you want them to appear in the <$numerics> section. You can remove the <$numerics> building block and replace it with a list of characters:

<$symbols>0 1 2 3 4 5 6 7 8 9 #<$alphabetics>

The result will be that the entries that begin with a pound sign (#) are sorted after the number nine and appear in the Numerics section.

Changing the group titles

When you generate the index, the alphabetic group titles are inserted automatically. For sorting purposes, the titles are treated just like regular index entries; however, they use the GroupTitlesIX paragraph tag instead of one of the LevelnIX tags. A common modification of the group titles is consolidating entries. For example, you might have an index that contains only two or three entries under X, Y, and Z. Instead of listing them separately, you could combine them under an X–Z group title. You can also create ranges using a comma separator to create an X, Y, Z group title.

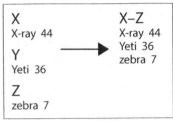

Using hyphens to create index entry ranges

The default GroupTitlesIX entry looks like this:

Symbols[\ ];Numerics[0];A;B;C;D;E;F;G;H;I;J;K;L;M;N;O;P;Q;R;S;T;U;V;W;X;Y;Z

To consolidate titles using hyphens, you must make two changes:

- Group the letters you want to combine.
- Force that title to appear at the beginning of the first letter (by customizing the alphabetization of that item).

To create an X–Z group title, modify GroupTitlesIX to look like this:

Symbols[\ ];Numerics[0];A;B;C;D;E;F;G;H;I;J;K;L;M;N;O;P;Q;R;S;T;U;V;W;X–Z[X]

Caution   If you omit the sort order, you may end up with unexpected results.

Using commas to create index entry ranges

The default GroupTitlesIX entry looks like this:

Symbols[\ ];Numerics[0];A;B;C;D;E;F;G;H;I;J;K;L;M;N;O;P;Q;R;S;T;U;V;W;X;Y;Z

To consolidate titles using commas, you simply replace the semicolon with a comma as needed. This collects the index entries in one group, and helps the reader understand that there are no missing letters in the index.

To create an X, Y, Z group title, modify GroupTitlesIX to look like this:

Symbols[\ ];Numerics[0];A;B;C;D;E;F;G;H;I;J;K;L;M;N;O;P;Q;R;S;T;U;V;W;X,Y,Z

Formatting the page number

The information presented in the page number is set by the IndexIX entry in the IX reference flow. By default, it is set to:

<$pagenum>

If you want to include a chapter number in your index page references, modify IndexIX to include the chapter number:

<$chapnum>-<$pagenum>

Tip Use a nonbreaking hyphen (ESC HYPHEN H) instead of a regular hyphen to prevent unattractive line breaks in the index.

You can also add the volume number (<$volnum>) and the paragraph number (<$paranum> or <$paranumonly>).

The formatting of the IndexIX paragraph tag controls the appearance of page numbers in the generated files, even though the page numbers end up being part of the LevelnIX paragraphs. If, for example, you want the page number to be italic while the index entry is regular text, you would define the Level1IX paragraph tag with the regular font and the IndexIX paragraph tag with the italic font.

Creating ranges automatically

FrameMaker has a little-known feature called autorange. Instead of creating matching startrange and endrange markers for each of your ranges (see page 363), you can let FrameMaker create the ranges automatically.

With autoranges activated, FrameMaker scans index entries for matching entries. When an entry occurs on two or more pages in sequence, a range is created in the index.

To activate autoranges, modify the IndexIX paragraph in the IX reference flow to contain the <$autorange> command:

<$pagenum><$autorange>

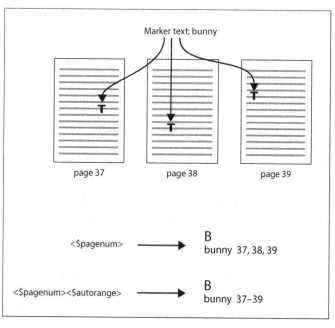

Autorange lets you avoid creating startrange and endrange markers. The disadvantage of autorange is that you must insert identical index markers on every page in the sequence. If the range is more than two or three pages, inserting all those markers will become quite tedious. Inserting a startrange and an endrange command is suddenly much more appealing.

Eliminating unwanted chapter numbers

Including the <$chapnum> building block in the IndexIX definition can cause a problem if your files are set up as follows:

- The main body text has multipart page numbers with chapter and page number (3-1, 3-2, and so on).
- The front matter (or back matter) uses a different numbering scheme without a chapter number (for example, i, ii, iii, and so on).

If you insert index entries into both numbering streams, you end up with the following numbers for the front matter entries: 1-iv, 1-x, and so on.

There are two solutions for this problem:

- Eliminate index entries from the front matter.
- Use a special marker type (for example, IndexRoman) for all of the index entries in the front matter.

For files needing Roman numeral entries, follow these steps:

1. Select **Special > Marker** to display the Marker pod.
2. In the Marker Type drop-down list, click Edit to display the Edit Custom Marker Type dialog.

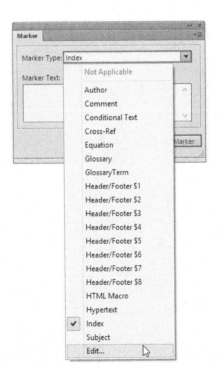

3. Type IndexRoman in the CustomMarkerType field.
4. Click the Add button.
5. Click the Done button (available after clicking Add) to close the Edit Custom Marker Type dialog box and return to the Marker pod. The new marker type is now shown in the Marker Type list.

Tip Custom marker definitions are stored in the document properties, so you can import formats from one file to another to copy over the new marker type.

6. Make sure that all of the index entries in the front matter use the IndexRoman marker type.
7. Go to your book file, click the index, then select **Edit > Set Up Standard Index**.
8. Add IndexRoman to the list of markers that are included when this file is generated.
9. Update the book.
10. Open the index and go to the reference pages. Locate the IndexRomanIX entry in the IX flow and make sure it includes only the page number:

   <$pagenum>

11. Update the book again to see the results.

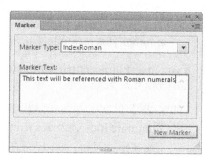

Chapter 20: Creating glossaries

There are two ways to approach glossaries:

- Manually maintaining a separate static FrameMaker file, updating whenever necessary.
- Placing markers on terms and definitions in the document and producing a generated file, similar to other generated files like table of contents and indexes.

Creating a manual glossary is fairly straightforward. You create a document, and type in the glossary terms and definitions. Add the glossary as you would any other file within your book, and update the glossary as needed, based upon the content of your book.

But a manual glossary has a significant drawback. Since the content is static, and not generated based upon the content of your book, it can easily become out of date, either with irrelevant entries, or entries that are missing altogether.

In this chapter, I'll cover creating glossaries for digital output, using the Glossary and GlossaryTerm markers. I'll also cover creating a marker for a generated FrameMaker index, suitable for print and PDF.

Here's a look at what's in this chapter:

Generated versus static glossary files	.374
Marking glossary definitions for digital publishing	.374
Marking a glossary definitions for print or PDF	.375
Marking glossary terms for digital publishing	.375
Create the generated glossary for print or PDF	.376
Modifying the reference pages	.376

Generated versus static glossary files

I'm not sure how many FrameMaker *authors* care whether a glossary is manually curated or is a file generated automatically from markers. The difference can be subtle, and each has its strengths.

However, starting with FrameMaker 12 you can do digital publishing directly from FrameMaker, in addition to the long standing ability to link FrameMaker content to RoboHelp projects. Both linked RoboHelp projects and direct digitally published projects can use markers to create interactive glossaries in various formats. See "Digital Publishing" on page 413 for more information.

Marking glossary content isn't difficult, but does require use of specific markers with specific attributes for the glossary to work properly in either FrameMaker generated documents or in digital output. The confounding part of the process is the labeling of the markers themselves, and the descriptions provided in the FrameMaker online help system.

Marking glossary definitions for digital publishing

To insert a marker for digital publishing, follow these steps:

1. If needed, create a single paragraph which contains only the definition of your glossary term.

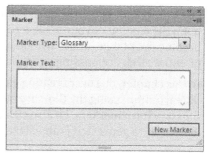

2. Place your cursor in the paragraph which is your term definition. I place this term marker at the beginning of the paragraph for the digital glossary definition so that I have consistency and findability when editing later.
3. Navigate to the Marker pod, and select Glossary from the list of available marker types.
4. For the marker text, type in the name of your glossary term. You can use an actual term consisting of multiple words here...you're not limited to a single word.
5. Select the New Marker button to create your marker.

   Tip Do not type the glossary definition into this marker's marker text. If you do, it will cause problems in your digital output.

While the FrameMaker help system does have information to help you figure out proper use of this marker, (see Create the glossary term definition in the FrameMaker help system) the explanation seems a bit circular, as the topic contains both term and definition in the heading and in the description.

Marking a glossary definitions for print or PDF

To insert a marker for print publishing, follow these steps:

1. Create a new marker type called GlossaryPrintDefinition.
2. I've asked the Adobe team to add this to FrameMaker's default marker list, so check to make sure it's not already in your document via a patch to FrameMaker 2015.
3. Place your cursor in the definition paragraph.
4. Navigate to the Marker pod.
5. In the marker text area, type in the term you want included in your print glossary. After the term, insert a semicolon and then type out the definition. The semicolon creates a second level index entry we'll use later in the Reference page section.
6. Insert the marker in the paragraph containing your glossary definition.

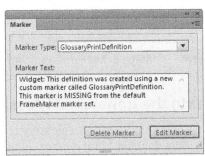

For consistency and findability, I put glossary markers at the end of paragraphs.

Marking glossary terms for digital publishing

Marking glossary terms is infinitely easier than what's required for the glossary definition(s).

To mark glossary terms, do the following:

1. Locate an instance of your glossary term in your text, and place your cursor at the beginning of the term.
2. Navigate to the Marker pod, change the marker type to GlossaryTerm.
3. Type the glossary term into the marker text and select New Marker

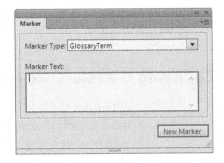

Tip If you highlighted the term in Step 1, the text may already be in the marker text area. If so, ensure the marker type is set to GlossaryTerm and select New Marker.

Generate your glossary for the document or for the book, and modify the formatting and reference pages as needed. See below for tips on modifying the reference page.

Tip While not technically necessary to enter the term into the GlossaryTerm marker (FrameMaker will use the word into which the marker is placed), for everyone's sanity, just use the glossary term as the marker text for both the glossary term and the glossary definition markers.

Chapter 20: Creating glossaries

4. For digital publishing output, use the Find/Change pod to search for instances of the glossary term.
5. Copy the glossary term and its marker to the clipboard.
6. Use Change by: Pasting to replace all instances of the term with the term+marker that you copied to clipboard in Step 6.

Note Steps 5-7 are optional, but ensure that all instances of your term will link to the glossary definition in the digital output.

Create the generated glossary for print or PDF

While you can set up a glossary for an individual file, I find it more common to create a glossary for a book, so the following describes that scenario:

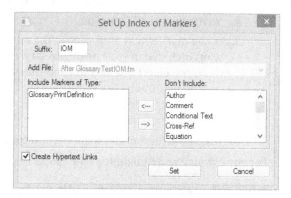

1. With the book file active, add an Index of Markers.
2. Include the GlossaryPrintDefinition marker in this generated file.

The Glossary and GlossaryTerm markers are used for creating digital output, and are not needed here. If you followed the model above, the GlossaryPrintDefinition marker already contains both the first second level index entries you need for your print glossary.

Modifying the reference pages

Use the following list as a guideline to setting up your print glossary FrameMaker file.

1. Remove the <$pagenum> building blocks from the Level1IOM and Level2IOM entries, leaving only an empty paragraph mark (a pilcrow, or ¶).
2. Remove the leading space from the IOM entry paragraph.

If you don't like the classic style of the first and second level index entries, try setting the Level1IOM paragraph to Run-In Head, setting default punctuation as needed. This gives you a more tabular look to the entries.

376

Chapter 21: Creating Other Generated Files

Tables of contents and indexes—discussed in Chapters 18 and 19, respectively—are the most common generated files, but you can also generate several other types of files. They are divided into generated lists and generated indexes.

Here's a look at what's in this chapter:

Examples of other generated files	.377
Creating paragraph-based lists	.380
List of figures	.381
List of tables	.381
List of paragraphs	.381
Alphabetical list of paragraphs	.382
Creating lists of other items	.382
List of markers	.382
Alphabetical list of markers	.382
List of references	.383
Generated indexes	.385
Index of authors	.385
Index of subjects	.386
Index of markers	.386
Index of references	.386

Examples of other generated files

Broadly defined, generated lists display information in the order that it occurs in the document; indexes list information in alphabetical order with separators (or "group titles") for each letter in the alphabet.

Just to confuse matters further, two lists are available in page order *or* alphabetical order.

The accompanying illustration shows how the same information is organized differently in a generated list and a generated index.

Like indexes and tables of contents, you can customize other generated lists by controlling the formatting and by using reference page flows to determine what information is included.

Each generated file has a unique suffix, which is appended to the file name, the paragraph tags in the generated file, and used as the reference page flow tag. For example, a list of markers uses the LOM suffix by default. If the book file you start from is named foo.book, the list of markers file will be fooLOM.fm, and the reference page flow tag is LOM. The paragraphs tags in the generated file will end with LOM (for example, if you include Cross-Ref markers, they will be formatted by the Cross-RefLOM paragraph tag).

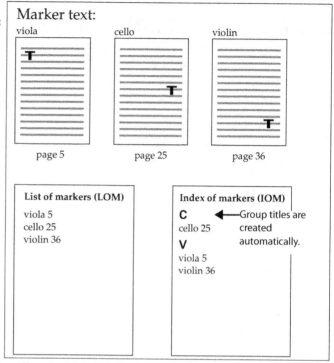

When the book file is active, you can add generated lists (List Of items) and generated indexes (Index of items) from the Add menu.

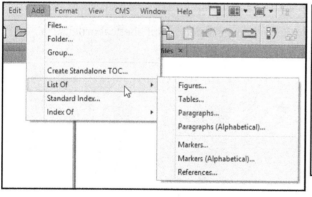
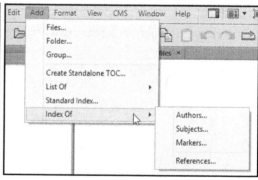

Table 21-1. Generated File Suffixes

Generated File	Suffix
Standalone TOC	TOC
List of figures	LOF
List of tables	LOT
List of paragraphs	LOP
Alphabetical list of paragraphs	APL
List of markers	LOM
Alphabetical list of markers	AML
List of references	LOR
Standard index	IX
Index of authors	AIX
Index of subjects	SIX
Index of markers	IOM
Index of references	IOR

The steps you follow to create a generated list or index are similar for each type of generated file.

To create a generated list or index, follow these steps:

1. Open the book or file for which you want to create a list.

    - If your starting document is a file, select **Special > List Of >** or **Special > Index Of >** and the appropriate item. Confirm that you want a standalone generated file by selecting Yes.

    - If your starting document is a book, select **Add > List Of** or **Add > Index Of >** and the appropriate item.

2. A setup dialog box is displayed for the list or index.

    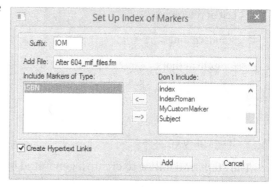

3. *(optional)* To create a hyperlinked generated list, confirm that the Create Hypertext Links checkbox is checked.

4. Select the settings you want for the generated file, then click the Add button to create it.

5. *(for book-level generated files)* Click the Update button to create the new file.

Tip If the item you want does not appear in the list, try updating the book.

The following sections discuss how to set up and customize each type of generated file.

Creating paragraph-based lists

Several of the generated lists are based on selecting items from a list of available paragraphs. They are as follows:

- List of figures
- List of tables
- List of paragraphs
- Table of contents

There is no functional difference between these four lists. In each case, you select various paragraph tags, and FrameMaker generates a list of those items in the order they occur in the documents.

Chapter 18, "Creating tables of contents," describes in detail how to format generated list files. You can apply the information provided there to format lists of figures, tables, and paragraphs.

Tip   Keep in mind that you can use the list of figures and list of tables for generated lists other than those implied by the titles. You could, for example, generate a list of syntax examples by creating a List of Figures that includes the CodeExampleTitle tag instead of the FigureTitle tag.

List of figures

When you create a list of figures, the Set Up List of Figures dialog box provides a list of paragraphs to select from.

To create a list of figures, select your figure caption paragraph tag (for example, figure) and generate the list.

To display the figure number along with the caption, format the CaptionLOF entry in the reference page flow similar to this example:

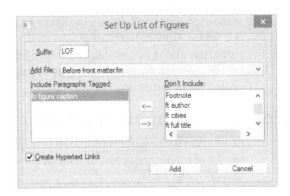

Figure <$paranumonly>:
<$paratext>\t<$pagenum>

Remember   You must define a tab stop in the Paragraph Designer for the tab.

List of tables

When you create a list of tables, the Set Up List of Tables dialog box provides a list of paragraphs to select from. The dialog box is identical to the Set Up List of Figures dialog box (see page 381), except for the window title and the default suffix.

To create a list of tables, select your table title paragraph tag (for example, TableTitle) and generate the list. To display the table number along with the caption, format the TableTitleLOT entry in the reference page flow similar to this example:

Table <$paranumonly>: <$paratext>\t<$pagenum>

Remember   You must define a tab stop in the Paragraph Designer for the tab.

List of paragraphs

The List of Paragraphs works just like the table of contents. The Set Up List of Paragraphs dialog box provides a list of paragraphs to select from. The dialog box is identical to the Set Up List of Figures dialog box except for the window title and the default suffix (see page 381).

To create a list of paragraphs, select the items you want to include and generate the list. The reference page LOP flow will provide entries for each included item, which you can format just like table of contents entries.

If you have reviewers that will work with the native FrameMaker files, the list of paragraphs can be useful for review comments. If you create and use a review comment paragraph tag, you can generate a hyperlinked list of all review comments in the file or book. Consider using the PDF Review function, however, unless you are doing major work on the files. See "Setting up a FrameMaker PDF review" on page 391 for more information.

Alphabetical list of paragraphs

The alphabetical list of paragraphs contains the same information as the regular list of paragraphs, but instead of listing the items in the order they occur in the book, they are listed in alphabetical order. To set up the list of paragraphs, specify the paragraphs you want to include in the Set Up Alphabetical Paragraph List dialog. Except for the dialog box name and the default suffix, the dialog box is identical to the Set Up List of Figures dialog box shown on page 381. Use this list to track down the paragraph styles used in your document, perhaps for updating and replacement purposes.

Creating lists of other items

In addition to the paragraph-based lists, you can create lists of markers and of references. Within the list of references, you can select from several items, including imported graphics, conditional text tags, and fonts.

List of markers

The generated list of markers is a page-by-page list of markers in a document or book. When you set up the list of markers, you can include one or all of the marker types, including custom markers, as shown in this example.

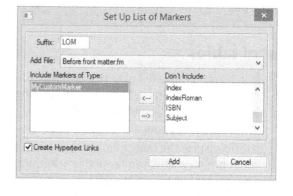

For each marker you select, the LOM flow on the reference page will contain a corresponding entry. For example, if you include the Index marker type, the LOM flow will contain an IndexLOM entry.

By default, the definition for each entry is:

<$markertext> <$pagenum>

The <$markertext> building block retrieves the information in the marker text field for each marker.

Alphabetical list of markers

Like the list of markers, the alphabetical list of markers lets you select which marker types you want to include. Except for the dialog box name and the default suffix, the dialog box is identical to the Set Up List of Markers dialog in the previous section.

The resulting list, however, is sorted alphabetically by marker text rather than in page order.

Note The only difference between the alphabetical list of markers and an index is how the entries are organized. The index provides group titles along with primary and secondary entries. See the generated lists example on page 378 for an example.

List of references

The list of references feature lets you create generated lists that include the following items:

- Condition tags
- External cross-references (that is, cross-references that point outside the current file)
- Fonts
- Imported graphics
- Text insets
- Unresolved cross-references
- Unresolved text insets

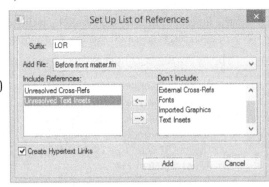

When you set up the list of references, you can choose to include any or all of these items in the Set Up List of References dialog.

For each item, FrameMaker creates an entry in the LOR flow on the reference pages. The entry's name is the included item with an LOR suffix; for example, Text InsetsLOR. The same building blocks are used for each item:

    <$referencename> <$pagenum>

The <$referencename> building block, however, results in different information for each type of reference.

Condition tags

Select condition tags to create a list of every occurrence of *shown* conditional text in the document. FrameMaker will ignore any conditional text that is currently hidden. The generated file lists the name of the condition tag and the page on which it is used, such as:

    Comment 47
    Draft 58

Tip   If you need a list of all conditional text, show all the conditional text tags before you create this list.

External cross-references

Select external cross-references to create a list of references from the current file to other files. If you create this list for a book, the list will include references from one file in the book to another as well as references to files outside the current book. The resulting list will show the name of the referenced file (including any path name that FrameMaker has stored) and the text of the Cross-Reference marker that identifies the cross-reference's target. For example:

    toc.fm (cn chapter_number: Chapter 5) 214
    index.fm (cn chapter_number: Chapter 9) 214

Lists of external cross-references are very useful for locating cross-references that point to local files instead of network files and the like.

Fonts

Select Fonts to create a list of all the fonts that occur in the document. FrameMaker lists the font, the font size, and the page on which the font occurs. The font list is based on the fonts that occur on each page, so you'll see the same font listed multiple times.

The following is a short excerpt from a generated list of fonts:

```
Univers 57 Condensed @ 12.0 pt 213
Palatino Linotype @ 75.0 pt 213
Univers Condensed @ 40.0 pt 213
Univers 57 Condensed @ 10.0 pt 214
Univers 57 Condensed @ 12.0 pt 214
Times @ 12.0 pt 214
```

Tip To create an alphabetical list of fonts, convert the listings to a table (each paragraph should be a cell) and sort the table. Better yet, generate an index of references for your fonts, which will automatically create one entry per font. See page 386.

Imported graphics

The list of imported graphics lists only graphics that are imported by reference—not graphics that were copied (embedded) into the document. For each graphic, FrameMaker lists the file path, the DPI setting being used (for bitmap images only), and the page number on which the graphic occurs, as shown in the following example:

```
graphics/para_cat_button.tif @ 48 dpi 6
graphics/status_bar_para.tif @ 96 dpi 7
graphics/formatting_bar.tif @ 125 dpi 8
graphics/override_para.tif @ 96 dpi 10
```

The list of imported graphics is useful for verifying any or all of the following items for a file:

- Checking DPI settings to ensure that graphics are sized consistently.
- Verifying file locations to ensure graphics are in the correct directory. (In the preceding example, all graphics are stored in a graphics subdirectory; an entry that doesn't begin with graphics/ would indicate a graphic that needs to be moved.)
- Checking file names to ensure that they follow naming conventions.
- Verifying that all graphics are copied into document instead of referenced. (If all graphics are copied and not referenced, the generated list of imported graphics should be empty.)

Text insets

Generating a list of text insets has a similar purpose as generating lists of imported graphics; it lets you verify that external files are being imported properly. For each text inset, the generated file lists the file name and path, and the page number on which it begins, as shown in the following example:

```
segment1.fm 222
segment2.fm 255
```

The list of text insets lets you quickly review a list of files that are being imported into the book or file you generated from.

Unresolved cross-references and text insets

When you generate a book file, the book error log creates a list of unresolved cross-references and text insets that occur in the documents. When you use the list of references to create a list of unresolved cross-references or text insets, you get the same information. If you created the list with hyperlinks (highly recommended), you can use the generated list to jump directly to the unresolved item and fix it.

Generated indexes

The various "Index Of" indexes are all quite similar to the standard index (IX) discussed in Chapter 19, "Creating indexes." In addition to the standard index, the following indexes are available:

- Index of authors
- Index of subjects
- Index of markers
- Index of references

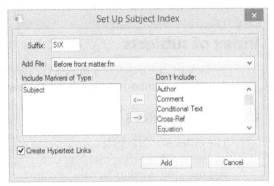

Each index creates an alphabetical list of entries, with matching entries grouped into a single index entry with multiple page numbers.

Although the standard index defaults to using only the Index marker type, you can use other (or additional) markers in the standard index. The index of authors and index of subjects default to including the matching index markers to create the index, but you can choose to include other marker types instead (or in addition to the default Author and Subject markers). In other words, you can use multiple marker types (possibly with certain different formatting attributes for each type) in a single generated "index" file.

Index of authors

The index of authors creates an index of the Author markers in the document or book. The index lets the reader look up information based on the original author. To set up an index of this type, you would first go through your document and identify the sections where different authors are involved. For each entry you want to create, set up an Author marker by following these steps:

1. Click in the text where the author's material occurs.
2. Create an Author marker (navigate to **Special > Marker** and make sure that Author is selected in the Marker Type drop-down list).

Note For detailed instructions on inserting markers, see "Creating index entries" on page 359.

3. In the marker text field, type in the author's name. For example:

   Sayers, Dorothy

4. Repeat this process for every item that you want to mark for a particular author. Then, generate the index of authors to see the result:

> **J**
> James, P. D. 35
> **M**
> Marsh, Ngaio 22
> Maron, Margaret 45
> **R**
> Reichs, Kathy 12
> **S**
> Sayers, Dorothy 100

Index of subjects

By default, the index of subjects creates an index from all of the Subject markers in the documents. You can customize them just like any other index entries.

Index of markers

The index of markers lets you create an index of any marker type. You might, for instance, create an index of hypertext markers to verify that the syntax and path names are correct for each. If you use spot cross-references extensively in a document, you could use an index of Cross-Reference markers to verify that each marker's text is unique.

Index of references

The index of references lets you create an index for each of the items that are available for the list of references. Instead of sorting items by page number, an index sorts them alphabetically, and matching items are grouped. That can make the index version much more manageable and usable than the list version. The index of fonts, for example, would look something like the following example:

```
Courier @ 9.0 pt 185–189, 192
Palatino Linotype @ 10.0 pt 178–193
Palatino Linotype @ 4.0 pt 178–180, 182–187, 189, 191–193
Palatino Linotype @ 75.0 pt 177
Palatino Linotype @ 8.0 pt 179, 182, 184
Times New Roman @ 2.0 pt 178–182, 184, 189–192
Univers @ 10.0 pt 178–180, 182, 184, 186, 189, 191
Univers @ 11.0 pt 186
Univers @ 16.0 pt 178, 181–183, 185, 187–188, 193
Univers 57 Condensed @ 10.0 pt 178–194
Univers 57 Condensed @ 12.0 pt 177–194
Univers Condensed @ 40.0 pt 177
ZapfDingbats @ 10.0 pt 180–181, 183, 187–188
```

You can quickly scan this list to find renegade fonts in your document.

Part V

Creating Output

Chapter 22: Print, PDF output, and package

You can create many kinds of output from FrameMaker documents. Print and Portable Document Format (PDF) files are common output formats, but tablet and mobile computing is rapidly gaining favor. When a document is an occasional reference, a PDF file, help, or a mobile app often suffices. These formats are less expensive to produce, and can be more rapidly updated than printed documentation. For an overview of other output options, see Chapter 23, "Digital Publishing."

While electronic output is useful, for longer documents (like this one) your audience may still prefer to read (and mark up) a printed document.

Here's a look at what's in this chapter:

- A comparison of print and electronic formats 390
- Setting up a FrameMaker PDF review 391
- Printing your documents 391
    - Printing an individual document 392
    - Printing from the book file 392
    - Specifying pages 392
    - Printing several copies 393
    - Changing the paper size 393
    - Printing double-sided documents 394
    - Skipping blank pages 394
- Advanced printing options 395
    - Printing thumbnails 395
    - Printing spot color in black and white 395
    - Printing low-resolution images 396
    - Printing registration marks 396
    - Changing the printer 397
    - About color separations 397
- Setting Adobe PDF document properties 398
    - Adobe PDF Settings sheet 398
    - Configuring PDF job options 399
- Creating PDF Files 401
    - PDF Setup dialog options 401
    - Generating PDF bookmarks 403
    - Generating tagged PDF files 405
    - Optimizing PDF files 407
    - Creating hyperlinks to other PDF files 409
- Adobe PDF printer setup 410
- Package 412
    - Choosing Package options 412
    - Settings 412

A comparison of print and electronic formats

Print and electronic formats both have advantages. The advantages to printed documents are as follows:

- Printed documents don't require access to a computer or a network. If the process documented is a computer task, the reader can simultaneously view the doc and the referenced software.
- During software installation, the installer application might use the entire screen. In many cases, the installer will not permit other applications to run, so the user cannot simultaneously read an electronic document. (Furthermore, the online books or online help included with the software won't be available until after the software has been fully installed.) Installation documents need to be available separately.
- Printed documentation can be an integral part of product marketing and sales. The documentation helps users understand the product and also enhances the quality of their experience with the product.
- Supplying printed material gives you control over the size of the printed document and the binding method. For example, a book designed for 5.5" x 8.5" output is easier to use as a small spiral-bound document than as letter-sized output.
- Users might prefer not to use their own paper and bindery services to print supplied electronic documentation.
- Some users may be likely to reach for printed documents than browse their computers or web sites for electronic files.

Advantages of PDF files include:

- Preservation of print formatting from the FrameMaker file, unlike formats like HTML or online help.
- PDF files (and other electronic formats) can provide clickable hyperlinks created from electronic links and references created in FrameMaker. Links in print can be tedious to follow.
- PDF files (and other electronic formats) can embed rich media such as interactive 3D files, video, and interactive SWF simulations that the user can manipulate and play.
- PDF versions of documentation (and other electronic formats) can be posted on an intranet or on the Internet, giving users and employees easy access to the information.
- PDF files (and other electronic formats) are easy to create and update.
- The user can perform text searches on the PDF file in Adobe Reader or Acrobat. In print, you have to rely on an index or table of contents to find information.
- The reader can conserve paper by printing only specific pages from a PDF instead of printing the entire book.
- PDF files eliminate (for the author) the time and expense associated with printing and distributing print documentation.
- It's easier and cheaper to update a PDF file (and other electronic formats) than a printed document.
- You can create tagged PDF files, which are useful for screen readers and small displays.

With FrameMaker's robust print, PDF, and online format options, you can choose the combination of outputs that best suit your organization and manage them all from a single set of FrameMaker documents.

Note In this chapter, the terms "print vendor" and "printer" have different meanings. A print vendor is a commercial printer that prints books on offset or digital presses. The print vendor may bind the books, or the binding may be outsourced to an external bindery. The term "printer" refers to a physical printer, such as the one in your office or home.

Setting up a FrameMaker PDF review

Importing of PDF comments has been around for a long time, but starting with FrameMaker 12 the features became more stable and are now part of every document review I do. In a nutshell, by utilizing PDF commenting tools on PDF produced by FrameMaker, I can use the Track Text Edits features to enable updating of my FrameMaker Content via editors who might only be using Adobe Reader, and might even be doing their edits on mobile devices.

For an overview of setting up a PDF review, see

http://www.techcommtools.com/mobile-pdf-commenting-workflow-framemaker-12/

Once you've made comments to the PDF, use the **File > Import > PDF Comments** command to bring the comments back into your FrameMaker document.

For more information about incorporating changes see "Tracking Changes" on page 55.

To download Adobe Reader for a specific platform, navigate to https://get.adobe.com/reader/ on that device.

Printing your documents

FrameMaker provides numerous printing options. You might use many of the basic features, such as outputting page ranges or printing several copies, every day. Other options, such as printing registration marks, are used when you prepare your document for final printing. You can print a single document, specific chapters in a book file, or an entire book. The print options for document files and book files are nearly the same. The printer you select provides additional features, such as support for special paper sizes, sophisticated color matching, and high-resolution printing. You should make sure the printer is set up correctly before printing. See your printer's documentation for details.

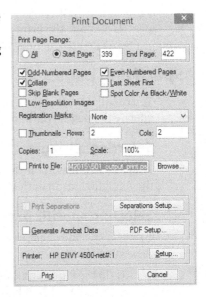

Tip Try producing PDF from FrameMaker files (using the Adobe PDF print driver), and *then* printing the resulting PDF to your printer. You'll be able to get back to FrameMaker faster, and EPS and PostScript code in the PDF can then be printed on non-PostScript printers.

Printing an individual document

To print a single document outside the book window, follow these steps:

1. Open the document you want to print, then select **File > Print** to display the Print dialog. (see page 391)
2. Change the printer options based on the following sections, then click the Print button to send the file to your default printer.

Printing from the book file

In a book file, you can print one file, specific pages in the book, or the entire book.

Tip   When printing from a book, first open all files in the book. This speeds printing and prevents missing font and other resource error messages from stopping the process. The file will print more quickly because FrameMaker doesn't have to open each file in the book. Before printing, you should also update the book ( ) to ensure accuracy of references and numbering. For more information, see "Printing a book" on page 341.

To print from the book file, follow these steps:

1. Open the book file, then specify the file or files you want to print by doing one of the following:
    - To print the entire book, select **File > Print Book**. The Print Book dialog box is displayed. The book and standard Print dialog boxes are identical (see page 391).
    - To print specific chapters in a book, CTRL+CLICK the files in the book window, then select **File > Print Selected Files**. The Print Selected Files in Book dialog box is displayed.
2. Change the printer options based on the following sections, then click the Print button. The book or selected documents are printed.

Specifying pages

You can print one page, a range of pages, or an entire document. This section describes how to specify the pages you want to print in a document or book.

Selecting all pages

To select all pages in the document or book, display the Print dialog box (see page 391), and click the All radio button in the Print Page Range section.

Selecting specific pages in a document

To select specific pages in a document, display the Print dialog box (see page 391), and follow these directions:

1. In the Print Page Range section, type the first page in the Start Page field and type the last page in the End Page field.
2. Click the Print button to print the selected range. (The current page number is displayed in both fields by default.) Note that you cannot specify multiple ranges in one step.

Printing your documents

Note You must specify numbers that match the file's numbering.

Selecting specific pages in a book

To select specific pages in a book, follow these steps:

1. CTRL+CLICK the file or files in which the pages occur.
2. Display the Print dialog box (see page 391).
3. Select **File > Print Selected Files**.
4. Type the first and last page in the Start Page and End Page fields, respectively.

If your document doesn't print

There are a few reasons why a document might not print. Before reprinting, verify the following:

- You could have inadvertently printed the document to file. In the Print dialog box, make sure the Print to File checkbox is unchecked before printing.
- If the document is long or graphics-intensive, check the print job dialog box to see if the document is displayed. The document might still be spooling to the printer.
- Look for problems with the printer. Check for a paper jam or empty paper tray.
- If you are seeing odd characters in the file, it may be that you are printing to a printer that does not include the correct fonts for your document. Select Save As PDF, then print the resulting PDF file to the printer.
- Shut down and restart your computer, then restart only FrameMaker.

Printing several copies

You can print several copies of one page, a range of pages, or the entire document, and you can print the copies one set at time. To specify copies, follow these steps:

1. Display the Print dialog box (see page 391), then indicate the pages you want to print. See "Specifying pages" on page 392 for details.
2. Indicate the number of copies to print by typing the number of copies to print in the Copies field.
3. To print one set at a time (for example, pages 1–15) instead of all copies of page 1, all copies of page 2, and so on, check the Collate checkbox.
4. If your printer places the printed page face-up, check the Last Sheet First check box.

Changing the paper size

You can change the paper size in your printer driver. You might print a document on a larger paper size to make room for trim marks. FrameMaker displays a warning if the paper isn't large enough for the text or marks to print completely.

To change the paper size, follow these steps:

1. Display the Print dialog box (see page 391).

393

2. Click the Setup button. Depending on your printer driver, the paper options might be displayed in this dialog.
3. If the paper options are not displayed, click the Properties button, then click the Advanced button. Click a different paper size in the Paper Size drop-down list.

Printing double-sided documents

Some printers have duplexing units that automatically print odd pages on one side of the paper and even pages on the other side. You can print a double-sided document on a printer that only prints single-sided by printing all odd pages first, flipping the paper in the printer, and printing the even pages. Before you print an entire document this way, test a few pages to make sure the paper is in the right direction. After printing one side of the document, you might need to print the other side in reverse order depending on your printer. Refer to your printer documentation for instructions on feeding the paper correctly.

To print a double-sided document on a printer without a duplexing unit, follow these steps:

1. Display the Print dialog box (see page 391), then check the Odd-Numbered Pages checkbox and uncheck the Even-Numbered Pages checkbox.
2. Click the Print button. The odd-numbered pages (1, 3, 5, 7…) are printed.
3. Flip over the paper in the paper tray. If you're printing to the manual feed paper tray, change the paper setting in your printer driver.
4. Select **File > Print** to display the Print dialog box again, then uncheck the Odd-Numbered Pages checkbox and check the Even-Numbered Pages checkbox.
5. If your printer requires it, check the Last Sheet First checkbox to print the other side in reverse order.
6. Click the Print button. The even-numbered pages (2, 4, 6, 8…) are printed on the back of the odd-numbered pages.

Skipping blank pages

You can skip printing blank pages. This is handy for printing a double-sided document as a single-sided document. Many double-sided documents have blank pages to force chapters to begin on a right page. When you print rough drafts, you can save paper by skipping the blank pages.

To suppress blank pages, display the Print dialog box (see page 391), and check the Skip Blank Pages checkbox. Another way to avoid printing blank pages is to convert a double-sided document to single-sided before you print, and then convert it back to double-sided. See "Switching from single- to double-sided pages" on page 253 for more information.

Note A page is considered blank if the main text flow is empty. The running headers and footers and other background objects don't count.

Advanced printing options

Printing thumbnails

You can print several pages as mini-images on a single page. This option is useful for checking the pagination of an entire document. You indicate the number of rows and columns you want printed (for instance, the 2-row by 4-column layout shown in the following illustration). The document is double-sided, so the thumbnails begin with the first right page.

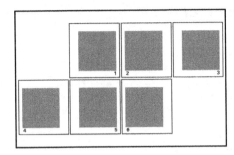

To print thumbnails, follow these steps:

1. Display the Print dialog box (see page 391), then check the Thumbnails check box.
2. In the Rows field, type the number of rows to print across the page.
3. In the Cols field, type the number of columns to print. Keep in mind the orientation of the paper. Refer to your printer documentation for details on specifying landscape pages.
4. Click the Print button.

Caution   Some printers don't support thumbnails. Refer to your printer documentation for details.

Printing spot color in black and white

You can save color ink by printing *spot color* as black and white. If your printer prints only in black and white, you may want to print spot color as black so that the areas that use spot colors print as a deep black instead of grayscale (for example, if you use spot color for the lines and arrows in your graphics).

When you send your book to the print vendor, spot color is premixed before the printing process and printed on one printing plate. *Process color* mixes shades of cyan, magenta, yellow, and black (CMYK), with each color printing on a different plate. To see which colors in your document are spot colors, select **View > Color > Definitions**, then click a color in the Name drop-down list. The setting is displayed in the Print As drop-down list. For more information, see Chapter 24, "Color output."

To print spot color in black and white, follow these steps:

1. Display the Print dialog box (see page 391).
2. Check the Spot Color as Black/White checkbox.

Printing low-resolution images

You can speed up the printing of graphic-intensive documents by printing the images at low resolution. The images print as gray boxes.

To turn on low-resolution printing, follow these steps:

1. Display the Print dialog box (see page 391).
2. Check the Low-Resolution Images checkbox.

Printing registration marks

Registration marks provide instructions for printing color separations, which are the individual pages that describe how to mix color for four-color printing. Information about the document is also displayed outside the registration marks—the date, time, file name, and page number. When you print registration marks in FrameMaker, *crop marks* also print in each corner of the page to indicate where to trim the paper after the book is printed.

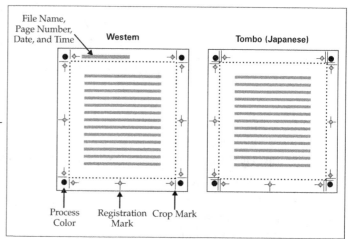

Registration marks are automatically added to color separations, and the color and halftone information is displayed outside the registration marks.

Caution To provide room for registration marks, you should print on paper that is at least one inch taller and wider than the paper size. If, for example, your document is 7 inches wide by 9 inches tall, you must use paper that's at least 8 inches wide by 10 inches tall to ensure that registration marks are displayed. In practice, this means that if your page size is a standard letter size (US Letter), you cannot print registration marks for your document—unless you output the pages onto larger paper.

Many print vendors require registration marks in the PDF file, but some printers add registration crop marks automatically. Be sure to consult with your print vendor. FrameMaker supports both Western registration marks and Tombo (Japanese) registration marks.

To select registration marks, follow these steps:

1. Display the Print dialog box (see page 391).
2. Click Western or Tombo in the Registration Marks drop-down list.

Caution To print registration marks in a PDF file, you must increase the page size in the PDF setup because the Adobe PDF Creation Add-On (or Acrobat Distiller) crops the page. See "Setting up registration marks in a PDF file" on page 402 for details.

Advanced printing options

Changing the printer

Initially, the printer displayed in the Print dialog box is the default printer for your system. You can, however, change the printer used in FrameMaker without changing the default printer. You may find it convenient to switch to the Adobe PDF printer to ensure that what you see on-screen matches what you will deliver in PDF files. Also, if some characters are not displaying correctly, selecting the Adobe PDF printer may fix the problem. If you print to a non-PostScript printer, any Encapsulated PostScript (EPS) images will print using the low resolution EPS "preview" image instead of the embedded PostScript data. If no preview exists, you may see a gray rectangle in place of the image. To print the EPS images correctly, print to a PostScript printer, or create a PDF of the file, then print the PDF on your device of choice. (See "Vector graphics" on page 271 for more information about EPS files.)

Tip It's good practice to keep Adobe PDF as your default printer because it is specifically optimized for PDF file creation. This will help special characters such as arrows and check marks to display correctly. You can install the free SetPrint.dll from Sundorne Communications to make the Adobe PDF printer the default printer for FrameMaker without impacting other Windows applications. For more information, see Appendix A, "Resources."

When you switch printers, a warning is displayed, indicating that the fonts for your system have changed. These fonts are the *printer-resident fonts* (described in "Printing fonts accurately" on page 399). If the required fonts are installed on your computer, the document should print correctly.

Caution Don't change drivers after you copyfit a book because the line breaks might change. If you must change the printer, check the pages for bad page and line breaks and regenerate the book file to update cross-references and generated files. Because the Adobe PDF printer driver has no settings for font substitution, leaving your default printer set to Adobe PDF can eliminate this problem.

To change the printer, follow these steps:

1. Display the Print dialog box (see page 391).
2. In the Printer section of the Print dialog box, click the Setup button to display the Print Setup dialog.
3. Click a printer in the Name drop-down list, then click the OK button to display the new printer. (If the resident fonts change, a warning is displayed. Click the OK button.)

About color separations

FrameMaker supports the creation of color separations. However, you don't need color separations if you deliver your documentation to your customers as PDF files or if your print vendor uses a digital printing process. If you do need color separations, your print vendor can create color separations from your PDF files. Discuss printing requirements with your print vendor before you produce and send the PDF files.

Setting Adobe PDF document properties

Starting with FrameMaker 8, Adobe no longer includes the full version of Acrobat Distiller with FrameMaker. Instead, FrameMaker includes the Adobe PDF Creation Add-On, which is a version of Acrobat Distiller that runs in the background when you select **File > Save As PDF**.

Note The Adobe Technical Communication Suite includes FrameMaker and Adobe Acrobat Professional, so the suite includes Adobe Distiller instead of the Adobe PDF Creation Add-On.

The biggest difference between the Adobe PDF Creation Add-On and Acrobat Distiller is that the add-on lets you create PDFs and enable online reviews only from FrameMaker (not from other applications). Also, Acrobat Distiller is part of the full Adobe Acrobat application, which lets you edit your PDF files, create forms, and more. Whether you use the Adobe PDF Creation Add-On or Acrobat Distiller, the process for creating PDF files from FrameMaker is the same.

Adobe PDF Settings sheet

The number one problem I've had with successfully producing PDF is with the Rely on system fonts only; do not use document fonts option. This option should be turned off.

To turn off this option, do the following:

1. Navigate to **File > Print Setup** to display the Print dialog.
2. Set your printer to Adobe PDF and select the Properties button to display the Adobe PDF Document Properties dialog.
3. Deselect the Rely on system fonts only; do not use document fonts option.
4. Click the OK button to dismiss the dialog.

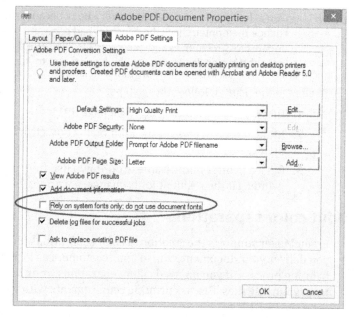

Configuring PDF job options

To access PDF Job Options, do the following:

1. Navigate to **File > Print Setup** to display the Print dialog.
2. Set your printer to Adobe PDF and select the Properties button to display the Adobe PDF Document Properties dialog. (see previous exercise)
3. Select the Default Settings which most closely matches your PDF output needs and select the Edit button.
4. Click each folder on the left to display its sheet. Change any options you like. Refer to the online help for details on the numerous settings.
5. Click the Save As button to display the Save Adobe PDF Settings As dialog.
6. Type a name for your job options file, then click the OK button.

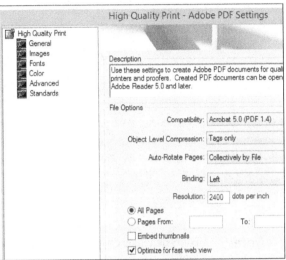

FrameMaker stores the custom job options file in your application data folder For example, in Windows 7, the file should be stored in

C:\Documents and Settings*user_name*\Application Data\Adobe\Adobe PDF\Settings

When you create a PDF file, you can choose your custom PDF job options file, as shown on page 402.

Tip   Ask your print vendor which PDF job options to use for best results. Some print vendors can send you a customized PDF job options file to make this easy.

Printing fonts accurately

Embedding fonts in your PDF files helps to avoid font substitution problems. By embedding fonts, for example, you can be sure that the print vendor has all of the fonts used in the document. To ensure that fonts print as you expect, do the following:

- Buy professional fonts whose licensing allows you to embed them in the PDF files. Set your expectations for your fonts based upon the vendor and price paid.
- Use the Adobe PDF printer driver (which should always be used for Save As PDF).

- In the PDF job options file, make sure that the option to embed all fonts is selected. To make the PDF smaller without loss of quality, select the option to subset the fonts when the percent of characters used is less than 100%.

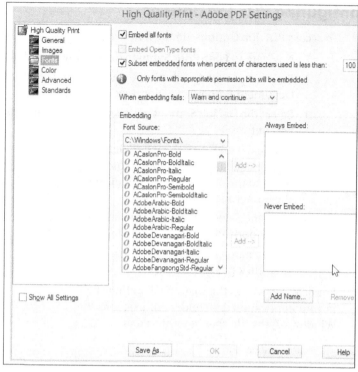

Note   Fonts generally do not occupy a great deal of disk space. I do not recommend subsetting fonts.

Note   If your PDF contains unrecognized characters (perhaps for fonts used in EPS files), try adding the problem font to the Always Embed area of the Fonts sheet. Every PDF created with this .joboption will be slightly larger, but the trade-off in recognized characters may be worth it.

Setting color mode in Adobe PDF document properties

While I've not experienced this problem, I've seen discussion of the Paper/Quality color setting in the Adobe user forums, so it's worth checking if you run into trouble.

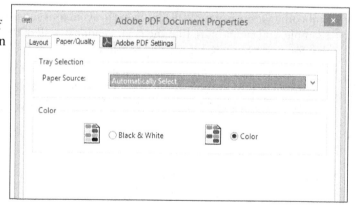

Creating PDF Files

You create a PDF file from FrameMaker using the Save As PDF function. This starts either the Adobe PDF Creation Add-On that is included with the FrameMaker product or the Adobe Distiller product that is included with the Technical Communication Suite, depending on what you are using.

To save a FrameMaker file as a PDF file, follow these steps:

1. Select **File > Save As PDF** to display the Save Document dialog.
2. *(optional)* Navigate to the directory where the PDF file will be created. By default, the PDF file is stored in the same folder as the FrameMaker file.
3. *(optional)* Specify the PDF file name.
4. Click the Save button. The PDF Setup dialog box is displayed (see page 402). The same options can be customized and stored by selecting **Format > Document > PDF Setup**.
5. Choose the settings you want in the PDF Setup dialog box, then click the Set button. You must click the Set button even if you don't customize the settings.

FrameMaker creates a temporary PostScript file (.tps) and starts the Adobe PDF Creation Add-On (or Adobe Distiller) to convert the file to a PDF file. A log file is placed in the same directory as the PDF file, with the same name as the PDF file but with a .log extension. You can open the log file in a text editor to check for any errors.

PDF Setup dialog options

Configuring the Settings sheet

This sheet lets you set the default PDF job options file used to save your PDF. It also lets you set options for the page the PDF opens on, and the paper size used for processing.

Choosing a PDF job option. From the PDF Job Options drop-down you can select from any PDF job options stored in the default location. See "Configuring PDF job options" on page 399 for more information.

Specifying the first page displayed and default zoom level. When the user opens a PDF file, the first page from the FrameMaker file is usually displayed. You might want another page displayed first if, for example, the user probably won't care about viewing the title page or front matter. You can also control the zoom level at which the PDF file is displayed. This zoom level overrides the default magnification in the user's Adobe Reader or Acrobat settings.

To specify the default zoom level and first displayed page, follow these steps:

1. Select **Format > Document > PDF Setup** to display the PDF Setup dialog.
2. Select the Settings sheet.
3. To change the first displayed page, type the page number in the Open PDF Document on Page field.
4. To change the zoom level, click an item in the Zoom drop-down list:

    - **Default:** Uses the Acrobat settings specified in the user's Adobe Reader or Adobe Acrobat program.
    - **Fit Page:** Displays the entire page in Adobe Reader or Acrobat.
    - **Fit Width:** Displays the width of the page.
    - **Fit Height:** Displays the height of the page.
    - **10% to 400%:** Specifies the magnification level.

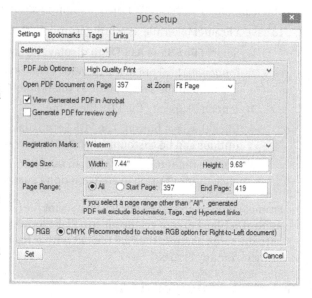

Specify View Generated PDF in Acrobat. Choose this option to automatically launch Acrobat after successful PDF conversion. This is handy, as it saves you from navigating to (or remembering) the directory into which you saved your file!

Generate PDF for Review Only. This option will prevent FrameMaker from producing separate review copies of the PDF so that you can retain an original uncommented copy on your drive. If you plan on performing a PDF review, see "Generating tagged PDF files" on page 405 for important information.

Setting up registration marks in a PDF file

By default, when you convert a document to a PDF file, the Adobe PDF Creation Add-On or Acrobat Distiller removes any extra space to maintain the page size of the FrameMaker file. Removing the extra space also removes registration marks because they fall into the area that's deleted. To prevent this, you must increase the page size in the PDF setup and uncheck the Generate Acrobat Data option in the Print dialog.

To set up registration marks in a PDF file, follow these steps:

1. Display the Print dialog box, then click Western or Tombo in the Registration Marks drop-down list.
2. Click the PDF Setup button. The PDF Setup dialog box is displayed (see previous section).
3. Click the Settings tab, then click the type of registration marks in the Registration Marks drop-down list.
4. Add 1 inch to the width and height, respectively, in the Page Size section.

Setting RGB/CMYK

I recommend setting your color mode to CMYK, and have had a few occasions where the RGB setting has caused PDF creation failures. However, starting with FrameMaker 2015, Adobe specifically recommends RGB color mode when producing PDF from RTL (Right to Left) languages. (see graphic on page 402 for example)

Generating PDF bookmarks

A *bookmark* is a link available in the left pane of the Adobe Reader or Acrobat window. The user can click the bookmark to view a specific topic in the PDF file. (Actually, bookmarks can also link to web sites, other files, or even perform actions—but when you set up bookmarks from the PDF Setup dialog, you are setting links to paragraphs within the PDF file.)

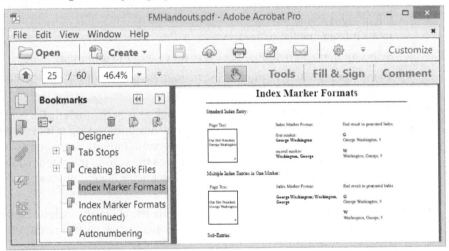

You can create bookmarks automatically from FrameMaker content by specifying which paragraph tags to include and their hierarchial relationship. Top-level bookmarks are displayed flush left. The second-level bookmarks are nested within first-level bookmarks, and third-level bookmarks are nested beneath second-level bookmarks, and so on. Since the nesting is hierarchical, if no second-level content exists, for example, the third-level bookmark nests inside a first-level bookmark, or flush left within the bookmarks pane if no first-level bookmark exists.

In the bookmark setup, you decide whether the bookmarks are expanded or collapsed by default when the user opens the PDF file. If all bookmarks are expanded, the user doesn't need to click each collapsed bookmark to display the list. But if the document has many bookmarks, and they're all expanded, the user will have to scroll down to search through all of the bookmarks.

To specify the paragraph tags you want to convert to bookmarks in a document, follow these steps:

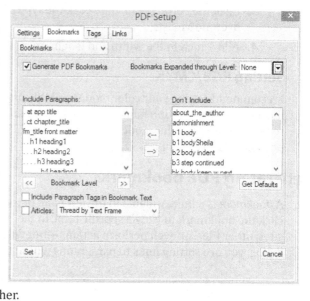

1. Select **Format > Document > PDF Setup** to display the PDF Setup dialog.
2. Click the Bookmarks tab.
3. Check the Generate PDF Bookmarks checkbox.
4. Move the items that you want to bookmark to the Include Paragraphs list on the left.

    You can move items by doing any of the following:

    - Select an item, then click the left or right arrow button to move it from one list to another.
    - Select an item, then double-click it to move it to the other list.
    - To move all the items from one list to another, press and SHIFT+CLICK the appropriate arrow button.

5. To specify the hierarchy of the bookmarks, click a tag in the Include Paragraphs column, then click the Bookmark Level (<< and >>) buttons until the tag is indented correctly. A dot is displayed next to the tag to indicate the level of indentation—one dot indicates the first indented paragraph tag, two dots indicates the second indented paragraph tag, up to the sixth level. Starting with the seventh bookmark level, a number is displayed instead of the dots.
6. To include the paragraph tag name in the bookmark, check the Include Tags in Bookmark Text checkbox. This option can help you verify bookmark levels in draft PDF files.
7. To expand or collapse bookmarks in the PDF file, do one of the following in the Bookmarks Expanded through Level drop-down list:
    - Accept Default to use the Acrobat settings specified in the user's Adobe Reader or Adobe Acrobat program.
    - Click None to collapse all bookmarks.
    - Click All to expand all bookmarks. A maximum of 25 bookmarks can be expanded.
    - Type a number to indicate the number of levels to expand. For example, to expand three levels of bookmarks, type 3.

Threading articles

Documents such as newsletters and magazine layouts often have articles that are in separate text flows. In a PDF file, you can thread articles by their text flow. When you reach the bottom of the article in flow A on page 1, the continued article at the bottom of page 2 is displayed.

In a single-column document, articles are usually threaded by text frame, so the cursor moves from text frame to text frame instead of from flow to flow. This is the default value in the PDF setup.

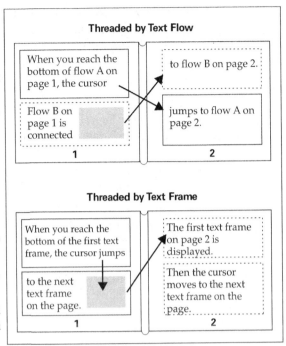

To indicate how the articles will be threaded, follow these steps:

1. Select **Format > Document > PDF Setup** to display the PDF Setup dialog.
2. Click the Bookmarks tab (see page 404), then do one of the following:
   - To read the articles in a multicolumn document, click Thread by Column in the Articles drop-down list.
   - To read the articles in a single-column document, click Thread by Text Frame in the Articles drop-down list.

Generating tagged PDF files

Tagged PDF files display a logical hierarchy of the document so you can repurpose the PDF file.

- **Tagged PDF is required in the FrameMaker/PDF review cycle**, and allows FrameMaker to import comments from a reviewed PDF for faster corrections using the Track Text Edits tools in FrameMaker.
- Tags allow for viewing of PDF on alternate displays, such as screen readers and small devices. For example, you can make sure any header or footer paragraph tags are not included in the tagged PDF file, so screen readers (such as the one built into Adobe Reader) know to skip them when reading the PDF file aloud.
- Tags provide targets which allow you set up links to specific locations within a PDF, not just to the PDF file itself.

The tree structure shows paragraphs in the document. The contents of anchored frames may be displayed if you assigned text attributes to the anchored frame object properties (see "Setting anchored frame object properties" on page 267).

To set up a tagged PDF file, follow these steps:

1. Select **Format > Document > PDF Setup** to display the PDF Setup dialog.
2. Click the Tags tab. The Tags sheet is displayed.
3. Check the Generate Tagged PDF checkbox.
4. Move the items that you want to include in the logical hierarchy to the Include Paragraphs list on the left.

   You can move items by doing any of the following:

   - Select an item, then click the left or right arrow button to move it from one list to another.
   - Select an item, then double-click it to move it to the other list.
   - To move all the items from one list to another, press and hold SHIFT, then click the appropriate arrow button.

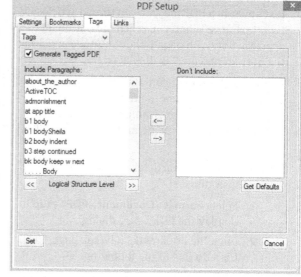

   It is not critical that you include the logical hierarchy if your primary concern is making the PDF file accessible for screen readers. Just make sure you include the paragraphs you want screen readers to speak aloud and omit those you don't.

Note   You must include at least one paragraph tag or the resulting PDF file will be untagged. If you include only one paragraph tag, that is the only paragraph that screen readers will read.

For information on describing the contents of anchored frames in tagged PDF files, see "Setting anchored frame object properties" on page 267.

Creating PDF Files

Optimizing PDF files

Using the Optimizing PDF Size command can reduce your FrameMaker file size, especially when the files were originally created in FrameMaker 6 or earlier. Additionally, if you choose not to create tagged PDF files (create named destinations for all paragraphs) you may find that some cross-document and cross-book hyperlinks in the PDF file don't work.

To set the optimization options, do the following:

1. Select **Format > Document > Optimize PDF Size > Options** to display the Optimization Options dialog.
2. *(optional)* Set the following options:

    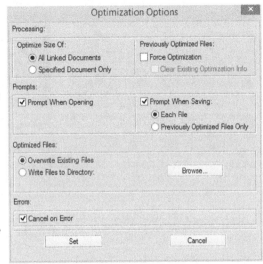

    - **Optimize Size Of.** Select whether you want FrameMaker to optimize only the file and all of the FrameMaker files it links to.
    - **Previously Optimized Files.** If the file has already been optimized, FrameMaker does not try to optimize it again. If you select Force Optimization, also select Clear Existing Optimization Info.
    - **Prompts.** FrameMaker displays a prompt before opening the linked files.
    - **Prompt When Saving.** If you selected Prompt When Opening, specify whether FrameMaker should display a prompt for each file or only for files that were previously optimized.
    - **Optimized Files.** Select whether FrameMaker should overwrite the existing files or write the files to a different directory.
    - **Errors.** Select whether you want FrameMaker to stop optimizing the file if an error occurs.
3. Click Set.

To optimize a file, do the following:

1. **Format > Document > Optimize PDF Size > Optimize File**.

407

2. Select the file you want FrameMaker to process and click the Select button.

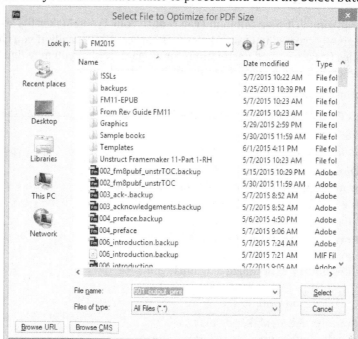

3. FrameMaker optimizes the file, and optionally, any files it links to. Progress and error messages are displayed in the console window.

Tip If you still have broken links in the PDF file, try creating named destinations, as explained in the next section. This makes the PDF file larger, but may fix unexpected broken links. For the best results, use the FrameMaker-To-Acrobat TimeSavers product (see "MicroType" on page 535), create all named destinations, and let TimeSavers select which named destinations to preserve.

Creating hyperlinks to other PDF files

You can hyperlink paragraphs from one PDF file to another by creating the PDF files with *named destinations*. This option lets you add cross-references or hypertext links from one FrameMaker document to another without resaving the documents as PDF files. Keep in mind that named destinations increase PDF size.

To generate named destinations, follow these steps:

1. Select **Format > Document > PDF Setup** to display the PDF Setup dialog.
2. Click the Links tab, then check the Create Named Destinations for All Paragraphs checkbox.
3. Click the Set button. FrameMaker will generate named destinations when you create the PDF file.

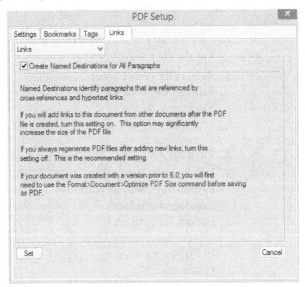

Adobe PDF printer setup

The Adobe PDF printer helps you easily specify PDF job options, or settings for PDF output. These settings include which version of PDF to produce (later versions let you create smaller files but users need an appropriate version of Adobe Reader), whether fonts are embedded in the PDF, how graphics are compressed, and many other settings. The Adobe PDF Creation Add-On includes several predefined PDF job options files that have names that indicate how they are intended to be used—High Quality Print.joboptions and Smallest File Size.joboptions, for example. You can also define your own PDF job options files.

To create your own customized PDF job options file, do the following:

1. Select **File > Print Setup**, then select Adobe PDF from the list.

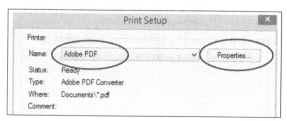

2. Click the Properties button to display the Adobe PDF Settings dialog.
3. Deselect the Rely on system fonts only; do not use document fonts option. Relying on system fonts may result in PDFs with missing resources.
4. From the Default Settings list, select the PDF job options that are closest to what you want (for example, High Quality Print).

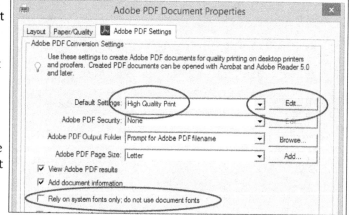

5. Click the Edit button to display the Adobe PDF Settings dialog box for the PDF job options file you selected.
6. Modify as needed and save in the default location. These settings will be available for any application (like Acrobat Distiller, or MS Word with the PDF Maker plugin) that uses .joboptions files.

If You Can't Create a PDF File

If you find that Save as PDF is not working as expected, here are some possible solutions:

- If the Adobe PDF Creation Add-On crashed, delete any .tps or .log files created by the software.
- Perform a MIF Wash. See "Performing a MIF wash" on page 565 for details.
- If you are using Save As PDF, create the PDF twice (same file name, same settings). It seems silly, but try it anyway.
- Open the properties for the Adobe PDF driver, then uncheck the "*Rely on system fonts only; do not use document fonts*" checkbox.
- If some text is not displaying in the PDF file, try deleting any temporary FrameMaker files. These may be named frm*.tmp.
- In Windows, use **Printer properties > Device Settings** to set the text and graphics grays options to *true postscript gray*.

See your network administrator for additional assistance.

Package

Starting with FrameMaker 2015, FrameMaker has a Package utility, available at **File > Package**. Using this feature, you can collect your book file, linked FrameMaker document and generated (TOC, IX) files, as well as files imported into your documents. This is a great way to create versioning points within your document, especially if you do not have a content management system in place. See "Using a Content Management System (CMS)" on page 527 for more information.

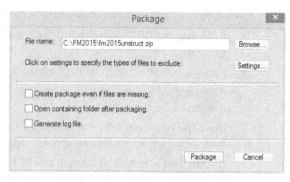

Choosing Package options

Consider placing your packaged project outside of your FrameMaker project directory. See more information about setting up a useful project directory

Settings

Chapter 23: Digital Publishing

I think direct digital publishing is the coolest feature in the last few FrameMaker releases. Using the Publish pod, you can produce content for just about any electronic device.

Here's a look at what's in this chapter:

- Accessing the Publish pod 413
- Exploring the Publish pod 414
  - Available formats 414
  - Pod controls 415
  - Style Mapping tab 415
  - Notes on style mapping 416
  - Outputs tab 416
  - Responsive HTML5 notes 417
  - Mobile App notes 418
  - WebHelp notes 420
  - ePub notes 421
  - Kindle notes 423
  - Microsoft HTML Help notes 424
  - Exporting XML output 424

Accessing the Publish pod

Access the Publish pod by selecting the **File > Publish** menu option.

Starting with FrameMaker 12, you can publish directly from FrameMaker to a series of online formats. FrameMaker 2015 adds to that list the ability to publish your content directly to mobile applications for iOS and Android devices.

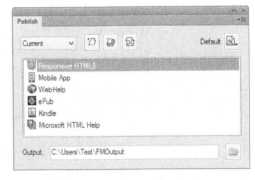

This chapter describes creating online output. (For details about PDF files, see Chapter 22, "Print, PDF output, and package.")

If your digital publishing needs go beyond what FrameMaker offers natively, then consider linking your FrameMaker project to Adobe RoboHelp for output. Requirements for more obscure online help output and a need to produce multiple independent versions of the content are two examples of where RoboHelp may extend beyond what FrameMaker can easily manage.

There are also the somewhat out of date options to Save as HTML and Save as XML. They give you "quick-and-dirty" conversions, but I don't recommend using those Save As options.. They are easy to set up but have limited control over the output.

For some specific needs you can also use other tools like RoboHelp and other non-Adobe software to convert your content. Those choices are more difficult to configure but may provide more specific customization options.

Note Avoid the Save as HTML and related Save As XML feature. See the sidebar "Exporting XML output" on page 424 for more information about the limits of Save As XML in unstructured FrameMaker. If you want to produce XML from FrameMaker, use structured FrameMaker.

Much like structured FrameMaker, FrameMaker digital publishing is a deep enough topic to warrant its own book. The default publishing options are fairly straight forward, but if you want to heavily customize the content, you'll benefit from some additional details. The Apple iOS publishing process, for example, requires a Developer license, as well as a series of control documents for processing.

To help you through these very specific processes, I've created both free and paid online courses available at training.techcommtools.com to help you out. These classes are a result of my work with clients who require things like EFB (electronic flight bag for aircraft pilots), application help, and delivery of content via membership websites and mobile (iOS and Android) applications.

Exploring the Publish pod

Available formats

Here is a list of the digital publishing formats available in FrameMaker 2015:

- Responsive HTML5
- Mobile Application
- WebHelp
- ePub
- Kindle
- MS HTML Help
- Adobe Digital Publishing System (DPS)

Pod controls

The Publish pod has the following controls available in the pod itself:

- Scope. You can choose to process the currently active file, or a specific opened file or book. Take care that this setting is correct whenever processing; I find that my work in FrameMaker changes my selected scope, resulting in a need to reprocess perhaps a third of my publishing attempts.
- Generate Selected Output. If you only need one format, selecting this button (or double clicking a specific option) is a fast way to start processing.
- Generate Multiple Outputs. This is a great way to queue final output or updates of a number of different formats.

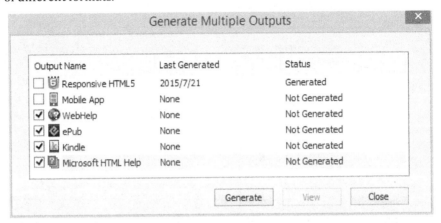

- View Output. Use this option to bring up either the output itself, or with Mobile App, to bring up the QR code you need to install the app on your device.
- Settings. Use the settings to control style mapping and specify details for your six different publishing formats. The Settings button will show the last loaded .sts file, and you should consider keeping a separate saved .sts file for each project requiring digital publishing. The .sts file can also be shared across devices, saving a great deal of time if you may publish from more than one computer.
- Available formats. As discussed above, there are five formats available in FrameMaker 12, and six formats available in FrameMaker 2015.
- Output. Each publishing format gets its own folder, so I recommend defining a digital publishing directory in your project folder where you can manage each of these folders.

Style Mapping tab

Within the Settings option is the ability to map FrameMaker styles to electronic CSS styles. If you need to match corporate branding standards, you will likely need to do this. Here is an example of a paragraph style mapping to the BodyLevel1 css style from the default FrameMaker mapping CSS file.

Note there are options for paginating content, as well as removing autonumbers from the content. You can map as much or as little of the content as you like, but I like the control I have over the output when I map just about everything to a style I can adjust from the CSS. You can access the CSS editor using the Edit Style button on this sheet.

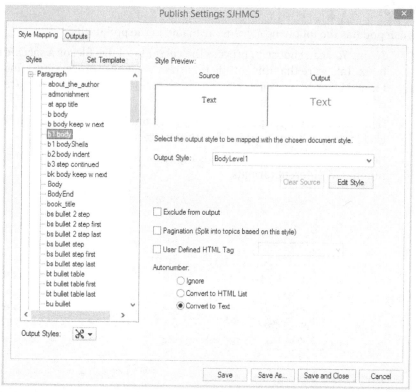

Fortunately, your mappings need only to be done once for each of the six available digital publishing outputs. Using the default mapping of [Source] will result in somewhat messy inline formatting in your HTML code. HTML code is basically what sits under all of the digital publishing formats.

Notes on style mapping

To set conversion formats via styles in a separate FrameMaker document, use the Set Template option.

If you prefer to edit CSS in a separate editor, use the Output Styles option at the bottom of this dialog.

The oddly named Pagination (Split into topics based on this style) option will split long documents into smaller chunks. Consider using it when mapping your larger section headings.

Section and reference numbering is not as important in online formats. Consider using Ignore for these types of numbers, and convert your ordered and unordered lists to HTML Lists.

Outputs tab

The other half of the Edit Settings dialog is the Outputs tab. Fortunately, the easiest format to configure is the Responsive HTML5, which is also one of the more commonly used formats. Responsive HTML5 allows content to be displayed on nearly any device, and to format content appropriately based upon screen width. It's worth noting that the Responsive HTML5 and Mobile App outputs allow dynamic content filtering, which lets the end user show or hide content based upon conditional text used in your documents.

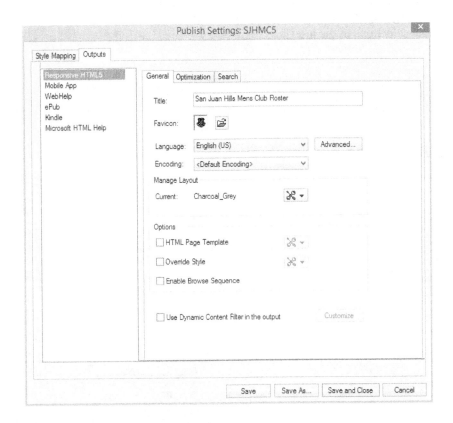

Responsive HTML5 notes

Refer to screen above for HTML5 options.

Title displays in the title bar of the user's browser window.

Favicon is either a PNG or ICO file, and can have a transparent background.

Labels and search options are available under the Language: Advanced button.

Manage Layout includes an Edit button that lets you choose and/or redefine your HTML5 browser skin. Use the most recent skins (Currently Charcoal Grey and Azure Blue) as starting points for the best results. Using older skins may disable more recent editing options.

To customize HTML5 content formatting, use the Overrride Style option.

The Use Dynamic Content Filter in the output option lets you map conditional text to user controlled content in the output. This is a nifty feature, and here's a look at what the interface looks like. You can create individual criteria or groups for display, and allow for single or multiple selections in the interface.

Mobile App notes

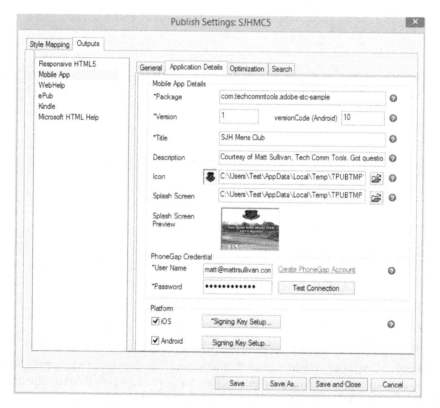

See "Responsive HTML5 notes" on page 417 for description of General tab items.

Package is required, and should follow the com.domain.appname convention.

Version is the actual version number of your app.

versionCode (Android) is your build number. This should be a whole number, and increment whenever you publish the app. In the example above, I have built the app ten times. If the next version were 2.0, then I would set the versionCode (Android) number to 11.

Title is the label used for your application on the home screen.

Description is shown when downloading, either manually, or on an app store.

Icon is the PNG file used with the Title on your device.

Splash screen shows while the application is loading on your user's device.

PhoneGap Credential is the Adobe ID and password you will use to access Adobe's PhoneGap:Build service. FrameMaker users have one PhoneGap application they can build and maintain for free. Beyond that, either a CC subscription or a PhoneGap subscription is required. Currently you are allowed to maintain up to 25 applications per month for a USD10 monthly fee.

Platform signing key setup (iOS). You must have an Apple Developers account to test and develop iOS apps (USD99 annually). Then you need to

- Create a Certificate Signing Request (CSR).
- Process the CSR into a Signing Certificate (CER).
- Process the CER into a P12 Certificate.
- Register your application.
- Register your Apple testing devices.
- Create a provisioning profile based on your CER, device(s) and application.

Once you have those things completed, you fill out the details in the Signing Key Setup.

Caution   Even after all that work you may not get the app to process properly at build.phonegap.com. You may need to upload your P12 and provisioning document manually, and unlock them on the PhoneGap server. If you're not attending my FrameMaker Digital Publishing course, do yourself a favor and keep an easily updatable checklist of all these steps.

WebHelp notes

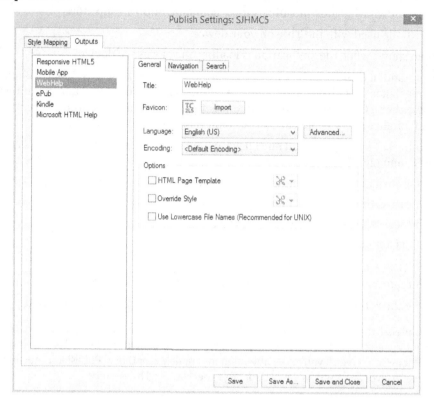

Title is shown in the user's browser title bar.

Favicon can be a small PNG or a from a favicon generator like

http://www.favicon-generator.org/

Override Style can be used to customize your WebHelp output beyond the style mapping applied to your other digital publishing formats.

Navigation allows you to select and customize WebHelp skins, as well as showing/hiding various controls in the WebHelp output.

ePub notes

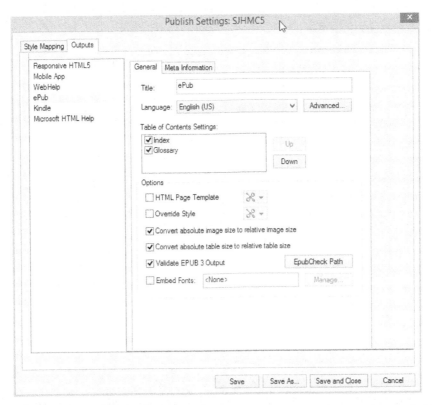

Use Override Style to customize ePub output beyond the formatting of your other digital formats.

I highly recommend the Validate EPUB3 Output button. Use the EpubCheck Path to identify/download the software.

Starting with FrameMaker 2015 you can embed the fonts used in your EPUB output.

- From the Outputs tab, choose ePub.
- In the Options area of the General tab, click Embed Fonts.
- In the Embed Fonts dialog box, select the fonts to embed in the eBook output.

You can choose the fonts currently used in the project. Alternatively, you can choose from all fonts available on your computer.

Caution   Make sure you have licensing rights to distribute fonts in your EPUB.

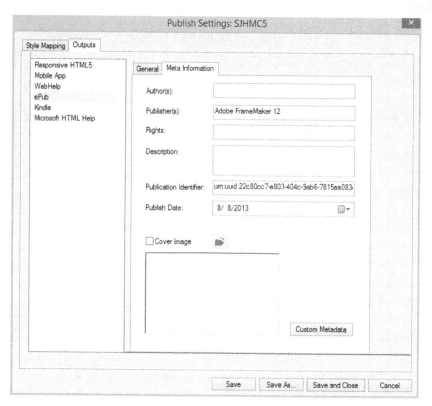

Use the Meta Information tab to define details surrounding publishing of your content.

Set your cover image here as well, and research .9.PNG to see if this is a proper format for your cover.

Kindle notes

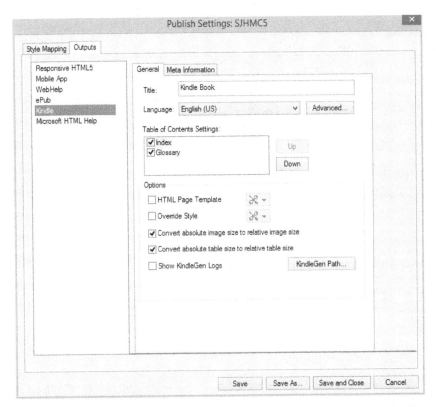

Use Override Style to customize Kindle output beyond the formatting of your other digital formats.

Use the KindleGen Path button do specify the location of and/or download the KindleGen software needed to produce Kindle and Kindle Fire output.

See ePub for a discussion of the Meta Information tab.

Microsoft HTML Help notes

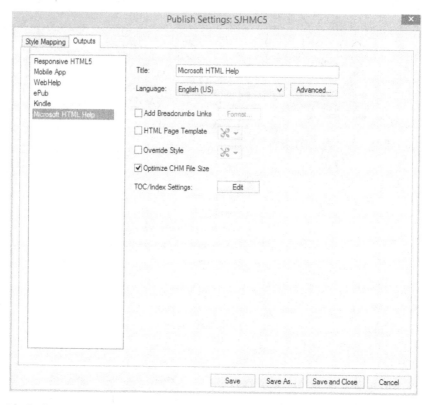

Use Override Style to customize MS HTML output beyond the formatting of your other digital formats.

There's not much else to discuss about MS HTML Help beyond what's been discussed in the other formats. If you have questions regarding this format email me directly, or check out the FrameMaker forum at forums.adobe.com.

Exporting XML output

Because of the simplistic nature of the **File > Save as XML** feature, it's best to only use that feature if you need to move content permanently to another editing system. Mapping paragraph and other tag names to element names does create *well-formed* XML files, meaning the XML has the necessary opening and closing XML tags. However, the XML files that it creates are probably not going to be very useful. Because the mapping moves data from a flat, unstructured file, the output has minimal hierarchy. In addition, unless you have chosen your FrameMaker tag names very carefully and with structure in mind, using tag names as element names will probably not convey the meaning of the information appropriately. To create *valid XML*, which has a hierarchy and can be validated against a document type definition (DTD) that provides the rules for the elements in the XML, you need to use structured FrameMaker to set up a structured application.

Chapter 24: Color output

FrameMaker lets you create full-color documentation using a rich set of tools. The available colors are stored in a color catalog, which is available through the Paragraph Designer, Character Designer, and many other dialog boxes.

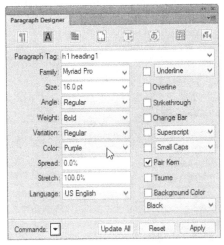

Here's a look at what's in this chapter:

    Advantages of defining custom colors 426
    Understanding types of color 427
        Understanding process color 428
        Understanding spot color 429
    Applying color to text and objects 430
    Managing color definitions 432
        Modifying existing colors 432
        Adding colors from a color library 434
        Creating custom colors 437
        Renaming colors 437
    Viewing colors 438
        Selecting a color view 438
        Setting up color views 439
    Deleting colors 439
    Controlling colors for output 440

425

Advantages of defining custom colors

You assign color to text or objects by applying tags that contain references to color definitions. For example, to create a purple first-level heading, you assign a color to the Default Font properties for the h1 heading1 paragraph tag.

To create blue, underlined cross-references that stand out in PDF files, you first create a character tag that applies blue and underlining.

Then insert the character tag in your cross-reference formats. In the PDF file, the hyperlink is underlined and displayed as blue text.

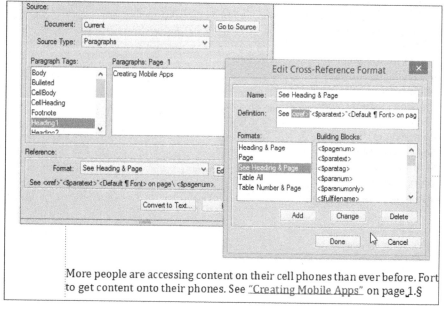

You can also assign color to some items, such as borders or FrameMaker drawings, by choosing a color in the Tools panel or object properties.

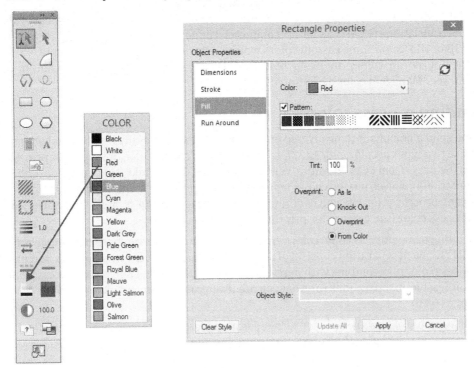

Understanding types of color

In FrameMaker, color definitions are based on one of three color models:

- **RGB.** Earlier versions of FramMaker define RGB colors in percentages of red, green, and blue. Starting with FrameMaker 2015, RGB colors are more classicly defined as 256 levels of each component. RGB is the color model that controls onscreen display.

- **CMYK.** Defines colors in percentages of cyan, magenta, yellow, and black inks (black is the *key* color, thus the "K"). CMYK is for color printing. Printed colors can be separated into percentages of the four component colors, thus the phrase "color separations" used in commercial printing. Because your computer monitor uses RGB to display colors, the colors you see onscreen do not match colors in print. Ask your print vendor to run a test document or get a swatch book to assist in proper color selection, and in setting expectations for printed output. CMYK is a subtractive color model, meaning the absence of all colors is equal to white.

- **HLS.** (Sometimes referred to as HSL.) Defines colors by hue, lightness, and saturation. The position along the edge of a color wheel (or the visible spectrum) determines the hue. The amount of black determines the lightness. Saturation refers to the amount of color, ranging from no color (white) to full color (fully saturated). HLS color is excellent for online formats and for digital photography professionals, although the color is translated to RGB or CMYK for output.

RGB, CMYK, and HLS colors are available *color libraries* in FrameMaker. If you plan to use a print vendor for output, your print vendor may recommend a color library based on paper stock and other printing factors, including press and printing method. Most color libraries will provide reasonable translations to CMYK color for commercial or desktop output; however, a few consist only of RGB or HLS color. Some custom colors may be outside the CMYK printing gamut, so if you have critical corporate colors, are using vibrant, vivid colors, or are using subtle pastels in your document, see "Managing color definitions" on page 432 for details.

Understanding process color

When you send a document to be printed on a printing press (generally *web* or *offset printing*), you can use two types of color: *process color* and *spot color*. Process color, or four-color printing, is mixed on the press from cyan, magenta, yellow, and black (CMYK) inks; spot color (described in the next section) involves mixing colors before going to press, much like buying paint for your house. Using process color lets you combine the inks on the press to create a wide variety of colors. Process color is necessary for printing photographs.

When printing using process color, colors are broken down into percentages of the four process colors (CMYK). Each separation consists of solid color areas, and areas where only a percentage of the process color may combine with percentages of the other inks. Four-color process printing requires at least four separations (more if you are using custom, or spot, colors in your document as well).

Here is an example of how the four process separation plates combine to produce the final printed image. This is the method of printing your desktop printer uses as well.

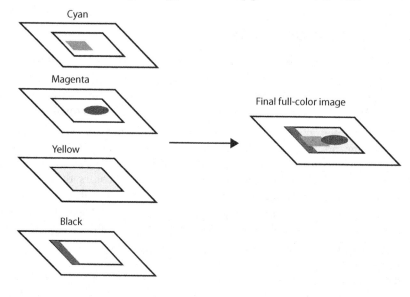

In commercial printing, separations are created for the individual plates. Each plate is used to print a single color onto a page. Registration marks allow the print technician to line up the separations correctly. Slight deviations in alignment can cause the color to print in the wrong place, so the dots on each separation are produced at different angles, and sometimes at slightly differing resolutions. See "Modifying existing colors" on page 432 for information on preventing registration errors. The following illustration shows what can happen (and is expected to happen) due to the registration limitations of a high-speed printing press.

Separations were accurately aligned.

Separations were misaligned, so there is a gap between the overlapping objects.

Your print vendor will likely address this problem by *trapping* your files prior to color separation. This is typically an automatic software process, and not something easily handled on the files and images that make up your FrameMaker project.

The color separations will be produced using a specific angle for each of the colors, which helps print color in a smooth *rosette* (named for its similarity to a flower) instead of in a jarring *moiré* pattern. Your print vendor will set up the *halftones* (dot patterns used in printing) during the separation process before going to press, so consult your vendor before changing these settings in FrameMaker.

Understanding spot color

In contrast to process color, spot color is mixed before your job goes to press. Instead of filling the ink wells on the press with CMYK, the press uses premixed inks. In theory, for print jobs that use only two or three colors, using spot colors can reduce the cost of printing (because printing three colors is less expensive than printing four colors—fewer separations, fewer plates, fewer layers to keep in registration). In reality, though, because of the automation available with process printing, you may spend more in setting up a spot color job than you save by reducing the colors involved. Additionally, if you have full-color images or photographs, spot color output is not an option.

Spot colors are also useful for printing colors that are outside the CMYK *gamut*; that is, colors that you cannot create by mixing CMYK inks. Metallic or fluorescent inks (neon colors) usually require a spot color because they are also outside the CMYK gamut.

If you are using one spot color (plus black) to reduce costs, you can create visual interest by using *tints*, or lighter versions of a base spot color. Instead of printing a color at full strength, you print only a percentage of the color. For example, 50 percent of a deep blue produces a light slate blue. Because tints use the same base color as the spot color, they are set up on the same separation and do not increase the cost of the print job.

If you're printing images with solid patches of color and can create the necessary color with a few inks, consider using spot color. If your document contains color photographs, though, you'll need to use process color. See "Understanding process color" on page 428 for more information.

Applying color to text and objects

You format text and objects with custom color by adding the color in the color catalog and assigning the color in one of the following items:

- Paragraph tags
- Character tags
- Table tags
- Condition tags
- Custom table ruling and shading
- Change bars
- Object properties
- Tools panel

Tip You can apply one or more of these properties using object styles. For more information, see Chapter 14, "Object styles."

You apply the appropriate tag, edit the object properties, or select the color in the Tools panel. If you modify a color in the color catalog, each item that uses that color is automatically updated. Suppose you create a custom color but decide to reduce the amount of red in the color. You update the color definition, and the color is updated globally—in existing text and objects and those you create later.

Table 24-1 summarizes assigning color to text and objects.

Table 24-1. Assigning Color to Text and Objects

Item	How to Assign Color	More Information
Assigning color to text		
Paragraph	Assign color in a paragraph tag.	"Modifying paragraph tags" on page 113
One or more words	Assign color in a character tag.	"Modifying character tags" on page 141
Conditional text	Assign color in the conditional tag definition.	"Creating condition tags" on page 452
Variable (all instances)	Type the character tag inside of brackets in the variable definition.	"Creating user variables" on page 201
Cross-reference style (all instances)	Type the character tag inside of brackets in the cross-reference format definition.	"Formatting cross-references" on page 180
Assigning color to objects		
One object	Select the color in the object properties.	"Modifying objects" on page 299
One or more objects	Select the color in the Tools panel.	"Modifying objects" on page 299
Table border/shading	Select the color in the custom ruling style definition or in the Table Designer.	"Changing ruling options" on page 164
Change bars	Assign color in the change bar properties, or select the change bar in a character tag.	"Using change bars" on page 58 and "Modifying character tags" on page 141

Managing color definitions

FrameMaker documents contain 16 basic colors by default. The color catalog lists them in the following order: Black, White, Red, Green, Blue, Cyan, Magenta, Yellow, Dark Grey, Pale Green, Forest Green, Royal Blue, Mauve, Light Salmon, Olive, and Salmon. (Color names are case sensitive.) You add more colors by doing one of the following:

- Modifying existing colors (except default colors)
- Creating custom colors
- Adding colors from a color library
- Renaming colors

Tip When you paste an object from one document to another, the color of the object is added to the color definitions if it doesn't already exist. Colors defined in imported EPS files are also added to the color catalog. If you import a PNG file that uses an optimized palette—which means the palette uses fewer than the maximum number of colors available—the color catalog adds the colors defined in the PNG. (Saving as MIF, reopening the file in FrameMaker, and saving as an FM file fixes this problem.) In addition, colors are imported when you open a Microsoft Word document in FrameMaker. The new colors are named "Color n," where "n" equals the location in the color catalog, but you can give the colors descriptive names. See "Renaming colors" on page 437 for details.

Modifying existing colors

To modify an existing color (other than the default colors), you move the sliders in the color definition or choose a color from a specific color library. Changes to the color definition or label will update throughout the document wherever the color is used.

To modify an existing color definition, follow these steps:

1. Select **View > Color > Definitions** to display the Color Definitions dialog.

2. In the Name drop-down list, select the color you want to modify. You cannot modify colors associated with base color models—the sliders are grayed out, and the color value fields aren't displayed.

3. To change how the color will be printed, click one of the following in the Print As drop-down list:

    - **Spot:** Specifies a custom ink for commercial printers, but will be converted to the CMYK equivalent for desktop output.
    - **Process:** Uses the CMYK values for all printed output.

- **Tint:** Prints a percentage of the selected color. (Be sure to test the tint. Lighter tints might not print predictably.)
- **Don't Print:** Prevents output of the color.

4. To change the type of color, click one of the following in the Model drop-down list:
   - **CMYK:** Defines each color in percentages of cyan, magenta, yellow, and black.
   - **RGB:** Defines the color using levels of red, green, and blue.
   - **HLS:** Displays the color in terms of the hue, lightness, and saturation, converts to RGB or CMYK as needed for output.

5. To adjust the color values, do one of the following:
   - Click a slider and drag it to the right or left until the color you need is displayed in the New swatch. Remember that the color will look different in print, so refer to a printed swatch book for color matching.
   - Change values as needed in the fields. Most values are percentages, but in the HLS color model, the Hue represents the angle of the color on a color wheel and RGB values are 0-255.

6. To indicate how FrameMaker prints the overlapping color of two objects, click one of the following in the Overprint drop-down list:
   - **Knock Out:** The overlapped portion looks white and doesn't print. Helps prevent the colors of a light object on top and a dark object on bottom from mixing, although the separations must be lined up correctly, or the knocked out piece may appear in the printed copy.

   The color of the top object is set to "knock out."

   The top object "knocks out" the color of the underlying object.

   "Leaks" may appear in the printed output.

   - **Overprint:** Both colors from the overlapped portion of the objects are printed. Helps prevent a gap from printing between objects; however, if the top object is lighter than the bottom object, the two colors might mix.

   The color of the top object is set to "overprint."

   The overlapped portion of the bottom object prints.

   Overprint can impact the color of the overprinting object.

   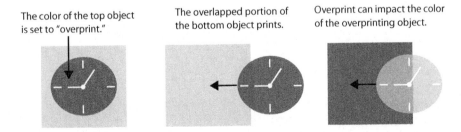

7. Click the Change button. The color definition is updated, and all items formatted with the color are reformatted. (If you modified the definition of a library color, a warning is displayed.

Unless you intended to change the library color, click the Cancel button, then click the color in the Name drop-down list to restore the definition.)

Caution Unless you use a library color as a basis for a new color, you shouldn't modify the definition of a library color. This changes the composition of the ink, so the color won't match the color you chose from the library. If you accidentally change a library color but don't save your changes, you can restore the original definition by clicking the color in the Name drop-down list. The original color definition and ink name are displayed.

> ## Naming colors
>
> When you add new colors, you should name the colors consistently. Most colors you choose from a library have an official ink name, but the name doesn't indicate that Pantone 252 CVU, for example, is a shade of pink. You can label the color before adding it to your color definitions, and the ink name is still displayed in the color definitions.
>
> Colors (and really all labels) are easier to manage if you name them based on their function, not their color. Your company's logo color is a good example. Instead of naming the color "Aqua," name it "Logo." If the color changes from aqua to teal, you can update the Logo color definition, import the new color definition into all your files, and the Logo color is instantly updated throughout your documents. If you had named the color "Aqua," you would need to rename the color after changing the definition. See "Renaming colors" on page 437 for details.

Adding colors from a color library

The color libraries provide predefined colors. Some libraries, such as the Pantone ones, may be updated after the FrameMaker release date, and thus may not contain all currently available colors. Take care when defining colors in FrameMaker to match Pantone colors in graphics created in applications such as InDesign, Illustrator, and Photoshop.

Tip You can add a color library to FrameMaker when stored in ASCII Color Format (.acf) version 2.1 or earlier, or in Binary Color Format (.bcf) version 2.0. Place the file in FrameMaker's \fminit\color directory, and restart FrameMaker. The library will be displayed with the default color libraries.

The following table describes the color libraries available in FrameMaker.

Table 24-2. Color Libraries in FrameMaker

Color Library	Description
RGB Color	
Common Color Picker	Lets you select a color using the operating system's color picker. You can also type specific RGB or HLS values.
Crayon	Developed by Adobe to provide access to common RGB colors using everyday names in alphabetical order. Do not use Crayon colors as spot colors.
MUNSELL High Chroma Colors and Book of Color	Provides RGB colors in terms of hue, value, and chroma. • **Hue:** Shades of red, yellow, green, blue, and purple in ten hue sectors: R, YR, Y, GY, G, BG, B, PB, P and RP (red, yellow-red, yellow, green-yellow, and so on). • **Value:** Lightness of the color from 0 (for pure black) to 10 (for pure white). • **Chroma:** Saturation level, or the degree of gray in the color, from 0 to infinity.
Online	Provides 216 "web-safe" colors that have a consistent appearance on all platforms when viewed with a web browser. The colors have a sequential number and a hexadecimal code.
CMYK Color	
DIC COLOR GUIDE SPOT	Provides spot colors; used mostly in Japan.
FOCOLTONE	Provides 860 process colors; helps prevent trapping and registration problems by showing the overprints that make up the colors.
Greys	Provides both process and spot shades of gray in 1 percent increments.
PANTONE Coated, PANTONE Uncoated, PANTONE ProSim, PANTONE ProSim EURO, PANTONE Process CSG, PANTONE Process Euro	Define ink colors in CMYK equivalents. • **PANTONE Coated:** For printing on coated paper. The paper coating affects how you perceive the color—color on coated paper looks lighter than color on uncoated paper. • **PANTONE Uncoated:** For printing on uncoated paper. • **PANTONE ProSim** and **PANTONE ProSim EURO:** For simulating Pantone spot colors onscreen. (ProSim stands for "process simulation.") Use the Euro library for documents printed in Europe. • **PANTONE Process CSG** and **PANTONE Process Euro:** For printing process colors. Use the Euro library for documents printed in Europe.
TOYO COLOR FINDER	Provides more than 1,000 colors based on the most common printing inks in Japan.
TRUMATCH 4-Color Selector	Provides more than 2,000 computer-generated colors that predictably match the CMYK color spectrum.

Chapter 24: Color output

To add a color from a color library, follow these steps:

1. Select **View > Color > Definitions** to display the Color Definitions dialog.
2. Click the Color Libraries button to display the Color Pickers pop-up menu.
3. Click the library you want to display, then do one of the following:
   - Type the first few characters of the color name or number in the Find field.
   - Click and drag the slider.
4. Click the color you need. The name or number of the color is displayed in the top field.
5. Click the Done button to display the Color Definitions dialog.
6. Click the Add button. The new color is added to the document's color catalog.

Caution Don't modify the definition of colors you choose from a library. This changes the composition of the ink, so the color won't match the color you chose from the library. If you accidentally change a library color, delete it and add it again from this dialog.

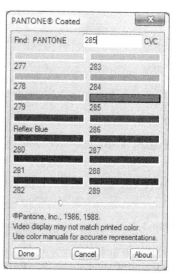

Creating custom colors

Instead of choosing a predefined color, you can create your own colors. To do so, follow these steps:

1. Select **View > Color > Definitions**. The Color Definitions dialog box is displayed (see page 432).
2. Do one of the following:
   - To base the color on an existing color, select the color in the Name drop-down list.
   - To create a unique color, select New Color in the Name drop-down list. Change the color's label to something more descriptive.
   - To choose from an RGB color picker, click the Color Libraries pop-up menu, then click Common Color Picker. Select from the displayed colors, type in the values in the RGB percentage fields, or click and drag the sliders, then click the Done button.

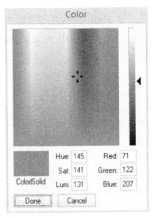

3. Type the name of the new color in the Name field. If you based the color on a default color, the Add button is no longer grayed out.
4. Click the Add button. The new color is added to the document's color catalog.

> **Caution** Do not click the Change button (which is displayed only if you based the color on an existing nondefault color), or you'll rename an existing color definition. See "Renaming colors" on page 437 for details.

Renaming colors

When you rename a color, the name is updated throughout the document. For example, if you rename "Ocean Blue" to "Ocean Green," the new name is displayed instead of the original name wherever you assign colors—paragraph tags, object properties, and the like.

You cannot rename the 16 basic colors; however, color names are case-sensitive, so you can create "green" in addition to the default "Green."

To rename a color, follow these steps:

1. Select **View > Color > Definitions**. The Color Definitions dialog box is displayed (see page 432).
2. In the Name drop-down list, click the color you want to rename.
3. Type a new name in the Name field, then click the Change button to display a confirmation dialog.
4. Click the OK button. The color is renamed throughout the document.

> **Note** If you tried to rename the color using the name of an existing color, that color is displayed. Redisplay the original color and then assign a name that's not in use.

Viewing colors

You can configure up to six color views that show and hide specific colors in your document. View 1, the default color view, displays all colors used in the document. You should not modify this view. Each of the remaining views displays specific colors and hides others. By assigning colors to content in your document, these views can be used as pseudo-layers.

For example, you can create an Instructor Notes color, then create a view that hides the Instructor Notes color, and use that view to output student versions with the exact same pagination as the instructor version.

In the following illustration, a different color view is displayed in each document:

View 1: All colors are displayed in this document.

View 2: The colors of the windowpane and sun are hidden.

View 3: The colors of the curtain and sun are hidden.

Selecting a color view

To select a color view, follow these steps:

1. Select **View > Color > Views** to display the Define Color Views dialog.

2. Click the radio button for the view you want to display, then click the Set button. The corresponding colors are displayed, cut out, or hidden in your document.

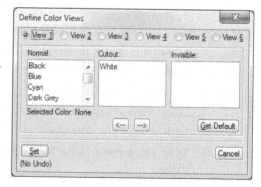

You can also toggle between the views using shortcuts (ESC V 1, ESC V 2, etc.).

Setting up color views

To set up color views, follow these steps:

1. Select **View > Color > Views** to display the Define Color Views dialog.
2. Click the radio button for the view you want to modify.
3. Click the color you want to move, then click the left arrow or right arrow button until the color is displayed in the correct section. Your choices are as follows:
   - **Normal:** The color prints as normal.
   - **Cutout:** The color prints as white on overlapping objects.
   - **Invisible:** The color doesn't print.
4. Click the Set button. The color view is updated.

Deleting colors

You can delete any color except the 16 default colors. If the color has been assigned in the document and you delete the color, a warning is displayed. You must substitute black for the deleted color or cancel the action.

To delete a color, follow these steps:

1. Select **View > Color > Definitions**. The Color Definitions dialog box is displayed (see page 438).
2. In the Name drop-down list, click the color you want to delete.
3. Click the Delete button. Do one of the following:
   - If the color is in use, a warning is displayed. Click the OK button to delete the color and reassign black to the items, or click the Cancel button if you don't want to delete the color.
   - If no warning is displayed, click the OK button to delete the color.

Controlling colors for output

You will likely produce print and a number of online formats (such as PDF, HTML, or HTML Help) from your FrameMaker documents.

Graphics and color definitions using the CMYK color model will print reliably, and maintain reasonable online representations, as the CMYK gamut is largely within the RGB gamut.

Graphics and color definitions using the RGB color model will produce accurately to online outputs, but some tones may shift when printed, as the RGB gamut is larger than the CMYK gamut.

Since some graphics (like screen captures) are RGB by nature, you won't see an improvement in color by converting them to CMYK in a graphics editor.

For colors defined in FrameMaker, determine which outputs are of primary importance for your organization (whether print or online), and use CMYK for optimal print accuracy or RGB for optimal online format accuracy. See "Understanding types of color" on page 427 for more information.

Note All default FrameMaker colors are RGB colors, and the Online, Crayon, MUNSELL, and Common Color Picker libraries also provide RGB colors.

The colors you specify in FrameMaker might not convert correctly. Some conversion processes alter color or replace it with a completely different color. Table 24-3 describes how to prevent color conversion problems.

Table 24-3. Common Online Color Problems

Problem	Solution
Windows converts CMYK color to RGB when the file is converted or printed.	• Apply CMYK colors in EPS graphics only, because EPS graphics preserve the original colors. • Check converted colors; if unacceptable, convert the color to RGB in FrameMaker.
Distiller might alter color during PDF conversion (for example, it can convert RGB color to sRGB, the RGB color standard optimized for onscreen display).	If you own Adobe Acrobat, create your own Distiller job option profile that leaves the color unchanged. Refer to the Adobe Acrobat documentation for more information.
Colors in different browsers don't match.	Apply colors from the Online color library. They're designed to display uniformly in web browsers.

Part VI

Advanced Techniques

Chapter 25: Setting up conditional text

Conditional text lets you identify content as belonging to a particular version of a document. By assigning condition tags to various text and graphics and then showing and hiding the condition tags in different combinations, you change the information that is displayed. You can maintain two or more versions of the same document in a single file. For example, an instructor could set up a file that includes the questions and the answers for an exam. In this scenario, the condition tag ExamAnswers is applied to the answers. The instructor hides the ExamAnswers information to print out the exam for students and then displays ExamAnswers to print out the answer key.

Exam

Question 1: What is Avogadro's number? What is it for?

Question 2: What is the Pythagorean theorem?

Answer Key

Question 1: What is Avogadro's number? What is it for?
Answer: 6.023×10^{23}
Extra credit for describing its use in chemistry. Extra extra credit for using the word "mole."

Question 2: What is the Pythagorean theorem?
Answer: In a right triangle, the square of the hypotenuse is equal to the sum of the squares of the other sides.

Here's a look at what's in this chapter:

    How conditional text works 444
        Notes regarding conditional text 444
        Strategies for using conditional text 445
    Conditional text examples 445
    Using conditional text in digital publishing 446
        Use conditions for multiple outputs 446
        Use conditions for personalized dynamic content filtering 446
    Applying condition tags 446
    Removing a single condition tag 448
    Removing all condition tags from text 449
    Using condition indicators 449
    Planning conditional text 450
        Common condition tags 451
        Alternatives to conditional text 451
    Creating condition tags 452
    Modifying condition tags 453
    Deleting a condition tag 453
    Showing and hiding conditional text 454
        Showing All Conditional Text 454
        Choosing the method for showing conditional text 454
        Showing or hiding conditional text as per condition 457
        Using conditional text expressions 458
        Modifying expressions 460
        Deleting expressions 460
        Examples of expressions 460

How conditional text works

When you show and hide conditional text, FrameMaker automatically repaginates the document. Any hidden conditional content is represented by a conditional text marker. The text that follows the conditional text automatically repaginates. When you show hidden conditional text, the text that follows the marker automatically adjusts to make room for the redisplayed information.

Conditional text is showing.

> **Building Your First Jet**¶
>
> Building a jet is fun and easy. You buy the kit for a few million dollars, then spend several years assembling it. This is much cheaper than purchasing those "ready-to-fly" jets from aircraft manufacturers.¶
> Joe, I think you might want to be a little more specific here. - Fred¶
> Begin by identifying the model you want to build. Then, you'll need some basic tools. Keep in mind that hydraulic work is a little more challenging.¶

Conditional text is hidden.

> **Building Your First Jet**¶
>
> Building a jet is fun and easy. You buy the kit for a few million dollars, then spend several years assembling it. This is much cheaper than purchasing those "ready-to-fly" jets from aircraft manufacturers.¶
> Begin by identifying the model you want to build. Then, you'll need some basic tools. Keep in mind that hydraulic work is a little more challenging.¶

Conditional text marker indicates hidden text.

Keep in mind that hiding conditional text affects the pagination and numbering of content in your document. If you want to hide specific information without impacting pagination and numbering, consider using custom colors and the **View > Color > Views...** command. For more information, see "Viewing colors" on page 438.

Caution Do not delete the conditional text marker; this will delete the entire section of hidden conditional text! Fortunately, FrameMaker warns you when you are about to delete a conditional text marker.

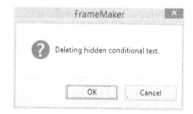

Notes regarding conditional text

Conditional text is an excellent way to create reusable content in FrameMaker. It works well with user variables and text insets. For complex content reuse, you may need a content management system (CMS), but conditional text can help you with basic reuse requrements.

Conditional text functionality was dramatically redesigned in FrameMaker 9. Although there are many improvements, the application and modification of conditional text settings is much more cumbersome than in FrameMaker 8 and earlier. Fortunately, starting in FrameMaker 11 the Smart Insert function streamlines conditional text assignment. The Smart Insert shortcuts (CONTROL-4, CONTROL-5 and CONTROL-6) are an invaluable part of the current conditional text workflow.

Strategies for using conditional text

There are two approaches to setting up conditional text:

- **Show as per Condition**. This approach identifies specific content to show or hide, and is the easiest to manage for simple tasks. This approach is discussed in the first portion of this chapter and is available in all versions of FrameMaker that support conditional text.
- **Show as per Expression**. This more sophisticated approach allows creation of Boolean expressions to determine the combination of content, and is discussed in the later parts of this chapter. This approach is available in FrameMaker 8 and later versions.

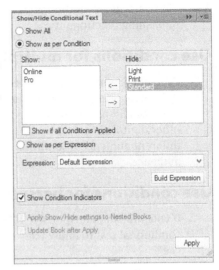

Conditional text examples

To create multiple versions of a document, you generally create two condition tags—one for information that's unique to version A and one for information that's unique to version B. (If there is only one type of unique content, you need only one condition tag.) Any information that's common to both versions is not tagged (that is, you leave it unconditional).

Caution A common mistake is to apply all available condition tags to information that's used in all versions. If the information is never hidden, it does not need a condition tag.

Because many writers use FrameMaker documents for print and online delivery, you might create two condition tags: Print and Online. Typically, the Print and Online condition tags are mutually exclusive—information existing in both conditions would remain unconditional.

When you create additional condition tags, the complexity of managing the versions increases. Consider a situation where you need to document two versions of a single product—a Light version and a Professional version. The two products are used by different people, so you've decided to create two books: *Widget Light User's Guide* and *Widget Professional User's Guide*. The content of the two books is quite similar; the Professional guide just contains more detailed information than the Light version. For this, you could create a single Pro condition tag. But in addition to that, you need to manage the printed and online information, so you also need Print and Online condition tags. To aid in switching between available combinations of tags, the Show Hide Conditional Text pod lets you display individual conditions independently. You can also create condition expressions, which are statements that include and exclude combinations of condition tags to create the intended output. (See "Using conditional text expressions" on page 458 for details on this feature.)

Using conditional text in digital publishing

There are two conditional text strategies to consider when using FrameMaker's digital publishing functionality.

Use conditions for multiple outputs

Print and Online conditions are common conditional tags in FrameMaker documents. When producing print, it's pretty easy to see that you don't want things to appear clickable. Additionally, print output may require higher quality images than you wish to deliver in PDF or other digital output. You may also need to produce different versions of your content for different audiences. Conditional tags let you display only the content needed for a specific audience.

Use conditions for personalized dynamic content filtering

If you publish to Responsive HTML5 or mobile apps for iOS and Android, you can also use conditional text to allow your users to selectively display content as they wish. For more detail, see

Applying condition tags

Although the feature is called conditional *text,* you can apply condition tags to many different items in your documents, including the following:

- Text in the main flow, in text frames, or inside table cells
- Paragraph characters
- Anchored frame anchors
- Table anchors
- Table rows
- Table columns (starting with FrameMaker 2015)
- Markers
- Cross-references
- Variables
- Equations
- Book components (starting with FrameMaker 2015)

A few items *cannot* be made conditional. They include the following:

- Text lines or text inside a text line

- Graphic (unanchored) frames
- Graphics that are not in frames
- Table cells which do not comprise an entire table row or column (but you *can* add conditions to text inside a table cell)

To apply a condition tag, follow these steps:

1. Select the information that you want to make conditional.
2. Select **Special > Conditional Text > Conditional Text**
3. ... or otherwise display the Conditional Text pod.

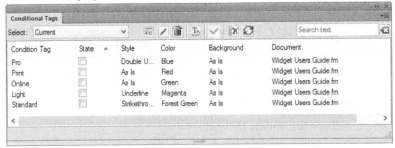

All of the condition tags that are defined in the document are listed, along with details about each tag, and controls for applying and managing them.

4. Select the condition tag you want to apply to the selected items.
5. Select the Conditional radio button.
6. Click the Apply (✓) button to tag the selected information with the specified condition tag.

You can verify this in three ways:

- Condition tags often have text colors, background colors, or special formatting (such as an underline or a strikethrough) associated with them (see page 452). If the formatting of the selected text changes, you know that you applied the condition tag successfully. This formatting change occurs only if your condition indicators (discussed on page 449) are turned on.
- The status bar indicates the condition tag for the selected text.
- The status column in the Conditional Text pod displays settings for the selected text.

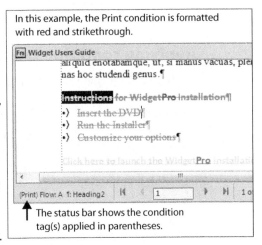

In this example, the Print condition is formatted with red and strikethrough.

The status bar shows the condition tag(s) applied in parentheses.

You can use keyboard shortcuts to apply condition tags. Here is an example:

1. Select the content that you want to make conditional.
2. Press CTRL+4 to display the Smart Insert for applying conditional text. Do one of the following:
   - Navigate to the desired tag using keyboard and/or arrow keys, then press ENTER.
   - Double-click on the desired tag.

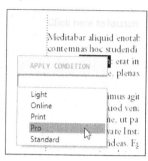

Removing a single condition tag

If you change your mind about a condition tag assignment, you can remove it. This process removes a single condition tag from the selected content. If you have several condition tags applied to the selected content and want to remove all of them at once, it's faster to remove all condition tags from the text, as described in the next section.

To remove a condition tag, follow these steps:

1. Select the content from which you want to remove the condition tag.
2. In the Conditional Text pod (see page 447) select the condition you want to remove from your content.
3. Select the Not In radio button.
4. Click the Apply (✓) button.

Use the CTRL+5 command to perform the same task via a keyboard shortcut.

Applying condition tags to tables and graphics

In tables, you can apply condition tags to entire tables (by applying conditions to the table anchor), table rows, or you can apply condition tags to text inside a table cell. You *cannot* apply condition tags to a cell or a group of cells in a table

For graphics, you can apply a condition tag to the anchor controlling an anchored frame, which makes the entire content of the frame conditional, and to text inside the anchored frame, provided that the text is in a text frame. You *cannot* apply condition tags to any of the following graphic elements:

- An actual graphic—you must apply the condition tag to the anchored frame that surrounds the graphic. If the graphic is not inside an anchored frame, you cannot make it conditional.
- A text line or part of a text line.
- (Unanchored) graphic frames.

Removing all condition tags from text

To remove all condition tags from text, follow these steps:

1. Select the text you want to make unconditional.
2. In the Conditional Text pod (see page 447) select the Unconditional radio button.
3. Click the Apply (✓) button.

Use the CTRL+6 command to perform the same task via a keyboard shortcut.

Tip If you want to remove *all* condition tags throughout a document, show all of the conditional text, and then select all the content in the document and press CTRL+6. If you do not display all of the condition tags, pressing CTRL+6 removes condition tags from all of the currently shown text without changing hidden conditional text. You can use this to quickly and completely remove one or more conditions from your content.

Using condition indicators

The style and colors that your condition tags use are called *condition indicators*. When you edit a file on screen, they make it easier to see which information uses one or more condition tags. When you output your content, though, you probably want conditional text to blend into the surrounding text.

You control display of the condition indicators using the Show/Hide Conditional Text pod. Turning off condition indicators *does not* affect your content, it only enables or disables the visual cues useful for editing.

To turn the condition indicators on or off, follow these steps:

1. In the Conditional Text pod (see page 447), click the Show/Hide Conditional Text ( ) button or otherwise display the Show/Hide Conditional Text pod.
2. Check the Show Condition Indicators checkbox to display your condition indicators.
3. Click the Apply button to apply your changes. Notice that although the text looks like regular text, when conditional content is selected, the status bar still indicates that it is conditional.

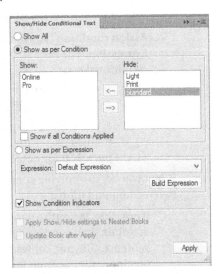

Planning conditional text

Before you create condition tags and start making your document conditional, consider doing some homework:

- **Analyze the information and identify the versions that are needed.** Determine which information is common to all versions (unconditional) and which information is needed in a specific version. Set a up a condition tag for each version.
- **Use distinct condition tag names.** Using names that are easily distinguished alphabetically makes applying tags via the keyboard easier. For example, instead of using "revision_3" and "revision_4," try "3rev" and "4rev." The latter two tags are much easier to distinguish in an alphabetical list. Even better would be something like "Dec2007" and "April2008."
- **Limit the number of condition tags to the bare minimum.** FrameMaker can handle a large number of tags, but it's difficult to create distinctive formatting for a large number of tags. When content has multiple condition tags applied, FrameMaker displays a mixture of the colors. Use background colors to help visually identify the condition. The limiting factor in creating condition tags is not FrameMaker; it's the user's ability to manage and apply the condition tags correctly.

Tip In a department or workgroup, avoid "tag creep" by ensuring that condition tags are set up and maintained as part of the template. Generally, you do not want personal tags—such as JoesFirstDraft—to remain in your documents.

You may need condition tags to specify which conditional text shows in the output. If you show by condition, each condition tag is either shown or hidden. If you show by expression, you can specify more complicated processing (for example, you might specify that content with the Online condition tag applied is shown only when the Pro condition is also applied). See "Choosing the method for showing conditional text" on page 454 for more information.

- **Assign condition tags based on whether you select Show as Per Condition or Show as per Expression.** The best way to tag your document varies depending on how you show the conditional text:
    - If you show by condition, avoid assigning multiple condition tags to the same information. Multiple condition tags are very difficult to maintain.
    - If you show by expression, assigning layers of all relevant tags to the content makes it much easier to use expressions for filtering. See "Examples of expressions" on page 460 for details on how layering condition tags affects expressions.

When you begin applying condition tags, try to keep the tagging as simple as possible. Here are two recommendations:

- **If possible, tag entire paragraphs as conditional.** Conditional paragraphs are much easier to maintain than conditional sentences or words. If you must tag inside paragraphs, tag entire sentences. If your content is translated, sentence structure varies across languages, so tagging sentence fragments can cause translation problems.
- **Tag consistently.** If you are tagging sentences, be consistent in whether you tag the space before and after the conditional part. If you

> If you need to transfer information from the database, save the information to a tab-delimited text file; then import that file into the new application. You can set several options, such as the line ending format, when you export the file. Just about every database should be able to accept tab-delimited text.

are not consistent, when you hide the conditional text, you will have difficulties with missing spaces or two spaces in a row in your document. The same is true when tagging individual words or phrases.

Common condition tags

Organization of information is an important part of deciding on how to set up condition tags. Here are a few common uses for condition tags:

- **Different product levels.** If your product has multiple versions (such as Light, Standard, and Professional), you can break out the differences with conditional text.
- **Information that's under development.** Use conditional text to hide information that isn't appropriate for the current release of the documentation but needs to be included in the next version.
- **Review comments.** You can embed queries to reviewers or comments from reviewers using conditional text. This lets you hide the information and print out a clean draft. (Another alternative is to use a paragraph tag that's hard to miss—such as 14-point bold italic blue text.) Consider using FrameMaker's Track Text Edits feature, which makes use of sophisticated built-in conditional text tags. See "Tracking Changes" on page 55.
- **Platform differences.** Use tags named iOS, Mac, Windows, and Android to manage platform-specific information. Note: These are conditional text labels for FrameMaker; FrameMaker is only supported on Windows.
- **Audience levels.** For a training manual, you might need Student and Instructor versions. Use condition tags to manage that information.
- **Delivery medium.** Conditional text is an important component of any single-sourcing effort. Use it to identify information that is delivered in a particular medium; you might have Print, PDF, and Online condition tags.

Alternatives to conditional text

In addition to conditional text, a few other techniques are available that let you reuse content inside a document. Consider using these instead of—or in addition to—conditional text:

- **Variables.** For company names, product names, and other short pieces of text that are repeated throughout the document, consider using a variable. See Chapter 10, "Storing content in variables," for details.
- **Marker types.** You can use custom marker types to store many types of information. You can then include or exclude those markers when you generate the index. For example, you could use the Index marker type for most index entries and then use a custom RangeIndex marker for ranges (so that you can exclude them from the online version of the index).

Remember If a marker is hidden as a result of conditional text settings, the information in that marker is not available. This means, for example, that an index entry is not generated for a marker inside hidden conditional text. It also means that a cross-reference to hidden content will result in an unresolved cross-reference only until the hidden text is again displayed.

Chapter 25: Setting up conditional text

- **Graphics.** If you need to create two different versions of each graphic, one approach is to create two anchored frames and make them conditional. Perhaps a more efficient idea is to set up two sets of source graphic files and swap out the directories as needed. To make this work, you might have the same file names and three graphics directories—the directory in use, the directory for Version A and the directory for Version B. To swap files, you overwrite the content of the "in use" directory with the files from Version A or Version B.

Creating condition tags

When you have analyzed your content and figured out what condition tags you need, you're ready to create those tags.

To add a condition tag, follow these steps:

1. In the Conditional Text pod (see page 447), select the Create New Tag ( ) button button to display the Add/Edit Condition Tag pod.
2. Type a name in the Tag Name area.
3. In the Style drop-down list, select a style. If you do not want to use a style, select As Is.
4. In the Color drop-down list, select a color. If you do not want to use a color, select As Is.

   The colors listed are the same as in your color catalog. If the color you want isn't available, click the New Color… option to add a new color to the color definitions, and then select it from the drop-down list for your condition tag.

   The colors and styles you assign to your condition tags can be hidden before you print the document. See "Using condition indicators" on page 449.

5. Click the Add button to add the new the condition tag.

After you create the condition tag, it is listed in the Conditional Text pod, and you can apply it to content in your document.

Modifying condition tags

To edit a condition tag, follow these steps:

1. Select the condition you want to modify in the Conditional Text pod (see page 447).
2. Click the Edit ( ) button. The Add/Edit Condition Tag pod is displayed with the selected tag in the Tag Name text area.
3. Modify the tag name, style, color and background as needed, then click the Edit button.

   If you modify the tag name, a message lets you know that conditional expressions are affected. (Expressions are not updated when you change a tag name, so if you have used the condition tag in expressions, you will need to update the expressions. See "Building expressions" on page 458 for more information.)

Deleting a condition tag

You can delete condition tags from your document. When you delete that condition tag, you can either delete the text with that condition applied (along with the condition tag itself) or make the text unconditional. If text uses more than one condition tag, the deleted tag is removed from your file, but the text stays in your file with the remaining condition tags still applied.

To delete a condition tag, follow these steps:

1. Select the condition you want to delete in the Conditional Text pod (see page 447).
2. Click the Delete ( ) button. A dialog box warns you that deleting tags can affect expressions. Ignore the typo and click the OK button to continue.

3. If your document contains text that uses the specified tag (and no other conditional tags), a dialog box is displayed. Do one of the following:

   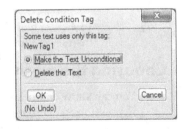

   - To delete the text that uses only this condition tag, click the Delete the Text radio button.
   - To remove the condition tag and make the text unconditional, click the Make the Text Unconditional radio button.

4. Click the OK button to delete the condition tag and modify the text.

Showing and hiding conditional text

After you have assigned condition tags in your document, you can show and hide each condition tag individually, or you can create expressions to show or hide sets of conditional text. FrameMaker dynamically repaginates your document whenever you show or hide information.

Caution If you hide a cross-reference marker and a cross-reference that points at the marker is still shown, the cross-reference becomes unresolved. To prevent this problem, be sure that the cross-reference and the cross-reference marker always have the same condition settings. You can also make all cross-reference markers unconditional. After changing show/hide settings, update the book so that the generated files are correct.

Showing All Conditional Text

To show or hide all the conditional text in your document, follow these steps:

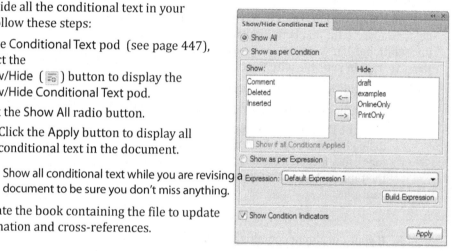

1. In the Conditional Text pod (see page 447), select the Show/Hide ( ) button to display the Show/Hide Conditional Text pod.
2. Click the Show All radio button.
    - Click the Apply button to display all conditional text in the document.

    Tip Show all conditional text while you are revising a document to be sure you don't miss anything.

3. Update the book containing the file to update pagination and cross-references.

Tip To show or hide all conditional text in a book, go to your book file, select all of the files in the book, then select **View > Show/Hide Conditional Text**.

Choosing the method for showing conditional text

You can show conditional text either by using the traditional Show as per Condition or by using expressions you create (Show as per Expression). The number of condition tags that you have in a document affects how you choose to show your conditional text, but the biggest factor is how many different groupings (or *dimensions*) of condition tags there are.

Here are some examples:

- One dimension: Each condition tag identifies a different product (for example, Light, Standard, and Pro).
- Two dimensions: The document has condition tags for both products (Light, Pro) and output types (Print, Online).
- Three dimensions: The document has condition tags for products, output types, and platforms (Windows, Mac, and UNIX).

Here is an example showing three dimensions of conditional text.

In general, Show as per Condition is easier if you have a single dimension of conditional text or only a few dimensions. Show as per Expression is probably worth the trouble if you have many dimensions to your conditional text. You can also use expressions to create new output from existing content that already has condition tags applied (although it won't be easy).

As you add other dimensions of conditional text, the number of condition tags you need increases, especially if you are showing by condition. For example, if you want to create a print version of the document for the Light product on the Windows platform, you may need extra conditions for LightPrint, LightWindows, and even LightPrintWindows. Using an expression, you would need only the Light, Print, and Windows condition tags. Even so, expressions are difficult to create, test, and maintain, so consider whether creating the extra tags and showing by condition may still be worthwhile.

Note   If you need to use complex combinations of condition tags, consider setting up a workflow in the structured interface of FrameMaker. Structured documents allow you to create versions based on attributes, which are easier to use and more powerful than expressions.

The number of tags you need to have (and the number you need to apply to the text) varies depending on the conditional text methods you use. The following sections provide information on implementing a Show as per Condition strategy.

Tags needed for Show as per Condition

If you are using the Show as per Condition method, you can apply layers of conditional tags to your content, or you can do the opposite and use layered tags. Both have advantages and disadvantages. Using layers of condition tags mean that a piece of content may have multiple simple tags applied. Using layered condition tags means that combinations of conditions are defined as separate condition tags. Using this method, content will contain only one tag.

Note   Discussions of conditional text become complicated quickly. To illustrate the concepts clearly, the examples that follow in this section assume that you always apply either a single condition tag or layers of condition tags. Of course, you could mix these methods to have some conditional text with a single tag and other chunks with layered tags.

Chapter 25: Setting up conditional text

Text tagged with a single condition tag.

When you apply condition tags to your content, the status line shows the names of the condition tags that are applied to the selected text. If you apply only one condition tag to a piece of content, the status line has enough space to display the name of that one condition tag. Also, the colors you define for your condition tags are easy to distinguish when you apply individual tags (as opposed to blended colors). Because it's easier to manage condition tags when you can see which ones are applied, the traditional advice has been to avoid layering condition tags.

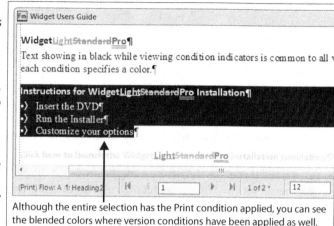

Although the entire selection has the Print condition applied, you can see the blended colors where version conditions have been applied as well.

The downside of avoiding layered condition tags is that you need more tags. For example, suppose that you have only the three dimensions discussed previously, and you don't want to assign multiple condition tags to any content. Assuming that there is no overlapping content within a dimension (for example, all content is unique among Light, Standard, and Pro, between Print and Online, and among Win, Mac, and UNIX), then you need 32 condition tags, as shown in the illustration.

Light			
Standard			
Pro			
Print	LightPrint	LightOnline	
Online	StandardPrint	StandardOnline	
	ProPrint	ProOnline	
Win	LightPrintWin	LightPrintMac	LightPrintUNIX
Mac	StandardPrintWin	StandardPrintMac	StandardPrintUNIX
UNIX	ProPrintWin	ProPrintMac	ProPrintUNIX
	LightOnlineWin	LightOnlineMac	LightOnlineUNIX
	StandardOnlineWin	StandardOnlineMac	StandardOnlineUNIX
	ProOnlineWin	ProOnlineMac	ProOnlineUNIX

If there is some text that is displayed across a dimension (for example, some text is displayed in both Light and Standard, you need tags like LightStandard, LightStandardPrint, LightStandardPrintWin, and so on), which means you need even more tags: up to 47 instead of 32. If you need other dimensions, the number of required condition tags is far more.

Text Tagged with Layered Condition Tags. Another choice is to apply layers of condition tags. In this case, you only need 18 tags, which doesn't seem bad in comparison to 32 or 47 tags. The condition tags are all hybrid tags such LightPrintWin, as shown in this illustration.

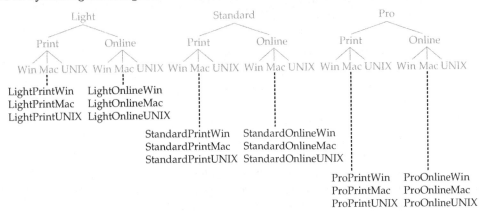

In the example on page 456, text that appeared in all *printed* outputs—but not in the *online* outputs—had a single condition tag (Print) applied. If you use layered condition tags, you must apply all nine condition tags that are used for the different printed versions of the document.

If you have a piece of text that is displayed in every version except, for example, LightPrintMac, you must apply all 17 of the other condition tags to exclude that one condition tag. To make sure you include all needed condition tags, display the Conditional Text pod instead of using the keyboard shortcuts like CTRL+4. You can't see that many condition tags displayed in the status line. Also, after you apply a few condition tags, the blended colors will look muddy, so you can't rely on color to reveal whether you have applied all the tags.

Tags needed for Show as per Expression

If you are using the Show as per Expression method to create output for the example shown, you again have the choice of applying either single or layered condition tags. In both cases, you need only eight condition tags.

| Light |
| Standard |
| Pro |
| Print |
| Online |
| Win |
| Mac |
| UNIX |

Note   Using expressions isn't necessarily easier than creating and applying numerous tags. "Using conditional text expressions" on page 458 provides detailed examples of expressions required when you apply single condition tags to content or layers of condition tags.

Showing or hiding conditional text as per condition

When you use Show as per Condition, you show or hide conditional text by selecting which condition tags to display. If you show a specific condition tag, all text marked with that condition tag is displayed, regardless of what other condition tags are applied to the text. For example, suppose that you have Windows, Mac, Print, and Online condition tags in your document. You want to create a Windows/Print version of the document, so you show the Windows and Print condition tags. The problem is that text that has the Windows condition tag applied could be tagged with any of the following combinations: Windows only, Windows and Print, or Windows and Online. If you show the Windows condition, then you could potentially display text that is intended for both the print and online versions. To solve the problem, create and apply additional condition tags as needed.

To show or hide conditional text by condition, follow these steps:

1. Select **Special > Conditional Text > Show/Hide Conditional Text** to display the Show/Hide Conditional Text pod.

2. To change which condition tags are displayed, do any of the following:

   - To show a condition tag, select it, then click the left arrow button to move it to the Show list on the left.

   - To hide a condition tag, select it, then click the right arrow button to move it to the Hide list on the right.

   - Select the Show If All Conditions Applied checkbox if you want text to show (as if it were unconditional) when every condition tag is applied to the text—even if you have moved all of the condition tags into the Hide list on the right.

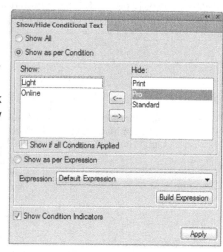

3. Click the Apply button. The document displays the condition tags you've specified.

4. Update any books containing the file to update pagination and cross-references.

Using conditional text expressions

To show or hide text using an expression, you must first create the expression from available condition tags using the Build Expression dialog.

Building expressions

Follow this example to create a conditional text expression:

1. In the Conditional Text pod (see page 447), select the Show/Hide Conditional Text ( ) button to display the Show/Hide Conditional Text pod.

2. In the Show/Hide Conditional Text pod, click the Build Expression button to display the Manage Conditional Expression dialog.

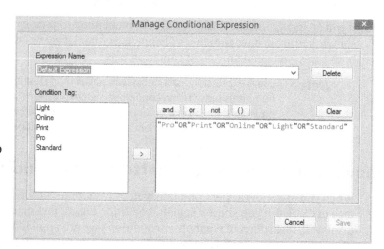

This dialog box lists the defined condition tags on the left. The operators (AND, OR, NOT) that let you combine condition tags are in the center. The result of your expression choices is displayed on the right.

Note Within the expression, condition tags are enclosed in straight double quotes, which ensures that condition tags with spaces in their names are processed correctly.

When you build expressions, you specify an operator or a condition tag, then provide additional operators and tags to create your expression.

3. To start a new expression, click the Clear button. (To edit an existing expression, see "Modifying expressions" on page 460.)
4. To build expressions, specify an operator or a condition tag and then provide additional operators and tags to create your expression:
   - To add a condition tag to the end of the expression, click the condition tag in the Condition Tag list on the left, then click the Arrow (>) button. Double-clicking the condition tag also works.
   - To add operators to the end of the expression, click operator buttons.

   The following table shows the available operators:

Operator	Usage
AND	The AND operator connects two or more two condition tags; both condition tags must be applied to the text for it to be displayed. For example, "Light" AND "Print" means that the content is shown when it has *both* the Light and Print condition tags applied. (Text that has only one of the condition tags applied is hidden.) The AND operator can connect more than two condition tags; for example, "Light" AND "Print" AND "Windows" means that all three condition tags must be applied to the text for it to be shown.
OR	Like the AND operator, the OR operator connects two condition tags; if either of the condition tags is applied, the text is displayed. For example, "Light" OR "Print" means that any content that has either of these condition tags applied is shown. OR is the default operator, so specifying only "Windows" means that all the text with the Windows condition tag is displayed. (Note that using Show as per Condition produces the same results as specifying an expression using only the OR operator.)
NOT	NOT excludes the condition tag that immediately follows it from what is shown. For example, NOT "Online" means that all conditional text that does not have the Online condition tag applied is displayed.
ANDNOT	Click the AND button and then click the NOT button to create the operator ANDNOT. The ANDNOT operator connects two operators to limit what is shown. For example, "Light" ANDNOT "Mac" shows all the content with the Light condition tag applied except the content that has both the Light and Mac condition tags applied.

5. Click the Set button.

Tip Expressions can be challenging. Create test documents that use all your defined condition tags and test creating all the output variations you will need with the expressions you create.

Applying conditional build expressions

To show or hide conditional text by creating an expression, follow these steps:

1. In the Show/Hide Conditional Text pod (see page 458) select from the available expressions.
   - For information about how to build expressions, see "Building expressions" on page 458.
2. Click the Apply button. The document displays the conditional text you specified.
3. Update any books containing the file to update pagination and cross-references.

Modifying expressions

To edit an existing expression, display the expression in the Build Expression dialog box and type your changes. You can use the buttons in the Build Expression dialog box to add the condition tags or operators only to the end of the expression.

You can type spaces between the condition tags and operators to make it easier to read the expressions (for example, "Light"AND"Online" produces the same result as "Light" AND "Online").

Often, it is easier to keep track of your expressions if you paste them from the Build Expression dialog box into a text document where you can add notes that explain how the expression is used. If you modify the expression and then paste text from that text document into the Manage Conditional Expression dialog, make sure that there is no space at the beginning of the expression or FrameMaker will reject the expression as invalid. Also, line breaks aren't accepted, so don't paste those in.

Be sure to use straight double quotes around the condition tag names. These are inserted automatically when you use the Build Expression dialog box to build the expression, but be sure you don't paste any curly quotes into the expression.

You cannot add parentheses to the expressions, so test thoroughly to make sure you are getting the results you expect.

Deleting expressions

As you are testing and creating new expressions, the list of expressions grows quickly. FrameMaker includes at least one expression by default (an OR expression including all tags and is equivalent to Show All). If you rename any condition tags, FrameMaker also adds some expressions to the list of expressions, but it doesn't update the expressions you have already created. If it's difficult to keep track of the expressions, delete the extraneous ones.

To delete expressions, you must save the file as MIF, search on Condition, find the expressions, and delete the ones you don't need. See Appendix F, "Maker Interchange Format." for information about editing MIF files.

Examples of expressions

As described earlier in the chapter, you can create condition tags for the different versions you want to produce. For the following examples, there are two dimensions of condition tags: product (Light and Pro) and output type (Print and Online).

Expressions are powerful, and you can use them to include or exclude complex combinations of condition tags. Depending on how the condition tags are applied in the document, the expressions you need to create can be very simple or very complex. In previous versions of FrameMaker, it was good practice to avoid using many layers of condition tags, but using multiple layers can greatly simplify the expressions you need. The following sections show two ways to apply the four condition tags (Light, Pro, Print, and Online) and to build the resulting expressions that produce identical results for Light/Print, Light/Online, Pro/Print, and Pro/Online output.

Example of layered condition tags and resulting expressions

For this example, there are four condition tags: Light and Pro are the tags for one dimension (product), and Print and Online form another dimension (output type). If you always layer at least two condition tags (one from each dimension) on all conditional text, the expressions you need are straightforward.

In the following sample document, none of the conditional text has fewer than two condition tags applied (Figure 25-1). For this example, some content applies only to the Light or Pro product. The print version includes pictures and figure captions that aren't included in the online version. The online version includes a list of related topics that doesn't appear in print.

Figure 25-1. All conditional text has condition tags from each dimension applied.

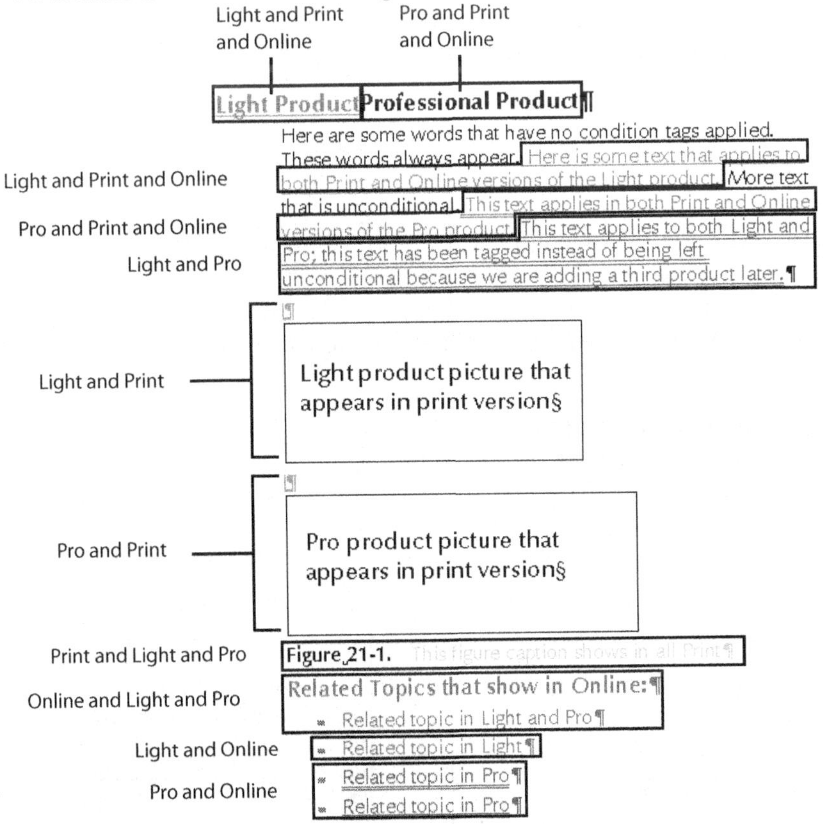

If you apply condition tags from each possible dimension to all the conditional text, you need to use only the AND operator when you build the expressions. For example, to create the Light product's print version, create an expression that shows all the text that has both the Light and Print condition tags applied:

"Light" AND "Print"

Here is an example showing the resulting output for Light/Print.

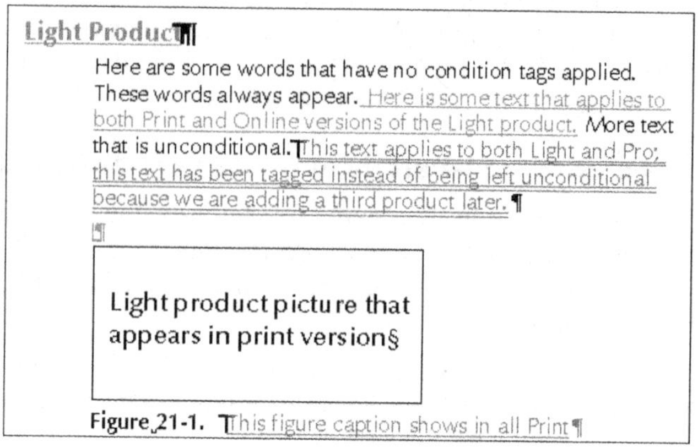
Light Product¶

Here are some words that have no condition tags applied. These words always appear. Here is some text that applies to both Print and Online versions of the Light product. More text that is unconditional. This text applies to both Light and Pro; this text has been tagged instead of being left unconditional because we are adding a third product later.¶

¶

Light product picture that appears in print version§

Figure 21-1. This figure caption shows in all Print¶

The conditional text can have condition tags other than Light and Print applied and that won't affect the output. The important thing is to make sure that both Light and Print are applied to all the text you want to show in this version.

The expressions for the other versions are just as simple. To create the online version for the Light product, build the following expression:

"Light" AND "Online"

Here is an example showing the resulting output for Light/Online.

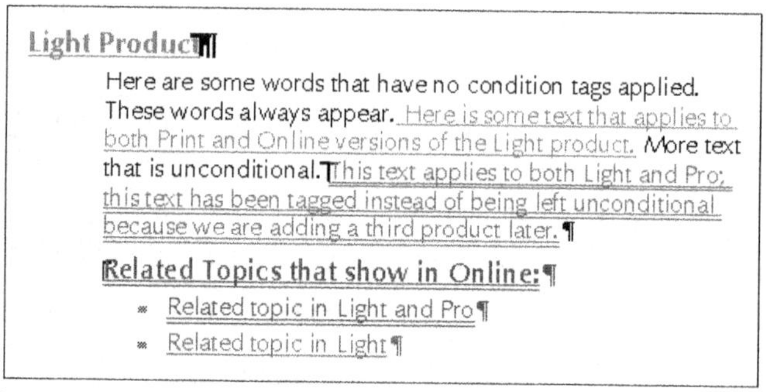
Light Product¶

Here are some words that have no condition tags applied. These words always appear. Here is some text that applies to both Print and Online versions of the Light product. More text that is unconditional. This text applies to both Light and Pro; this text has been tagged instead of being left unconditional because we are adding a third product later.¶

Related Topics that show in Online:¶

- Related topic in Light and Pro¶
- Related topic in Light¶

To create the print version for the Professional product, build the following expression:

"Pro" AND "Print"

Here is an example showing the resulting output for Pro/Print.

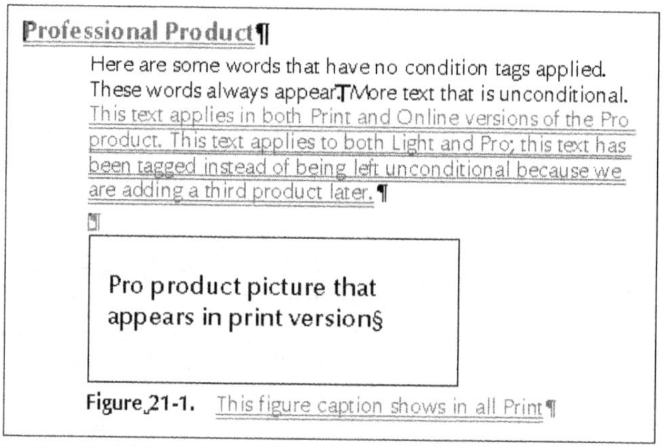
Professional Product¶

Here are some words that have no condition tags applied. These words always appear. More text that is unconditional. This text applies in both Print and Online versions of the Pro product. This text applies to both Light and Pro; this text has been tagged instead of being left unconditional because we are adding a third product later.¶

¶

Pro product picture that appears in print version§

Figure 21-1. This figure caption shows in all Print¶

To create the print version for the Professional product, build the following expression:

"Pro" AND "Online"

The following example shows the resulting output for Pro/Online.

> *Remember* Applying multiple condition tags to all conditional text isn't required, but it allows you to use simpler expressions.

Example of traditional condition tagging and resulting expressions

When possible, try to avoid layering condition tags. For example, if a piece of content appeared in all print versions, writers would apply the Print condition tag. You can still write expressions to process content that is tagged this way, but the expressions can become extremely complicated, because the expression must exclude any content that has the Print condition tag applied but doesn't belong in the output you are trying to create.

The sample document shown in Figure 25-2 is similar to Figure 25-1 on page 461, but the conditional text doesn't always have layers. The conditional text sometimes has only a single condition tag. Also, the conditional text may not include condition tags from both the product dimension (Light or Pro) and the output type dimension (Print or Online).

Figure 25-2. Much of the text in this example does not include condition tags from all dimensions.

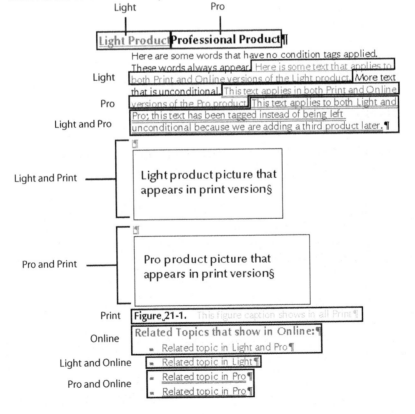

Chapter 25: Setting up conditional text

This following expression produces the same Light/Print output shown previously:

"Light" ANDNOT "Online" OR "Light" AND "Print" OR "Print" ANDNOT "Pro"

Logically, you can break the expression at the OR operators to separate the expression into understandable pieces (but omit the line breaks when you build the expression).

Here is an explanation of what this expression is doing:

"Light" ANDNOT "Online"	Shows all the conditional text that has the Light condition tag applied *except* for the content that has both the Light and Online condition tags applied. Because this expression produces Light/Print version, we want to exclude any content that is intended only for the Light/Online version.
OR "Light" AND "Print"	Shows conditional text tagged with both Light and Print. This part of the expression might not be needed, but if there are pieces that have Light, Print, and Online condition tags all three applied, this line includes those.
OR "Print" ANDNOT "Pro"	Shows conditional text with only the Print condition tag applied, but excludes content that has both the Print and Pro condition tags applied. If you apply only the Print condition tag to content, you are logically marking that content as being Print, but not Light, not Pro, and not Online. You need this line to process any content that is marked with only the Print condition tag.

The expressions that create the other outputs are similar:

- **Light/Online.** This expression produces the output shown on page 462:

    "Light" ANDNOT "Print"
    OR "Light "AND" Online"
    OR "Online "ANDNOT" Pro"

- **Professional/Print.** This expression produces the output shown on page 462:

    "Pro" ANDNOT "Online"
    OR "Pro" AND "Print"
    OR "Print" ANDNOT "Light"

- **Professional/Online.** This expression produces the output shown on page 463:

    "Pro" ANDNOT "Print"
    OR"Pro" AND "Online"
    OR"Online" ANDNOT "Light"

As you can see, conditional text lets you maintain multiple versions within a single document, but implementing conditional text can get messy. If you need to use conditional text extensively, consider setting up structured documents instead.

Chapter 26: Automation with ExtendScript

ExtendScript is a scripting environment built into FrameMaker 11. Adobe added ExtendScript support to FrameMaker starting with version 10. ExtendScript is essentially Adobe's version of the JavaScript language for application automation. It is extended to include features that go beyond the core JavaScript language. ExtendScript is also available in other Adobe applications including RoboHelp, InDesign, Illustrator, Photoshop, and Acrobat.

Here's a look at what's in this chapter:

Advantages of using ExtendScript	.465
A brief history of FrameMaker automation	.465
Getting started with scripting	.466
Interacting with FrameMaker ExtendScript	.466
Using the Script Library pod	.467
Creating and editing scripts	.469
Creating an alert box	.470
Automating text entry	.470
Objects, Properties, and Methods	.472
Finding the correct objects, properties, and methods	.473
Principles of successful script writing	.476
Start small	.477
Use functions	.477
Working with selections	.478

Advantages of using ExtendScript

Scripting is all about automation and making things faster. There are three major reasons to use scripting in FrameMaker:

- **Automate tedious, repetitive tasks.** For example, you could use a script to perform find/change operations on marker text.
- **Extend FrameMaker functionality by adding new features.** For example, you could have a script automatically set File Info fields based on document content.
- **Make impractical tasks practical.** Later in the chapter, we will develop a script that adds "highlighting" to table cells that contain certain part numbers.

A brief history of FrameMaker automation

FrameMaker has long provided options to automate and extend its functionality. The Frame Developers Kit (FDK) is a free download from Adobe that allows you to write API (Application Programming Interface) clients to extend FrameMaker. The FDK is a series of C/C++ libraries and header files that allow you to build compiled "plug-ins" for FrameMaker. These plug-ins are .dll files that are installed and activated when you start FrameMaker. Some of FrameMaker's core features are actually implemented as API clients, including the **Table > Sort** command and the **Format > Page Layout > Apply Master Pages** command.

In 1997, Finite Matters Ltd. released FrameScript for FrameMaker. FrameScript is a scripting environment for FrameMaker similar to ExtendScript. Its proprietary language is similar to Visual

Basic and JavaScript. FrameScript is an excellent, stable product, but unlike the FDK and ExtendScript, it is not free.

Getting started with scripting

While scripting with ExtendScript is generally more "user-friendly" than using the FDK, it is important to realize that you are still writing computer programs. There is no facility for "recording" actions and playing them back as you can with Microsoft's Visual Basic for Applications. Even so, you can learn to write scripts with little or no programming background. There are three basic requirements for successfully writing scripts:

- **Understand the ExtendScript language "syntax."** Computer language syntax is similar to the spelling and grammar rules in a spoken language, but is more rigid.
- **Understand FrameMaker's "object model."** This is how objects such as documents, paragraphs, and formats are represented in ExtendScript.
- **Have the ability to develop "algorithms" for solving FrameMaker problems.** An algorithm is a basic approach for solving a programming problem and is usually independent of the language you are using.

Interacting with FrameMaker ExtendScript

The ExtendScript environment is integrated with FrameMaker 11. You can access the scripting options with the **File > Script** menu item.

Here is a description of the Script submenu commands.

- **Run**. Displays the Choose Script dialog box in which you can select and run an existing script.
- **New Script**. Launches the ExtendScript Toolkit in which you can create a new script. The ExtendScript Toolkit is a standalone application that is installed with any Adobe application that supports ExtendScript. You can run scripts directly from the ExtendScript Toolkit.
- **Catalog**. This command opens the Script Library pod, in which you can manage and run your scripts.
- **History**. Lists recently run scripts.

Using the Script Library pod

Select **File > Script > Catalog** to display the Script Library pod. The Script Library pod provides three views: Favourite, for frequently used scripts; Autorun, for scripts that must run when starting FrameMaker; and Registered, for scripts that respond to FrameMaker events. There are some common elements to each view:

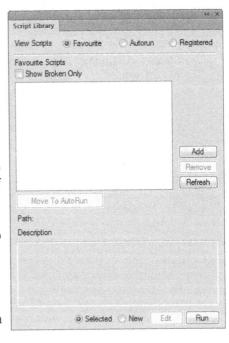

- **Show Broken Only** checkbox. Check this to see only scripts that cannot be located. Broken scripts are displayed with an asterisk at the end of their names.
- **Edit** button. Select a script, then click the Edit button to open the selected script in the ExtendScript Toolkit. If the **New** radio button is selected, the Edit button will be disabled.
- **Run** button. Select a script, then click the Run button to execute that script. To execute a script that is not in the list, select the New radio button, then click Run. The Choose Script dialog box prompts you to select a script.
- **Refresh** button. Updates the list if scripts are moved on your computer.
- **Path**. Shows the absolute path to the selected script.
- **Description**. Shows a JavaScript block comment if one exists at the top of the selected script. Block comments are discussed later in the chapter.

Favourite View

The **Favourite** (current spelling in FrameMaker at time of publishing) view lets you store frequently used scripts so you can easily run and edit them from the Script Library pod.

Click the **Add** button to select scripts from your computer and add them to the list.

To remove a script from the list, select it and click the **Remove** button.

Some scripts need to run when FrameMaker starts. For example, a script that adds menus or commands to FrameMaker must run when FrameMaker starts. If you have a script like this in the Favourite view, select it and click the **Move To AutoRun** button. The script is copied from its current location to the startup folder. Autorun scripts are shown in the Autorun view of the pod.

Autorun view

Autorun scripts run automatically when you start FrameMaker. Autorun scripts are stored in a startup folder located at:

C:\Users\<UserName>\AppData\Roaming\Adobe\FrameMaker\11\startup

where <UserName> is your Windows login name.

Note The exact path may vary, depending on installation choices and whether you installed FrameMaker as part of the Adobe Technical Communication Suite.

There are three ways to add a script to the Autorun view of the Script Library pod:

- Select a script in the Favorite view and click the **Move To AutoRun** button.
- Click the **Add** button in the Autorun view and select a script.
- Use Windows to copy a script to the startup folder. When you click the **Refresh** button, the script will show up in the list.

To remove a script from the list and delete it from the startup folder, click the **Delete** button.

Caution Make sure you have a backup copy of your script. Clicking the Delete button removes the script from the list and from the file system.

Registered view

Some scripts respond to FrameMaker *events*. Events are FrameMaker actions such as saving a document, updating a book, or copying text to the clipboard. You can write scripts that are triggered by one or more of these events. For example, RunaroundNone is an ExtendScript script (used for the production of this book and commercially available at www.frameexpert.com) that sets an imported graphic's Runaround property to Don't Run Around as soon as the graphic is imported or pasted from the clipboard. This script automatically responds to FrameMaker's import file and paste events. These types of scripts are known as notification scripts.

Notification scripts use a Notification function to tell ExtendScript to "listen" for particular FrameMaker events to take place. For example, the following command:

Notification(Constants.FA_Note_PostImport, true);

Creating and editing scripts

tells ExtendScript to notify the script after the user selects **File > Import > File** and imports a graphic or text. If a graphic was imported, the script sets the graphic's Runaround property to Don't Run Around.

When you run a script with one or more Notification commands, it is listed in the **Registered** view of the Script Library pod.

To cancel a script's notifications, select the script in the list and click the **Unregister** button. This does not delete or move the script from its current location—but it cancels its notifications. If the script is an Autorun script, it will run next time you start FrameMaker and reappear in the Registered list.

Creating and editing scripts

The ExtendScript Toolkit application is included with your FrameMaker installation. This standalone application lets you write and edit your scripts. To launch the Toolkit from FrameMaker, select **File > Script > New Script**.

Because the ExtendScript Toolkit is a standalone application, you can also start it from the Windows Start menu. A third option is to select a script in any of the Script Library pod lists and click the Edit button at the bottom of the pod.

The Toolkit is a full-featured ExtendScript editor and debugger with a lot of useful features. You can get comprehensive information by choosing **Help > JavaScript Tools Guide**.

You can use the ExtendScript Toolkit with other Adobe applications that you may have installed. As a result, you need to tell the Toolkit application which Adobe application you want to "target" with each script. The target application is indicated by the drop-down list in the upper left corner of each document window, next to the chain-link icon. In this screenshot, you can see that Adobe FrameMaker 11 is the target application. This is the default when you launch the Toolkit from the **File > Script > New Script** command. The screenshot also shows some of the other targets you may see if you have other Adobe applications installed.

469

When you launch the Toolkit directly from Windows, the default target is the ExtendScript Toolkit. You can change the target from the drop-down list. To force a script to always target FrameMaker, start the script with the following line:

#target framemaker

Creating an alert box

Type the following in the Toolkit window and click the **Run** button ( ) on the toolbar (or press F5).

#target framemaker
alert ('Hello world!');

You will see an alert dialog box appear on-screen:

Automating text entry

Open a new, untitled FrameMaker document, type the following code in the Toolkit window, and run it.

```
#target framemaker
// Make a variable for the active document.
var doc = app.ActiveDoc;
// Make a variable for the first paragraph in the flow.
var pgf = doc.MainFlowInDoc.FirstTextFrameInFlow.FirstPgf;
// Make a text location at the beginning of the paragraph.
var textLoc = New TextLoc (pgf, 0);
// Add text to the text location.
doc.AddText(textLoc,"Hello world!");
// Get the Title paragraph format.
var pgfFmt = doc.GetNamedPgfFmt("Title");
// If the Title paragraph format exists, apply it to the paragraph.
if (pgfFmt.ObjectValid()) {
    pgf.SetProps(pgfFmt.GetProps());
}
```

Note Lines beginning with // are comments. They help you to document or annotate your script, and have no effect on the code within the script itself.

You should see the Hello world! text in the document's first paragraph, and the paragraph should have the Title paragraph format applied to it. If there are errors in the code, the script stops and the offending line is highlighted in the ExtendScript window. The cause of the error is shown in the lower left corner of the script window. Click the Stop button ( ) on the toolbar to stop the script. Correct any errors and run the script again.

Creating and editing scripts

Here is an explanation of how the code works:

#target framemaker

This command directs ExtendScript to execute the script in FrameMaker and not another application.

// Make a variable for the active document.
var doc = app.ActiveDoc;

The two forward slashes identify the first line as a comment. Comments are ignored by ExtendScript, but give you a way to document your code. Commenting your code helps you and others remember what your code does. Comments starting with two forward slashes are called *line* comments because anything on the line after the forward slashes is ignored by ExtendScript.

ExtendScript also supports *block* comments, which are delimited by /* and */ characters. Block comments allow to have comments that span multiple lines, like this:

/* This is a block comment

It has multiple lines, which are comments.

*/

Note   A block comment at the top of a script in the Script Library pod will show in the Description area of the pod.

The second line sets a *variable* to FrameMaker's active document (**doc**). A variable in ExtendScript (and in other computer languages) is a "placeholder" that holds an object or property value. When you create a variable, you use the **var** keyword and a name to identify the variable. You can choose any name for your variable as long as it starts with a letter or an underscore and is not the same as an ExtendScript or JavaScript reserved word. It is best to use short, descriptive names so that it is evident what the variable represents.

// Make a variable for the first paragraph in the flow.
var pgf = doc.MainFlowInDoc.FirstTextFrameInFlow.FirstPgf;

Here we are setting another variable for the first paragraph in the document's main flow (**pgf**). An important principle here is that we are using "dot notation" to navigate FrameMaker's object model. The object model is how FrameMaker represents its internal objects. FrameMaker's objects are organized in a somewhat intuitive hierarchy, with **app** as the top-most object. The **doc** object contains flows (**MainFlowInDoc**); a flow contains text frames (**FirstTextFrameInFlow**); and a text frame contains paragraphs (**FirstPgf**). You use the dot notation to move down, and sometimes up, the hierarchy.

There is really no limit to how far you can navigate the object model with dot notation. In our example script, we could have skipped the doc variable and done this:

var pgf = app.ActiveDoc.MainFlowInDoc.FirstTextFrameInFlow.FirstPgf;

Or, we could have used more variables and only moved a single step on each line:

var doc = app.ActiveDoc;
var flow = doc.MainFlowInDoc;
var textFrame = flow.FirstTextFrameInFlow;
var pgf = textFrame.FirstPgf;

How do you know which approach is best? In general, it is better if you only declare variables that you need to access directly in your code. In our example, we do not need to capture the flow and text frame in variables, because we are not going to access them elsewhere in the script. It is customary to capture the document object, because we often perform multiple tasks on a document in a script.

Referring to our earlier example script, this command creates a *text location*, or position, at the beginning of the paragraph:

// Make a text location at the beginning of the paragraph.
var textLoc = New TextLoc (pgf, 0);

This command adds the text to the text location:

// Add text to the text location.
doc.AddText(textLoc,"Hello world!");

These lines get theTitle paragraph format object and, if it exists, applies its properties to the paragraph object:

// Get the Title paragraph format.

var pgfFmt = doc.GetNamedPgfFmt("Title");

// If the Title paragraph format exists, apply it to the paragraph.
if (pgfFmt.ObjectValid()) {
   pgf.SetProps(pgfFmt.GetProps());
}

Objects, Properties, and Methods

When scripting, you are mainly working with three items: *objects*, *properties*, and *methods*.

An object is a FrameMaker object, such as a document, paragraph, or text frame. Your scripts will be accessing and manipulating objects. Make an effort to know the proper names of FrameMaker objects, as you will often need them in your scripts. For example, **Doc**, **Pgf**, **TextFrame** are the internal names for documents, paragraphs, and text frames, respectively.

Objects have one or more properties associated with them. A property is a characteristic of an object. For example, a **Pgf** object has a **Name** property, which indicates the paragraph format that is applied to the paragraph. Put your cursor in the first paragraph of the document, type the following in an empty script and run it:

#target framemaker

var doc = app.ActiveDoc;

// Get the paragraph at the insertion point.

var pgf = doc.TextSelection.beg.obj;

alert (pgf.Name);

In this instance, the result shows that the **Pgf** object represented by the **pgf** variable has a **Name** property of "Title." Most of your scripts will change the properties of objects.

Many properties are actually objects that have properties of their own. For example, a **Pgf**'s **Color** property is actually a **Color** object. To get the Color name, we have to get the Color's **Name** property.

```
#target framemaker
var doc = app.ActiveDoc;
// Get the paragraph at the insertion point.
var pgf = doc.TextSelection.beg.obj;
alert (pgf.Color.Name);
```

Methods perform an action on an object. Previously, we used a method to add text to the document:

```
// Add text to the text location.
doc.AddText(textLoc,"Hello world!");
```

Typically, methods are attached to objects. In this example, the **Doc** object uses an **AddText** method to add text to the document. Methods require trailing parentheses, where you add any necessary parameters. A parameter is a specific piece of information that a affects the function of the method. The **AddText** method requires two parameters: a text location (**TextLoc**), and a string containing the text to be added. Some methods don't require any parameters, but you still must add the trailing parentheses. For example:

```
// Refresh the document's display.
doc.Redisplay();
```

ExtendScript also has "global" methods that are not attached to objects. For example, to create a new document, you use the global **NewDoc** method:

```
var docPath = "E:\My Documents\FrontMatterTemplate.fm";
var doc =  NewDoc(docPath,false);
```

Finding the correct objects, properties, and methods

A major part of learning to write scripts is figuring out the appropriate objects, properties, and methods that you need to complete the task at hand. The Adobe ExtendScript documentation helps you learn your way around the FrameMaker object model. You can access the latest FrameMaker ExtendScript information at Adobe's web site: www.adobe.com/devnet/framemaker.html.

The ExtendScript Toolkit has an Object Model Viewer that allows you to see the object model of installed Adobe applications that support ExtendScript. Select **Help > Object Model Viewer** (or press **F1**) to display it.

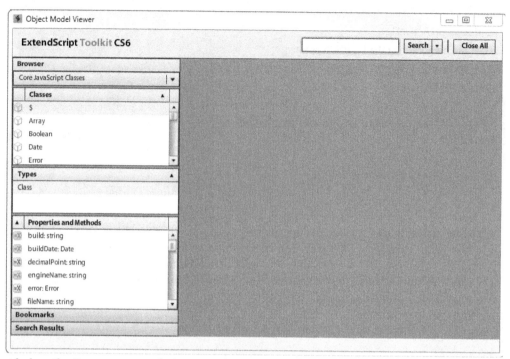

Just below the Browser control, there is a drop-down list where you select the application that you want to browse. Select Adobe FrameMaker 11 Object Model from the list. FrameMaker's classes will appear in the Classes section of the browser.

Note   We have been referring to classes as objects and will continue to do so because, conceptually, objects are a little less abstract.

When you click on an item in the Classes section, information about that object is displayed in the Types and Properties and Methods sections below it. This information helps you navigate the object model in your scripts. Here is an exercise using the previous example script which illustrates the Object Model Viewer.

1. In the ExtendScript Toolkit, select **Help > Object Model Viewer**
2. Select Adobe FrameMaker 11 Object Model from the Browser drop-down list.
3. Click the app object in the Classes section.
4. Locate the ActiveDoc property in the Properties and Methods section and click it. Notice that its Data Type is Doc.
5. Click the Doc link to highlight the Doc object in the Classes section.
6. Locate the MainFlowInDoc property in the Properties and Methods section and click it. Notice that this is a Flow object.
7. Click the Flow link to highlight the Flow object in the Classes section.
8. Locate the FirstTextFrameInFlow property in the Properties and Methods section and click it. Notice that this is a TextFrame object.
9. Click the TextFrame link and the TextFrame object will highlight in the Classes section.

10. Locate the FirstPgf property in the Properties and Methods section. Notice that this is a Pgf object.

When you are done, the Object Model Viewer should look like this:

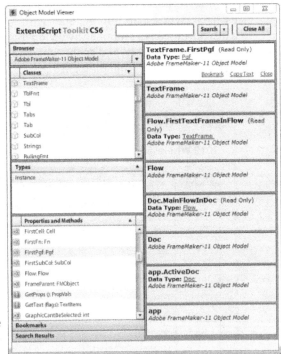

Admittedly, it's easier to work backward from an example script. In practice, you will use the Object Model Viewer to help you figure out the code as you write it. The Object Model Viewer will help you with the correct spelling and hierarchy as you write your scripts. As you become more familiar with the object model, it will get easier to work with.

The Object Model Viewer is convenient because it is built into the ExtendScript Toolkit. However, it is usually easier to use the *Adobe FrameMaker Scripting Guide*, which is a PDF file available from adobe.com. This document is light on examples, so you should also download and install the Frame Developers Kit (FDK) from adobe.com. The FDK documentation contains more example code. The examples will help you develop your solutions, and you can translate the example code from the C language to ExtendScript by using the ExtendScript documentation.

Scripting Resources

Because this is an introduction to ExtendScript, we won't go into a lot of detail about the ExtendScript syntax. You can use the Object Model Viewer, ExtendScript documentation, and example scripts to help you with the language syntax. Since ExtendScript is essentially JavaScript, you can also use non-FrameMaker resources for learning the language.

Chapter 26: Automation with ExtendScript

Principles of successful script writing

To illustrate basic principles of successful script writing, we will finish the chapter with an example script. Our client has a parts list in a FrameMaker table. They are going to discontinue each part whose part number ends with "-10". They want to "highlight" each table cell containing a discontinued part number by giving it a yellow background. The "-10" parts are scattered throughout the table, and as the table is almost 50 pages long we don't want to do this by hand. Here is an excerpt from the table:

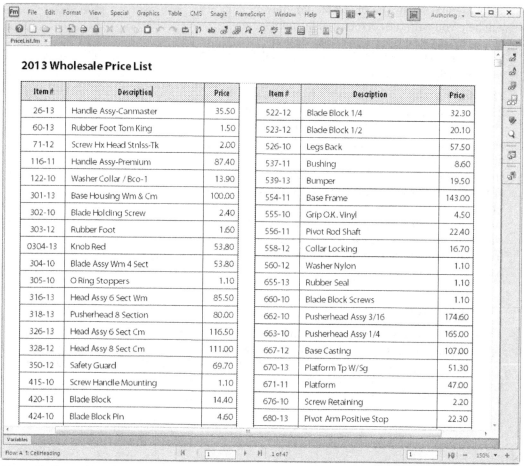

Our main focus here is on good principles for developing scripting solutions. If you start with a solid foundation, you will find it much easier to apply your knowledge as you learn the language.

Start small

The first principle of script writing success is to start with the smallest possible task and work your way outward to the complete script. In our example, we want to highlight table cells that contain a particular part number pattern. Here is the suggested approach to writing the script:

- Identify the desired pattern in a single part number, returning true or false
- Highlight a single table cell by making its color yellow
- Navigate through all of the cells in a table, and identify cells needing highlighting
- Navigate through all of the tables in a document, and identify tables that should be processed
- Navigate through all of the documents in a book

The list goes from specific tasks that are unique to this script (the first two) to more general tasks that could be reused in other scripts (the last two). The middle task is a hybrid; navigating table cells is a general task, but identifying which cells to process is determined by the requirements of this particular script.

There are two advantages to starting with the smallest tasks. First, these are the specific tasks that central to the overall solution. Once these are done the bulk of the work is finished. The rest of the code is usually boilerplate code that can be copied and reused from other scripts. Second, isolating the smallest tasks allows us to test and troubleshoot the code in the smallest possible context, giving us confidence that it will work in the overall script.

Use functions

The second principle is to split tasks into functions. A function is a reusable block of code that can be called from anywhere in your code. Most beginning script writers write their scripts in one long chunk of code. As your scripts get more involved, they become more difficult to understand and troubleshoot. Functions are useful because they isolate scripting tasks into small units that are easier to test, troubleshoot, and reuse. Understanding and using functions right from the beginning will give you a head start on being a productive script writer.

Here is how we can organize our example script using functions. We will just show the function names with comments on what each function will do.

```
function isSpecialCell (cell) {
    // Determines if the cell contains a special part number.
    // Returns true or false.
}

function highlightCell (cell, doc) {
    // Colors the table cell with a yellow background.
}

function navigatePartNumberCells (tbl, doc) {
    // Navigates the part number cells, and check for cells to highlight.
}

function processTables (doc) {
    // Processes all of the tables in the document, and test for parts tables.
}
```

Chapter 26: Automation with ExtendScript

```
function processBook (book) {
    // Processes all of the documents in the book.
}
```

In addition to these functions, we will have a few "utility" functions that perform common tasks. For example, we will need a function that will get all of the text in a table cell, and we will need functions for opening, saving, and closing each document in a book.

Working with selections

When starting with the smallest possible task or tasks, it is useful to work with selections in your document. You use code to select an object in the document, and then you develop code to modify the selected object. For example, if your script is designed to modify paragraphs in the document, you can use the currently selected paragraph to develop and test your code. Here is how to get the paragraph containing the insertion point:

#target framemaker

var doc = app.ActiveDoc;

// Get the paragraph containing the insertion point.

var pgf = doc.TextSelection.beg.obj;

// Display the paragraph's format name.

alert (pgf.Name);

Now you can develop your production code by working with the selected paragraph. After it is sufficiently tested, you can expand the code to work with other paragraphs in the document.

In our example script, we need to work with table cells. We can use code to isolate the table cell that contains our cursor, and use this cell object for developing and testing the code that will work on table cells in the main script.

#target framemaker

var doc = app.ActiveDoc;

// Get the paragraph containing the insertion point.

var pgf = doc.TextSelection.beg.obj;

// Get the cell object that contains the paragraph.

var cell = pgf.InTextObj;

// Display the cell object.

alert (cell);

Highlight a single cell

Now that we have a cell object to work with, use the function for highlighting a cell. You will also add the code for calling the function on the selected cell.

#target framemaker

var doc = app.ActiveDoc;

// Get the paragraph containing the insertion point.

var pgf = doc.TextSelection.beg.obj;

// Get the cell object that contains the paragraph.

```
var cell = pgf.InTextObj;
// Highlight the selected cell.
highlightCell (cell, doc);

function highlightCell (cell, doc) {
    // Colors the table cell with a Yellow background.
}
```

This gives a platform for developing the highlighting code. We have a single cell containing our cursor, so we will get immediate feedback on our highlighting code. We will know if it works for before trying to use it on a bunch of other cells in the table or document. Here is the finished function and its call; try it on a table in a sample document. Make sure your cursor is in a single cell paragraph before running it.

```
#target framemaker
var doc = app.ActiveDoc;
// Get the paragraph containing the insertion point.
var pgf = doc.TextSelection.beg.obj;
// Get the cell object that contains the paragraph.
var cell = pgf.InTextObj;
// Highlight the selected cell.
highlightCell (cell, doc);
// Refresh the screen.
doc.Redisplay();

function highlightCell (cell, doc) {
    // Colors the table cell with a Yellow background.
    // Get the Yellow color object from the document.
    var color = doc.GetNamedColor("Yellow");
    // Change the cell's color to Yellow.
    cell.CellOverrideShading = color;
    cell.CellUseOverrideShading = true;
    cell.CellOverrideFill = Constants.FV_FILL_BLACK;
    cell.CellUseOverrideFill = true;
}
```

The code that makes the cell Yellow is a bit cryptic, but as you gain experience with the ExtendScript syntax, it will become easier to work with the objects and properties that are required for each task. You now have one of your tasks completed, tested, and isolated in a function. Save this chunk of code so that it can be copied into the main script.

Determining if a cell should be highlighted

We have another small task to complete: we want to decide whether a cell should be highlighted based on the content of that cell. Here we can use a general-purpose function that gets all of the text in a text object. Text objects include paragraphs, table cells, text frames, and text lines, etc. Type this code into a new script.

```
#target framemaker
var doc = app.ActiveDoc;
// Get the paragraph containing the insertion point.
var pgf = doc.TextSelection.beg.obj;
// Get the cell object that contains the paragraph.
var cell = pgf.InTextObj;
// Get the cell's text and display it.
var cellText = getText(cell);
alert (cellText);

function getText (textObj) {
    // Gets the text from the text object.
    var text = "";
    // Get a list of the strings in the text object.
    var textItems = textObj.GetText(Constants.FTI_String);
    // Concatenate the strings.
    for (var i = 0; i < textItems.len; i += 1) {
        text += (textItems[i].sdata);
    }
    return text; // Return the text
}
```

Now that we can get the text from a table cell, we can test it to see if it matches our criteria. We can use a regular expression to test the part number and see if the cell should be highlighted. We will put this code into a function.

```
function isSpecialCell (cell) {
    // Determines if the cell contains a special part number.
    // Returns true or false.
    // Get the cell's text and see if the part number matches.
    var cellText = getText(cell);
    if (/-10$/.test(cellText) === true) {
        return true;
    } else { // Does not match.
        return false;
    }
}
```

We want to highlight cells where the part number ends with -10, and that is what the regular expression /-10$/ is looking for. If the cell text contains this, the exec function will return true; otherwise it will return false. If it returns true, then we return true from the function; otherwise, we will return false. Save this chunk of code, along with the getText function.

Navigating Among Cells in a Table

Our first two tasks are completed. Now that we can identify parts that match our spec, and we know how to highlight their cells, we need to be able to navigate from cell to cell, looking for matches. To keep things simple, we will work with a single table. ExtendScript can get the table object when the cursor is in the table, or any of its cells are selected. Click in a table and run this code:

```
#target framemaker
var doc = app.ActiveDoc;
// Get the table containing the insertion point.
var tbl = doc.SelectedTbl;
alert (tbl);
```

Once you have the table object, there are several ways to navigate through the table. You can go row-by-row, and for each row, cell-by-cell:

```
#target framemaker
var doc = app.ActiveDoc;
// Get the table containing the insertion point.
var tbl = doc.SelectedTbl;
// Navigate row-by-row.
var row = tbl.FirstRowInTbl;
while (row.ObjectValid()) {
    // For each row, navigate cell-by-cell.
    var cell = row.FirstCellInRow;
    while (cell.ObjectValid()) {
        // Do something with the cell here.
        // ...
        cell = cell.NextCellInRow;
    }
    row = row.NextRowInTbl;
}
```

In our table, the part numbers are in the first column so we can move down the column cell-by-cell.

```
#target framemaker
var doc = app.ActiveDoc;
// Get the table containing the insertion point.
var tbl = doc.SelectedTbl;
// Get the first cell in the first row.
var cell = tbl.FirstRowInTbl.FirstCellInRow;
while (cell.ObjectValid()) {
    // Do something with the cell here.
    // ...
    cell = cell.CellBelowInCol; // Move down to the next cell.
}
```

When we outlined our script with functions, we wanted one that would navigate through a table and search for matching part numbers. If we convert the code above to a function, we have what we need.

```
function navigatePartNumberCells (tbl, doc) {
    // Navigate the part number cells, and check for cells to highlight.
    // Get the first cell in the first row.
    var cell = tbl.FirstRowInTbl.FirstCellInRow;
    while (cell.ObjectValid()) {
        // See if the cell's text matches the required pattern.
        if (isSpecialCell (cell) === true) {
            // Call the function to highlight the cell.
            highlightCell (cell, doc);
        }
        cell = cell.CellBelowInCol; // Move down to the next cell.
    }
}
```

Save this function before moving to the next step.

Processing all of the tables in the document

Next, we will need to navigate through all of the tables in the document, testing for parts tables. We will put this code into a function.

```
function processTables (doc) {
    // Loop through the tables in the main flow, looking for parts tables.

    // Get all of the tables in the document's main flow.
    var textList = doc.MainFlowInDoc.GetText (Constants.FTI_TblAnchor);
    for (var i = 0; i < textList.length; i += 1) {
        var tbl = textList[i].obj;
        // Test for a parts table.
        if (tbl.TblTag === "PartsTable") {
            navigatePartNumberCells (tbl, doc);
        }
    }
}
```

Save this function. We can now use the functions to assemble the completed script. We will also need an entry point to the functions at the top of the script. We will test for an active document, and if there is one, call the processTables function.

Here is the code along with an outline of each function.

```
#target framemaker

// Test for an active document.
if (app.ActiveDoc.ObjectValid()) {
    processTables (app.ActiveDoc);
} else {
    alert ("There is no active document.");
}

function processTables (doc) {
...
        navigatePartNumberCells (tbl, doc);
...
}

function navigatePartNumberCells (tbl, doc) {
...
        if (isSpecialCell (cell) === true) {
            highlightCell (cell, doc);
        }
...
}

function highlightCell (cell, doc) {
...
}

function isSpecialCell (cell) {
...
    var cellText = getText(cell);
....
}

function getText (textObj) {
...
}
```

Combine the actual code into a single script file and test it on a sample document. If you have tested each function as you developed them, and if the author has properly applied the PartsTable tag to the content, the finished script should work properly.

Processing all of the files in a book

A final step would be to expand the script to work on an active book as well as the active document. Keep in mind that once you develop the book code, it can be reused for any script that you want to work on all the files in a book.

Chapter 27: Creating Interactive Content with Hypertext

Hypertext refers to electronic links in online documents. In a printed document, you can instruct readers to "see page 88." The reader then turns to page 88 to find the relevant information. In online documents, hypertext lets you create a live link; to go from one topic to another, the reader clicks on the link and immediately jumps to the new information.

The term *hypertext* also describes the features that make documents interactive (such as cross-references and links). FrameMaker provides hypertext commands—for things like links, pop-up alerts, and image maps—that let you create interactive online documents

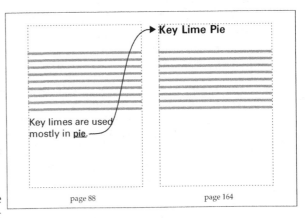

Here's a look at what's in this chapter:

    Advantages of hypertext interactivity 486
    Setting up a basic hypertext link 487
        Creating Link Destinations 488
        Creating a link to a destination 489
        Creating the active area in text 490
        Activating links 490
    Creating hotspots 490
        Using hotspot properties to create links .491
        Using hotspot mode to create links 491
        Using a text frame as a hotspot 491
    Creating a text link to a web address 492
    Creating an email link 492
    Using markers to create notes in PDF files 493
    Locking and unlocking view-only documents 494
    Hypertext Command Reference 495

485

Advantages of hypertext interactivity

Interactivity becomes increasingly important as we deliver more documents on the web and on mobile. Starting with FrameMaker 2015, you can also deliver your documents as mobile applications on the iOS and Android platforms, in addition to PDF, responsive HTML5, and other formats. See "Digital Publishing" on page 413 for more information.

Hypertext links are necessarily static in print but active when you convert FrameMaker files to online formats. Several of the FrameMaker hypertext commands work only when accessed in native FrameMaker documents, and are generally refinements of commands that work in other media For example, the "jump to link" command works in all media but the "jump to link and open in a new window" works only in native FrameMaker files.

FrameMaker hypertext is an important part of a single-sourcing strategy because you can embed commands that are ignored in the print version but make the online files interactive.

Hypertext uses markers to identify the source and destination of a link. One marker will identify a target, or *named destination*, with a keyword. The other marker identifies the named destination and provides the link (or jump) to that spot.

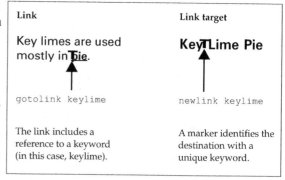

Documents for online distribution often contain a navigation bar, where users might click a button in the bar to go to the next page in the document. You can mimic this in PDF by setting up hypertext links that perform a nextpage command on the master pages of your document. Because the command is relative, it will work across all the pages (except the last) of your PDF. On the body pages, the result is a link that takes the reader to the next page. When clicked on page 4, the link goes to page 5; from page 88, it goes to page 89. However, as we move to more web-friendly online formats like responsive HTML5 and mobile apps, these features are less important because many online formats include their own navigation features that are more sophisticated.

This chapter describes how to create interactive hypertext features in your documents.

Cross-references vs. hypertext

Cross-references are a specialized form of hypertext. When you create a cross-reference, FrameMaker automatically creates a link for you, which is preserved in electronic outputs. Often cross-references and hypertext accomplish the same goal, and you can use both features in the same document.

Here are some things to consider when deciding which feature to use for a specific link:

- You can automatically format cross-references by embedding character tags in the cross-reference definitions. For hypertext, you must apply character tags separately.
- Hypertext lets you make any area of text (or a graphic) a link. With cross-references, you use the text of the destination or a page reference.
- Cross-references are easier and faster to create than cross-references.

As a rule, use cross-references whenever you can and reserve hyperlinks for situations that need the flexibility that hypertext links provide. For example, assume that you have some information about rowing a boat. You can quickly create cross-references that look like this:

- see "Rowing a boat" on page 77
- see page 77
- Rowing a boat

You need hypertext commands to create links with wording that differs from the heading text:

- Before you begin, make sure you know how to row.
- Learn all about rowing.

Setting up a basic hypertext link

A hypertext link lets the reader click an item and jump to another document or a different part of the current document. To create a hypertext link in FrameMaker, you need to complete three steps (described in detail in the sections that follow):

- Identify the target of the link
- Create a link to the target
- Create the link's active area within your document

Hypertext links in generated files

When you set up indexes, tables of contents, and other generated files, you can specify that entries become hypertext links by checking the Create Hypertext Links checkbox. (For complete instructions, see "Setting up table of contents file" on page 345.) After creating the generated file, the reader can jump from a page number in the file to the referenced item like any other hypertext link.

Creating Link Destinations

To identify a target, you can create a hypertext marker that creates a *named destination*. A named destination serves as a unique label or identifier for a location in the document.

Note Cross-references and generated docs like TOC and IX create their own destination markers automatically.

To create a named destination for a link, follow these steps:

1. Place your cursor at the target or ending location.
2. Select **Special>Hypertext** to display the Hypertext pod.
3. In the Command drop-down list, click Specify Named Destination. This inserts the newlink command into the text field.

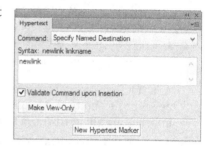

   Tip You can type **newlink** into the text field instead of selecting the command, but the syntax shown can help avoid errors in your markers.

4. Type a name for this destination after newlink. To make maintenance easier, use simple phrases, with alphanumeric, lowercase characters. I recommend avoiding spaces and keeping the names as short as possible. Each named destination within a FrameMaker file must have a unique name. The marker text should look like this:

   newlink *linkname*

5. By default, the Validate Command Upon Insertion checkbox is checked. We recommend leaving it checked; FrameMaker will check the syntax of your hypertext command when you create the marker. If you create the gotolink command before the newlink command, validation fails because the destination doesn't exist yet. This results in a stern warning from FrameMaker.

Caution Clicking the Make View-Only button changes the document to view-only. This might be useful if you want to check hypertext links, but to change the document back to an editable state, you must select the Make Editable button, or type ESC SHIFT-F L K.

6. Click the New Hypertext Marker button to insert the new named destination. A new hypertext marker is created at the cursor location. The marker stores the named destination information for you.

Creating a link to a destination

After identifying the link's destination, you create the link itself. You can create links that point to your current file or to another FrameMaker document. You can create the link to a destination before creating the destination marker, but if you have the Validate Command upon Insertion option selected, FrameMaker will produce an error message when you select the New Hypertext Marker button. If this happens, dismiss the dialog and proceed to create the named destination you need.

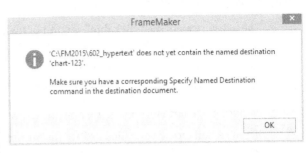

If you create cross-file links, all of the files must be available for the link to work in FrameMaker. If FileA links to FileB, you must ship FileA and FileB to your editors. If you forget to send FileB, or if the online formats are in locations other than specified, readers will encounter a broken link.

To create a hypertext link, follow these steps:

1. Position your cursor in the text where you want place the link. If the files are going to be sent out for translation, or delivered electronically (other than PDF) place marker at the beginning or end of a word instead of in the middle of a word.

2. Select **Special > Hypertext** to display the Hypertext pod.

3. In the Command drop-down list, click Jump to Named Destination. This inserts the gotolink command into the text field.

   Tip You can type gotolink into the text field instead of selecting the command. Spaces count! Spaces, spelling and punctuation must match exactly.

4. Type in the destination after gotolink. The *linkname* corresponds to the keyword you set up in the preceding section with the newlink command.

   For links within a file, use this syntax:

   gotolink *linkname*

   For links to another file in the same directory, use this syntax:

   gotolink *file.fm:linkname*

   For links to a file in a different directory, use this syntax (be sure to use forward slashes and not backward slashes in the path name):

   gotolink *path/file.fm:linkname*

5. Click the New Hypertext Marker button to insert the hypertext marker.

Creating the active area in text

The active area is the "clickable" section of text that lets you use the hypertext link.

When you create a hypertext link, FrameMaker creates an active area in the paragraph where the link occurs. Often the active area is the entire paragraph containing the marker. However, if you have any character tag formatting inside the paragraph, the active area stops where the character tag changes. If you want an entire paragraph to be active, the paragraph cannot contain any character tags or formatting overrides.

If you're delivering your files as PDF files for online viewing, consider setting up a special Hyperlink character tag, to help indicate the active area to readers.

To specify an active area for a hyperlink, follow these steps:

1. Select the text you want to serve as the clickable area. The selected area must contain the hypertext marker you set up in the preceding section.

2. Apply a character tag to the text. If you do not want the appearance of the text to change, create a character tag in which every attribute is set to As Is and apply that.

Tip If the marker is inside text formatted with a character tag, then the formatted text is the active area. If the marker is the only thing in a text frame, the entire frame is the active area. If you want to create active areas on a graphic, see "Creating hotspots."

Activating links

Test links to verify they are working. To do so, press and hold CTRL+ALT and move the cursor over the link. The cursor turns into a finger pointer. Click the link.

> Key limes are used mostly in pie.
>
> The word "pie" is the active area.

If you need to test a large number of links, you may want to convert the document to a view-only document. In view-only documents, you click on links without holding down additional keys. To convert a document from editable to view-only, use the button in the Hypertext pod or press ESC SHIFT-F L K. Click the corresponding button in Hypertex or repeat the key sequence to convert from view-only back to a regular, editable document.

Creating hotspots

You can use frames and graphics as *hotspots* in your document, or set up active areas over a graphic. Hotspots create links to different locations based on where the reader clicks. Starting with FrameMaker 11 there is a Hotspot command (**Graphics > Hotspot Properties**) and the graphics tools include a Hotspot Mode ( ) that lets you draw hotspots on graphics using any of the FrameMaker shape tools.

You can use a hotspot for a web link, or you can link to a destination in a FrameMaker document (defined with a newlink command as described on page 488). Then, you create a hotspot which links to the destination.

Using hotspot properties to create links

Starting with FrameMaker 11, you can now define an graphic or frame as a hotspot. To do this, follow these steps:

1. Select the graphic or frame you want to act as a hotspot.
2. Select the **Graphics > Hotspot Properties…** command to display the Hotspot Properties dialog.
3. Do one of the following:
   - Select a document from the drop-down list and choose from the available markers.
   - Select the URL radio button and enter the link in the text box.
4. *(optional)* Enter the tooltip text.
5. Click the Save button.

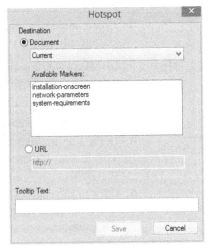

Using hotspot mode to create links

Starting with FrameMaker 11, if you want multiple hotspots on an image (often called an *image map*) or if you only want part of an image to become a link, you can use the Hotspot Mode to create those areas using the FrameMaker graphic tools. To define a hotspot with the graphics tools, do the following:

1. Select the Hotspot Mode ( ) button in the Graphics Tools pod. (see page 294) to display the Hotspot dialog.
2. See "Using hotspot properties to create links" for details on choosing options.

Using a text frame as a hotspot

To use a text frame as a hotspot, follow these steps:

1. Draw a text frame on top of the graphic. The dimensions of the text frame will become the active area on the graphic.
2. Position your cursor inside the text frame.
3. Create a hypertext marker in the text frame as described in "Creating a link to a destination" on page 489.

Note   If using text frames as hotspots, you insert only the hypertext marker in the text frame. If the marker is the only content in a text frame, the entire frame is the active area.

Chapter 27:   Creating Interactive Content with Hypertext

Creating a text link to a web address

You can create text links that navigate to a web address. When you activate the link in FrameMaker (CTRL-ALT click), it opens using your default web browser. In electronic output formats, the link may open into the same window, or open up a new window.

To create a web link, follow these steps:

1. Create an active area for the link by applying a character tag. If no tag exists, the paragraph will be your active area.

Tip To create a web link on a graphic, see "Using hotspot properties to create links" and "Using hotspot mode to create links" sections.

2. Position your cursor in the active area.
3. Select **Special > Hypertext** to display the Hypertext pod.
4. In the Command drop-down list, click Go to URL. This inserts the message URL command into the text field.
5. Click to the right of the space following message URL and insert your web address.
   For example:

   message URL http://www.mattrsullivan.com

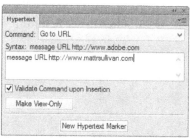

6. Click the New Hypertext Marker button to insert the marker.

Creating an email link

You can add links that create an email message. The link opens the user's default email program when activated in an electronic file.

To create an email link, do the following:

1. Create a Go to URL hypertext link, as described in steps 1 through 4 of the previous procedure, "Creating a text link to a web address."
2. Instead of the web address, type a *mail URL* so that the marker text is, for example:

   message url mailto:matt@mattrsullivan.com

You can also type extra information to populate the message. For example, you can add a predefined subject:

   message url mailto:mailto:matt@mattrsullivan.com?subject=FM2015%20book%20Comments

Note Some punctuation marks have to be specified as hex values (for example, a space is %20). See www.blooberry.com/indexdot/html/topics/urlencoding.htm for more information about using hex values in HTML.

3. Click the New Hypertext Marker button to insert the marker.

Using markers to create notes in PDF files

FrameMaker provides the Alert and Alert with Title hypertext commands, which convert to notes in PDF files. For example, you could include alert messages in a draft document where information is not yet available, or where you wanted to flag something in the PDF file.

Caution Alert messages are dropped when content is converted to HTML. Converted Alert boxes initially appear on top of the text in your PDF output, so consider whether this is acceptable before using the Alert commands.

To create an alert message, follow these steps:

1. Position your cursor where you want the note in the PDF file. You do not need to create an active area for an alert.
2. Select **Special > Hypertext** to display the Hypertext pod.
3. In the Command drop-down list, click Alert. This inserts the alert command into the text field.

Tip The Alert with Title command is only partially supported when you create a PDF file. The title portion is embedded in the PDF note, and not rendered as a title.

4. Type the alert message in the marker text field. Alerts are limited to 1023 characters. For example, to create an alert with the message "Verify the day", use this syntax:

   alert Verify the day

5. Click the New Hypertext Marker button to insert the hypertext marker.

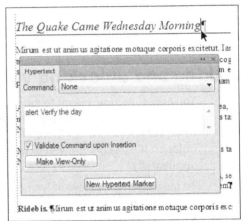

In the PDF file, the alert becomes a note.

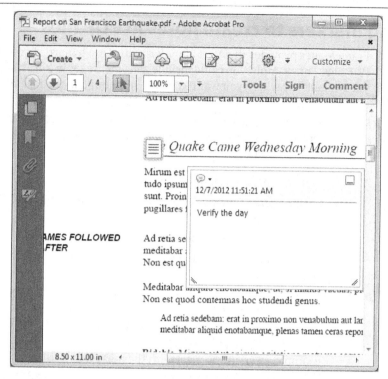

Locking and unlocking view-only documents

You may sometimes want to make a FrameMaker file view-only so it can't be changed automatically or so you can test hypertext links. More often, you may need to unlock a file has been saved to the view-only format accidentally.

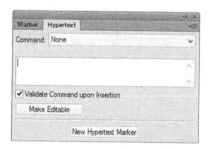

To toggle between view-only and editable formats in an open document, use the Hypertext pod press ESC SHIFT-F L K.

Hypertext Command Reference

Table 27-1 lists the available hypertext commands and specifies whether the commands work in PDF files, HTML output, and view-only FrameMaker files. Commands marked "Obsolete" apply to view-only FrameMaker files and have no effect on PDF or HTML output.

Note If you are creating HTML output, double-check the resulting HTML code after conversion and modify your FrameMaker marker text accordingly.

Table 27-1. Hypertext Commands

Command	PDF	HTML	View-Only FrameMaker	Description and Syntax
Alert	•		•	Displays an alert box. alert *message of up to 1023 characters* See "Using markers to create notes in PDF files" on page 493 for details.
Alert with Title	•		•	Displays an alert box (and a title in FrameMaker view-only documents). alerttitle *title text:message of up to 1023 characters* Title is pushed into note body in PDF files. The Alert command is preferred. See "Using markers to create notes in PDF files" on page 493 for details.
Specify Named Destination	•	•	•	Defines a name used to create a link to the marker location. newlink *linkname* newlink *filename:linkname* See "Creating Link Destinations" on page 488 for details.
Jump to Named Destination	•	•	•	Displays content that contains the named link. gotolink *linkname* gotolink *filename:linkname* See "Creating a link to a destination" on page 489 for details.
Jump to Named Destination & Fit to Page	•	•	•	Displays destination (and fits to page in FrameMaker view-only documents). gotolinkfitwin *linkname* gotolinkfitwin *filename:linkname* Does not fit to page for PDF and HTML files. The Jump to Named Destination command (gotolink) is preferred.
Jump to First Page	•		•	Displays first page of specified document. gotolink firstpage gotolink *filename:*firstpage

Table 27-1. Hypertext Commands (Continued)

Command	PDF	HTML	View-Only FrameMaker	Description and Syntax
Jump to Last Page	•		•	Displays last page of specified document. gotolink lastpage gotolink *filename*:lastpage
Jump to Page Number	•		•	Displays specified page. gotolink *pagenumber* gotopage *filename:pagenumber* Page numbers are hard to maintain. The Jump to Named Destination command (gotolink) is preferred.
Jump to Previous Page	•		•	Displays previous page viewed. previouspage
Jump to Next Page	•		•	Displays next page of current document. nextpage
Jump Back	•		•	Displays location last viewed. previouslink
Jump Back and Fit to Page	•		•	Displays location last viewed (and fits to page for FrameMaker view-only documents). previouslinkfitwin Does not fit to page in PDF files. Jump Back command (previouslink) is preferred.
Open Document	•	•	•	Displays another page of current or different document. openlink *linkname* openlink *filename:linkname* The Jump to Named Destination command (gotolink) is preferred.
Open Document & Fit to Page	•	•	•	Displays another page of current or different document and fits to page in FrameMaker view-only documents. openlinkfitwin *linkname* openlinkfitwin *filename:linkname* Does not fit to page in PDF or HTML files. The Jump to Named Destination command (gotolink) is preferred.

Table 27-1. Hypertext Commands (Continued)

Command	PDF	HTML	View-Only FrameMaker	Description and Syntax
Open Document at First Page	•	•	•	Displays first page of current or different document. openlink firstpage openlink *filename*:firstpage The Jump to First Page command (gotolink firstpage) is preferred.
Open Document at Last Page	•	•	•	Displays last page of current or different document. openlink lastpage openlink *filename*:lastpage The Jump to Last Page command (gotolink lastpage) is preferred.
Open Document at Page Number	•	•	•	Displays specific page number of current or different document. openpage *pagenumber* openpage *filename pagenumber* Page numbers are hard to maintain. The Jump to Named Destination command (gotolink) is preferred.
Open Document as New			•	Obsolete. Opens file as untitled FrameMaker document from FrameMaker view-only document. opennew *filename*
Popup Menu			•	Obsolete. Displays a pop-up menu containing hypertext links as choices in the FrameMaker view-only file. popup *flowname*
Button Matrix			•	Obsolete. Allows reader to select from a matrix to run hypertext commands in FrameMaker view-only document. matrix *number_of_rows number_of_columns flowname* To create an image map instead of a button matrix, see "Creating hotspots" on page 490.
Go to URL	•	•	•	Opens a web address. message URL *web_address* See "Creating a text link to a web address" on page 492 and "Creating an email link" on page 492 for details.
Message Client	•		•	Starts other programs. message openfile *filename*

Chapter 27: Creating Interactive Content with Hypertext

Table 27-1. Hypertext Commands (Continued)

Command	PDF	HTML	View-Only FrameMaker	Description and Syntax
Close Current Window			•	Obsolete. Closes current viewer window in FrameMaker view-only document. quit
Close All Hypertext Windows			•	Obsolete. Closes all windows in FrameMaker view-only document. quitall
Exit Application			•	Obsolete. Exits application in FrameMaker view-only document. exit
Multimedia Object Link	•	•		Controls placed multimedia files. See "Placing rich media" on page 283 for more information. multimedia *objectId type command value*

Chapter 28: Writing equations

There are a few options for placing equations in your FrameMaker content:

- Import an equation graphic as with any other vector image. (see page 255)
- Use FrameMaker's native equation editor.
- Place MathML objects. (starting with FrameMaker 12)

Here's a look at what's in this chapter:

Using native FrameMaker equation editor	.499
Understanding the Equations pod	.500
Inserting equations	.502
Using the Equations Pod	.503
Typing an equation	.504
Selecting equations and math elements	.504
Navigating through equations	.505
Moving equations	.506
Modifying equations	.506
Moving text using keyboard shortcuts	.506
Moving text using the equations palette	.507
Deleting equations	.508
Formatting equations	.509
Changing equation fonts	.509
Changing equation font sizes	.509
Inserting automatic line breaks	.510
Changing the equation size	.510
Applying character tags	.511
Evaluating equations	.511
Working with MathML	.511

Using native FrameMaker equation editor

In FrameMaker, you can write complex equations, such as those found in scientific or mathematical documents. Equations consist of operators, alphanumeric characters, symbols, text strings, and other math elements.

$$(x \times y)^2 = y \qquad area = \pi r(\ )^2 \qquad \frac{(9+3)}{6}$$

To write an equation, type the math elements or use the Equations pod to insert the items. For example, consider the following equation:

$$\log_a \Delta = \frac{\ln \Delta}{\ln a}$$

Chapter 28: Writing equations

To write this equation, you first create a high-level approximation of the equation, as shown here:

$$? = \frac{?}{?}$$

You then replace the question marks with symbols, functions, and other math elements.

Equations are displayed in anchored or graphic frames. You can move an equation inside the frame by selecting the equation and using keyboard shortcuts or commands in the Equations palette. You can also shrink-wrap anchored frames to remove extra space between the frame and equation.

Understanding the Equations pod

Symbols, operators, delimiters, and other math elements are displayed in the Equations palette. Select **Special>Equations** to display the Equations pod. The palette is divided into nine sheets for different types of math elements or commands. Click one of the buttons to display the appropriate page, and then click a button in the palette.

Table 28-1 describes each sheet in the Equations palette.

Table 28-1. Equations Palette Sheets

Category	Description	Sheet
Symbols	Greek characters, atomic symbols, diacritical marks, and strings	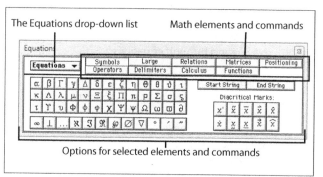
Large	Sums, products, integrals, intersections, and unions	

500

Table 28-1. Equations Palette Sheets (Continued)

Category	Description	Sheet
Relations	Equal, less than, greater than, similar to, subset of, and proportional to symbols	
Matrices	Matrices and matrix commands	
Positioning	Micropositioning, alignment, line breaks, and spacing options	
Operators	Roots, powers, signs, subscripts, superscripts, and logic symbols	
Delimiters	Parentheses, brackets, braces, and substitution symbols	

Table 28-1. Equations Palette Sheets (Continued)

Category	Description	Sheet
Calculus	Integrals, derivatives, partial derivatives, gradients, and limit symbols	
Functions	Trigonometric, hyperbolic, and logarithm functions; commands for evaluating expressions and for creating and applying rules	

Inserting equations

Equations inserted in empty paragraphs, as shown in the following example, are called *display equations*.

> Figure the square footage by solving the following equation:
> l × w = h

When inserted in sentences, as shown in the following example, equations are called *inline equations*.

> Figure the square footage by calculating l × w = h.

To insert an equation, follow these steps:

1. Do one of the following:
   - To create a display equation within an anchored frame, position your cursor in an empty paragraph.

     > Figure the square footage by solving the following equation:
     > Cursor → |

   - To create an inline equation within an anchored frame, position your cursor where the equation should begin.

     > Figure the square footage by calculating |
     > ↑
     > Cursor

   - To create an equation in a graphic frame, select **Graphics > Tools** to display the Tools panel, click the Graphic Frame icon ( ), then draw a frame on the page. Select the frame before proceeding.

Inserting equations

2. Select **Special > Equations** to display the Equations pod.
3. In the Equations drop-down list, click the size of the equation you want to insert. The equation size is determined by the default font size and spread in the Equations Sizes dialog. See "Changing equation font sizes" on page 509 for details.

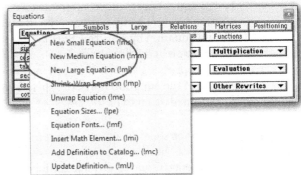

Note You can change the size of the equation later in the object properties if you don't like the size you selected. See "Changing the equation size" on page 510 for details.

An *equation object* is displayed as a highlighted question mark. (?)

Do one of the following:

- Click math elements and commands in the Equations palette to replace the equation object with your base equation.
- Type the equation in place of the equation object.

The next sections show how to write the following equation using each method:

$$? \longrightarrow \frac{8}{2} + \frac{10}{5} = 6$$

Using the Equations Pod

To use the Equations palette to write the preceding equation, follow these steps:

1. CTRL+CLICK the equation object, or press the SPACEBAR. The equation object is selected. (?)

2. Click the Operators button in the Equations pod, then click the [?=?] button. The object $? = ?$ is updated.
3. With the question mark on the far left selected, click the [?+?] button. The object is $? + ? = ?$ updated.
4. With the question mark on the far left selected, click the [?/?] button. The object $\frac{?}{?} + ? = ?$ is updated.
5. Press the TAB key until the next single question mark is selected, then click the [?/?] button. The object is updated, and the final form of the equation object is $\frac{?}{?} + \frac{?}{?} = ?$ displayed.
6. Type the numbers in the equation; press the TAB key to move between objects. $\frac{8}{2} + \frac{10}{5} = 6$

Note Though you must calculate equations, FrameMaker can transform selected expressions to alternate formats. See "Evaluating equations" on page 511 for details.

7. For equations in anchored frames, remove extra space between the frame and equation by selecting Shrink-Wrap Equation in the Equations drop-down list or by pressing ESC M P. If you change the size of the equation fonts, the shrink-wrapped equations are automatically shrink-wrapped again to fit the new font sizes.

Note When you shrink-wrap an anchored frame, the anchoring position changes to At Insertion Point. You might need to reposition the frame based on where you want the equation displayed. See "Anchoring Graphics" on page 256 for details.

Typing an equation

If you're writing a simple equation or know the keystrokes, you might want to type the equation instead of using the Equations palette. To do so, follow these steps:

1. CTRL+CLICK the equation object to highlight it.
2. Type **8/2** to display the fraction.
3. Press the SPACEBAR twice to select the equation, then type **+10/5**. The object is updated.
4. Press the SPACEBAR twice to select the equation, then type **=6** to display the final equation.

Tip To type text in an equation, click the Symbols page in the Equations palette, and then click the Start String button. Quotes are displayed, and when you type text, the text replaces the quotes. Click the End String button to close the text string.

Selecting equations and math elements

You can select an entire equation as you do other objects, or you can select specific math elements. In individual equations, the location of the insertion point determines which characters are selected. For example, if a fraction is selected, and you press the DOWN ARROW key, the numerator in the equation is selected.

To select one or more characters in an equation, do one of the following:

- To select an equation, CTRL+CLICK the equation. The selection handles are displayed around the object.
- To select part of an equation, drag the cursor across the text, or use one of the shortcuts in Table 28-2.

Caution You must use your cursor to select a text string; the keyboard shortcuts don't work.

Inserting equations

Table 28-2. Selecting Items in Equations

Keystroke	Description
SPACEBAR	Expands the selection.
LEFT ARROW	Selects the next character on the left.
RIGHT ARROW	Selects the next character on the right.
DOWN ARROW	Selects the numerator in a fraction when the whole fraction is selected.
CTRL+A	Selects all objects in an anchored or graphic frame. Before using this shortcut, you must select the frame or click an item inside the frame; otherwise, you'll insert a new equation object above the insertion point.

Navigating through equations

To move the cursor between math elements in an equation, you use the arrow keys. Table 28-3 shows how the insertion point moves when you press the arrow keys.

Table 28-3. Moving the Cursor Through Equations

Key	Description
LEFT ARROW	Moves the cursor to the left.
RIGHT ARROW	Moves the cursor to the right.

Chapter 28: Writing equations

Table 28-3. Moving the Cursor Through Equations (Continued)

Key	Description
DOWN ARROW	Moves the cursor from the side of a fraction to its numerator. $$\left|\frac{10}{2}+\frac{6}{3}=7 \qquad \frac{\lfloor 10}{2}+\frac{6}{3}=7\right.$$
UP ARROW	Moves the cursor from the numerator to the entire fraction. $$\frac{\lfloor 10}{2}+\frac{6}{3}=7 \qquad \left|\frac{10}{2}+\frac{6}{3}=7\right.$$

Moving equations

After you select an equation, you can move it on the page by one of three methods. To move an equation, select the equation, and then do one of the following:

- Drag the selected equation to a new location.
- Press ALT or SHIFT+ALT while pressing one of the arrow keys.
- Right-click the equation, then select **Graphics > Object Properties** to display the Object Properties dialog. Change properties as needed, then click the Apply button.

For more information on modifying the position and alignment of objects, see "Rearranging objects" on page 310.

Caution   You can move the equation outside the anchored frame; however, the equation is no longer linked to the paragraph, and text added or deleted above the equation may be displayed on top of the equation.

Modifying equations

After you write an equation, you can modify the spacing and alignment in several ways. For example, you can tighten the spacing or add a line break. You should wait to modify spacing and alignment until the equation is final, or you might need to make additional adjustments.

Moving text using keyboard shortcuts

You can move selected text up and down, left, right, and diagonally using keyboard shortcuts. To do so, select the text you want to move, and use keyboard shortcuts in the following table to microposition the text. (The point values apply to documents displayed at 100 percent.)

Direction	Shortcut
Left one point	ALT-LEFT ARROW
Right one point	ALT-RIGHT ARROW
Up one point	ALT-UP ARROW

Modifying equations

Direction	Shortcut
Down one point	ALT-DOWN ARROW
Left six points	ALT-SHIFT-LEFT ARROW
Right six points	ALT-SHIFT-RIGHT ARROW
Up six points	ALT-SHIFT-UP ARROW
Down six points	ALT-SHIFT-DOWN ARROW

Note Some of the keys on your computer may be specially programmed. If so, one or more keyboard shortcuts may not work as described. Consult your system documentation for help.

Moving text using the equations palette

In addition to using keyboard shortcuts, you can reposition an equation using commands in the Positioning page.

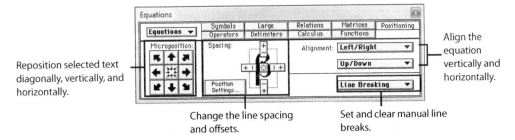

The following list describes some of the most common ways to modify equations using the Equations palette:

- **Adjusting the spread.** Add or remove space between characters in an equation by selecting the characters and clicking the appropriate Spacing arrow. (The middle button resets the alignment.)

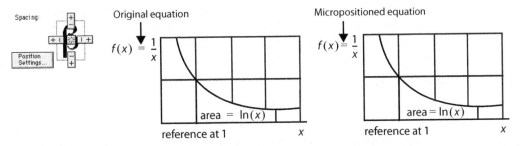

- **Adjusting the line spacing.** Add or remove space between lines of a multi-line equation by editing spacing in the Position Settings dialog.

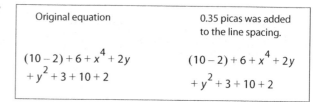

- **Repositioning text.** Move selected text up and down, side-to-side, and diagonally by clicking one of the Microposition arrows. (The middle button resets the alignment.)

Caution If you add math elements after using the Microposition commands, the repositioned text will move back to its original location.

- **Aligning.** Line up equations at specific points (for example, along the equal signs or centers of the equations).

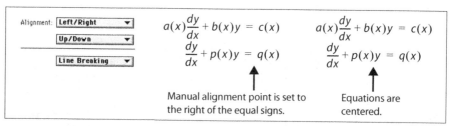

- **Adding a line break.** Sets a manual line break in an equation. (You can set up an automatic line break in the equation's object properties.)

$x - 3y + (10 \times 350) + 9 - 2 \longrightarrow x - 3y + (10 \times 350) + 9 - 2$

For more information on modifying the spacing and alignment of equations, see the FrameMaker online help.

Deleting equations

You can delete specific math elements in an equation or the entire equation. When you delete a math element, a question mark is displayed in place of the element. You can insert another item in place of the question mark or continue to delete more of the equation.

To delete items in an equation, follow these steps:

1. Do one of the following:
    - Select the item you want to delete.
    - Position the cursor on the right side of the item.
2. Press the BACKSPACE key. The item is deleted, and the question mark is displayed in its place.
3. Repeat to delete additional items.

To delete an entire equation, select the equation, then press the DELETE key.

Formatting equations

FrameMaker lets you set the global font properties for equations. You customize equation fonts by assigning a character format. For example, you might create a character tag with a sans-serif typeface and assign the format in the Equations palette. You can also modify the font sizes in small, medium, and large equations and adjust the default spread of each equation. When you modify global font settings, new and existing equations are updated, and shrink-wrapped equations are rewrapped, if necessary.

You reformat individual equations in the equation's object properties. In addition to modifying the typical object properties—color, tint, overprint, and angle—you set up automatic line breaks or change the size of the equation. To reformat certain characters in an equation, you must use character tags.

Changing equation fonts

In equations, alphanumeric characters, including functions, are formatted with the Times or Times New Roman typeface. You change the font properties by assigning a character tag. The five types of equation characters—math symbols, functions, numbers, strings, and variables—may use different character tags.

To modify default equation fonts, follow these steps:

1. Create a character tag for each kind of formatting. For example, you can create a separate tag for functions and numbers. See "Creating character tags" on page 143 for details.
2. Display the Equations palette, then click Equation Fonts in the Equations drop-down list to display the Equation Fonts dialog.
3. To change the math symbol font, click a font in the Math Symbols drop-down list. Only math fonts are displayed.
4. To apply a character tag to other math elements, click the character tag in the appropriate drop-down list.

Caution The EquationVariables character tag is assigned to variables to italicize the characters. Don't delete this character tag from your template if you insert equations in your document.

5. Click the Set button. The default font properties in new and existing equations are updated.

Changing equation font sizes

When you insert an equation, you choose a small, medium, or large equation. The default font sizes determine the size of the equation. Certain types of characters within each equation also differ in size. For example, integral and sigma symbols are larger than other characters, and there are three font sizes for different levels of alphanumeric characters. You can change the default font sizes for equations, and all existing equations are also updated.

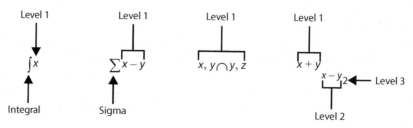

In addition to editing font sizes, you can change the equation's vertical or horizontal spread. The spread setting functions the same as spread in a paragraph tag. See "Default Font sheet" on page 117 for more information.

To change the default font sizes and spreads, follow these steps:

1. Display the Equations pod, then click Equation Sizes in the Equations drop-down list to display the Equation Sizes dialog.
2. To change the font sizes for equations, type new values in the Small, Medium, and Large fields.
3. To change the spacing between characters, do any of the following:

   - To modify space horizontally, type new values in the Horizontal Spread values.
   - To modify space vertically, type new values in the Vertical Spread values.

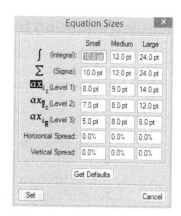

Note To restore the default font sizes, click the Get Defaults button. (The defaults are displayed in the previous graphic.)

4. Click the Set button to update the default font sizes in new and existing equations are updated.

Inserting automatic line breaks

You can set up automatic line breaks in individual equations. When an equation reaches a certain length, the equation wraps to the next line. To insert an automatic line break, follow these steps:

1. Select the equation, then select **Graphics > Object Properties** to display the Object Properties dialog. (see page 300)
2. Check the Automatic Line Break After checkbox, type the length at which the line break occurs, then click the Set button. The equation properties are updated.

Changing the equation size

After you insert an equation, you can change the size. For example, you might insert a small equation and decide the font is too small. Instead of modifying the size of the font globally, you can assign a different size to the equation. To change the equation size, follow these steps:

1. Select the equation, then select **Graphics > Object Properties** to display the Object Properties dialog. (see page 300)

2. Click a size in the Size drop-down list, then click the Set button. The equation size is updated.

For details on changing equation sizes globally, see "Changing equation font sizes" on page 509.

Applying character tags

You can reformat specific characters in an equation by applying character tags. This overrides the global character formatting specified in the Equation Fonts dialog. For details, see "Changing equation fonts" on page 509.

To apply a character tag, select the text, and click a tag in the Character Catalog. You cannot use the keyboard shortcut for applying character tags because the shortcut is mapped to one of the operators in the Equations palette. For more information, see "Applying Character Tags" on page 140.

Evaluating equations

FrameMaker can evaluate a selected expression and either display an alternate format or correct the syntax. The following example shows an expression before and after evaluation. In the numerator of the first expression, the parenthetical values were transformed to display the values in the correct order. The denominator was converted to its simplest expression. The second expression was also converted to its simplest value.

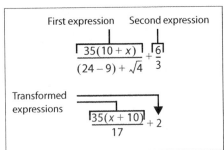

To evaluate an equation, follow these steps:

1. Display the Equations pod, then click the Functions button to display the Functions page.
2. Select the expression you want to evaluate.
3. Click the Evaluation drop-down list, then click the appropriate command for your expression. Your equation is evaluated, and the alternate values are displayed.

For more information on evaluating equations, see the FrameMaker online help.

Working with MathML

You can use MathML in both structured and unstructured FrameMaker.

Starting with FrameMaker 12, you can use Design Science's MathFlow application to insert complex equations. The MathFlow Style and Structure Editors are built into FrameMaker and XML Author as 30-day trial software that can be unlocked at any time by purchasing a key from Design Science. See http://www.dessci.com for more information.

See the preferences to set how equations are handled in both structured and unstructured FrameMaker. Your FrameMaker install comes with a 30-day trial of MathFlow to make it easier to try the product.

Equations created in MathFlow can be placed and managed, giving you control beyond what FrameMaker's native equation editor will allow. Because the equations are managed using MathML,

Chapter 28: Writing equations

a form of XML, you have control over the equation using an application like MathFlow, or using just a text or XML editor to make changes.

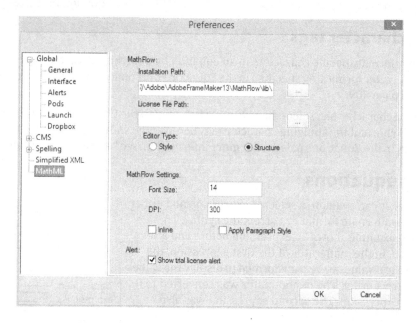

Chapter 29: Content reuse with text insets

Most desktop publishing applications let you link graphics—you set up a pointer from a document to a particular graphic file. If that file is updated, the graphic in the document is updated automatically. FrameMaker supports linked graphics (they are discussed in Chapter 13, "Importing graphic content"), but in addition to graphics, FrameMaker also lets you link text into a document. Instead of importing a graphic, you import a file that contains text. If you update the file, its content is also updated where it is linked.

Imported-by-reference text fragments are called *text insets*. A development process based on assembling text fragments into larger documents is referred to as creating *modular documentation* or modular text—your final document is made up of smaller text chunks or *modules*. The composite document that's assembled from modules is called a *container document*.

Tip Text insets are used most commonly to import one FrameMaker file into another, but you can also import other supported file formats, such as Word files, as text insets. This works best if the file is relatively simple and if the styles used in Word match paragraph tags in the FrameMaker file.

Here's a look at what's in this chapter:

    Advantages of using text insets 513
    A text inset example 514
    Working with text in insets. 517
    Considering text insets and other reuse options 518
    Planning modular text. 518
        Breaking text down into modules 518
        File storage 519
        Information retrieval 519
        Controlling formatting in text insets 519
    Creating a text inset 521
    Managing text insets. 522
        Opening the source file 524
        Converting text insets to text 524

Advantages of using text insets

Breaking up documents into modules lets you create small, reusable topics. This modular approach makes sense if you must write documentation for several related products. By dividing information in small chunks, you can maximize reuse of common information and write separate modules where the products differ.

Consider the example shown here. You have two installation documents, one for Product A and one for Product B. The two products have identical minimum system requirements and licensing agreements, but information in the installation instructions is different. To take advantage of reuse, you created a minimum system requirements document (reqs.fm) and a licensing agreement document (license.fm) and imported these as text insets into the install.fm files in both the productA and productB books. Any updates to the inset files will be reflected in both installation documents. Because you only have one copy of the license.fm and reqs.fm files, it doesn't make sense to put them in a specific product's directory; instead, many authors set up a shared directory, as shown in example shown in this example.

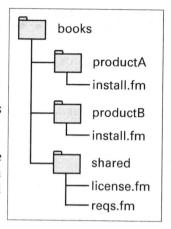

A text inset example

Although FrameMaker creates text insets via the **File > Import > File** command, creating a text inset isn't exactly the same as importing a file.

When you create a text inset, you import a specific *flow* from a file; most often, the main body flow (flow A). You can set up a document that contains several different flows : for example, a series of standard warnings that you need to repeat throughout a book. Instead of copying and pasting, you can create text insets in the book to ensure consistency.

In the following example, the author has created a document with separate text flows for each warning.

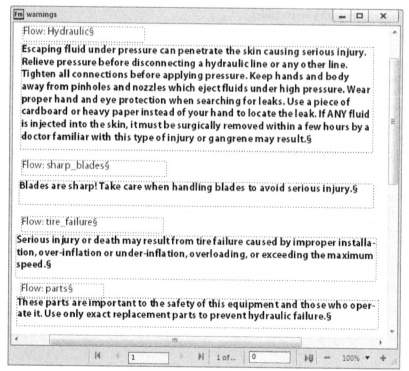

When you specify that you want to import text from the warnings file (**File > Import > File**), you are prompted to choose a flow. In this example, there are four separate body page text flows to choose from.

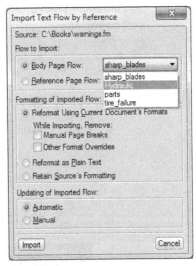

After you select a flow, that information is imported into the container document. Notice that the text inset has been reformatted using the current document's paragraph catalog.

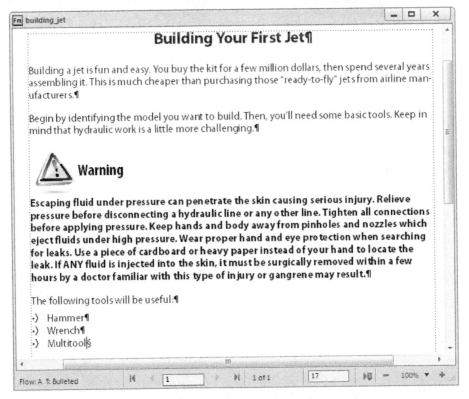

Note   In the example shown here, formatting of the warning paragraph is quite different in the source (warnings.fm) than it is when used in a container. The paragraph tag definition in the container (destination) document is overriding the information provided in the source document. You can control this with the Formatting of Imported Flow section when you import the text flow.

You cannot edit text insets in the container document. Selecting them results in the entire text inset being selected, just like a cross-reference or a variable (except that the text inset could be many pages long).

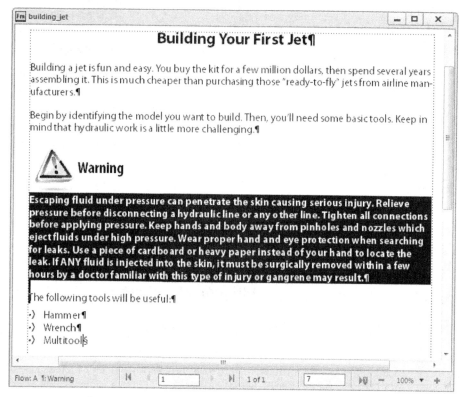

Instead, you must open the source document and make any changes there. After you save the file and update your text insets, the new information is displayed wherever that text inset is used.

Working with text in insets

For the most part, content inside a text inset behaves like regular text. In some cases, though, there are differences between regular text and text in a text inset:

- **Adding cross-references.** If you want to add a cross-reference to a heading that is part of a text inset, there is an extra step that isn't obvious. Before you cross-reference anything in the text inset, open the text inset and add a cross-reference marker to the heading you want to cross-reference. Save the text inset. To add cross-references to headings in the text inset, point to the container file, not the text inset.

Tip If you add a cross-reference to a container file without first making sure the text inset has the necessary cross-reference marker saved in it, the cross-reference breaks when you update the cross-references. The cross-reference often looks fine (although it won't be updated if the text or page number changes), but the broken cross-reference message keeps reappearing. It's not unusual for authors to redo the cross-references repeatedly and then finally give up on text insets in frustration.

- **Finding unresolved cross-references.** Text insets are checked for unresolved cross-references when you open the container file.
- **Finding and replacing information.** The Find/Change command does not work inside text insets when searching a container file. To search inside a text inset, you must open the text inset file.
- **Spell-checking.** When you spell-check a container file, its text insets are not spell-checked. You must open the text inset file to spell-check it.
- **Variable definitions.** If you use a variable inside a text inset, the container file may display the variable definition from the text inset or the variable definition from the container file, depending on your settings. If you specify that you want to retain the formatting from the source file, the variable definition from the text inset file is used. If you specify that the text inset should be reformatted using the container document's settings, the variable definition from the container document is used, *provided that the variable is defined in the container document*. If it is not, the original value from the inset file is used.
- **Conditional text settings.** Show/Hide settings and conditional text definitions (color and style settings) are handled the same way that variable definitions are. If you retain formatting from the source document, settings from the text inset are used. If you reformat with the container document's settings, the container document overrides tags with matching names.
- **Generated lists.** When you create a generated list from the container file, information from the inset file is included.
- **Separating stacked text insets.** After a text inset, make sure you insert a blank paragraph (as regular text in the container document) with a space in it before beginning the next text inset. Stacking text insets without the "separator" paragraph causes formatting problems. The second text inset takes on the formatting used by the first line of the first text inset. (You can define the separator paragraph with a two-point font size and negative space above or below so that it doesn't take up any space on the page.)
- **Converting text with insets to HTML.** When you convert to HTML, text insets are treated like regular text.

Considering text insets and other reuse options

Text insets are not the only way you can reuse information. As you begin to plan your approach, consider using any or all of the following techniques:

- **Cross-references.** With cross-references, you can link to a paragraph of text. This works well for standard, repeated information, such as a copyright statement or standard warnings. However, you cannot create a cross-reference that picks up more than a paragraph of text at one time, so if you need to insert more than one paragraph of text, cross-references won't work.
- **Conditional text.** Conditional text lets you label information as belonging to a particular version of a document. You can use conditional text to embed two (or more) versions of the same document in a single file. Text insets, by contrast, let you reuse the same information in multiple files.
- **Multiple books.** You can create two versions of a document by creating two book files. You include the files that the two versions have in common in both books; the unique files are in only one book. The main disadvantage to this approach is that each file must start on a new page. If your product differences are at the chapter level, this approach usually works quite well. Text insets do not have to start a new page.
- **Text insets.** Lets you embed chunks of text from one file in another. Chunks can include any type of content—paragraphs, character tags, anchored frames, equations, and so on—and can be a paragraph or several pages long. When you create a text inset, the information is integrated into the container file without requiring a page break. To create different versions of content, you must create two or more container files and import different sets of text modules into those files.

Planning modular text

Like conditional text, text insets are much easier to create than to plan. Assigning a condition or importing a text inset is the easy part; deciding what conditions to use and what information should go in which text inset is much more difficult. Without careful planning, you're likely to encounter serious frustration with text insets.

Breaking text down into modules

Breaking down information into modules can be a bit of a black art. Headings are a logical starting place. Some very hierarchical documents are easy to break up. For example, an alphabetical list of programming commands is easy to break into modules where each module contains one command along with its explanation, code example, and the like. For task-oriented user's guides, a logical starting point for creating modules would be procedures. The most difficult part in creating modular procedures is the transition information from one procedure to the next. Instead of trying to write generic transitions in the text inset, consider inserting that content in the container document as you assemble the documents.

File storage

File storage requirements will depend on the number of modules. If you are sharing just a few modules across a few books (such as a copyright page, system requirements list, and document conventions), you may be able to create a simple file structure with a shared folder for the modules.

As the number of modules increases, however, it becomes more difficult to manage the module files. If you have hundreds or thousands of module files, you will probably need to set up a content management system (CMS), which lets you check in and check out files.

Information retrieval

Setting up a storage system for files can be complicated, but establishing a system that ensures you can find relevant modules is an even bigger challenge. To locate files, you need to track some or all of the following information about each module:

- Main topic/title
- Keywords
- File location
- Author
- Last revision date
- Used in which books

For a small group of modules, a table or spreadsheet could suffice:

Topic	Keywords	File	Author	Revised
copyright	copyright, legal statements, trademarks	shared/copyright.fm	SSO	1/30/2013
system requirements	system requirements, hardware, software, RAM, memory	shared/reqs.fm	SAL	1/12/2013
document conventions	boldface, italic, conventions, menu selection	shared/conventions.fm	ASP	1/16/2013

If you have hundreds or thousands of modules, consider setting up a database or content management system to keep track of them.

Controlling formatting in text insets

Like any other FrameMaker file, a file used as a text inset contains tagging information. The text inset file also includes formatting information that specifies how a paragraph tag should be displayed.

When you import a file as a text inset, you can use the formatting specified in the text inset file. But more often, you specify that the container file's formatting should override the information in the text inset. This results in a situation where the appearance of information in the text inset file may be significantly different from the appearance in the text inset because the formatting templates are different.

The following example shows our original example warning in the original file, but with different formatting. Notice the autonumbering string ("WARNING:") and the hanging indent.

> WARNING:) Escaping fluid under pressure can penetrate the skin causing serious injury. Relieve pressure before disconnecting a hydraulic line or any other line. Tighten all connections before applying pressure. Keep hands and body away from pinholes and nozzles which eject fluids under high pressure. Wear proper hand and eye protection when searching for leaks. Use a piece of cardboard or heavy paper instead of your hand to locate the leak. If ANY fluid is injected into the skin, it must be surgically removed within a few hours by a doctor familiar with this type of injury or gangrene may result.§

After that text flow is imported into a container document, the appearance of the warning changes significantly. The warning paragraph tag in the destination document drops the autonumbering string ("WARNING:) and instead uses a frame above the paragraph (see page 124) to insert the exclamation mark icon and the word warning. The font is different, and there is no hanging indent.

> ⚠ **WARNING**
> Escaping fluid under pressure can penetrate the skin causing serious injury. Relieve pressure before disconnecting a hydraulic line or any other line. Tighten all connections before applying pressure. Keep hands and body away from pinholes and nozzles which eject fluids under high pressure. Wear proper hand and eye protection when searching for leaks. Use a piece of cardboard or heavy paper instead of your hand to locate the leak. If ANY fluid is injected into the skin, it must be surgically removed within a few hours by a doctor familiar with this type of injury or gangrene may result.¶

Tip In addition to modifying paragraph styles, the inset items also inherit other settings, such as variable definitions and conditional text settings. That means you can set a variable for a product name in the text inset, and display different product names depending on the variable definition in each container document. You do not have to update the text inset's variable definition. The inset is displayed in different containers with different values for the variable.

If you are overriding formatting in the source files, the actual paragraph tag settings in the source files are irrelevant. It is important, though, to ensure that each text inset uses the correct set of tags. When you update your template, you must import those formats into each text inset file. To help make this task easier, consider creating a book to manage text insets. The book contains a link to each of the text inset files, so you can update formats definitions for each text inset by importing formats to all of the files in the book.

Using a dummy book also lets you perform global search-and-replace operations across text insets—remember that search and replace on a container file does *not* look inside the inset files.

Creating a text inset

To create a text inset, you need at least two documents: the module file and the container file into which you plan to import the module.

To set up the text inset, follow these steps:

1. Open the container document.
2. Position your cursor where you want the text inset to start.

Tip Use a blank paragraph before the inset to ensure that the text inset doesn't get mashed into the previous paragraph.

3. Select **File > Import > File** (just as if you were getting ready to import a graphic) to display the Import dialog.

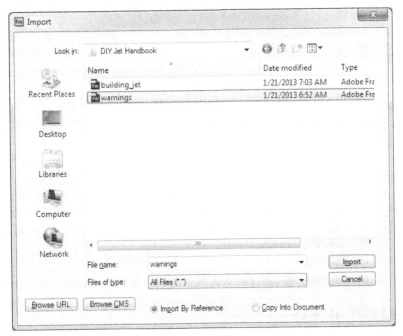

4. In the Import dialog box, locate and select the file you want to use. Make sure that the Import by Reference radio button is selected.

5. Click the Import button to display the Import Text Flow by Reference dialog.

6. Select a flow to import. The main body page text flow is usually A.

7. In the Formatting of Imported Flow section, specify whether you want to apply the current document's template. Your choices are as follows:

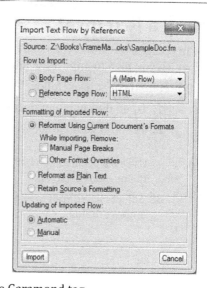

- **Reformat Using Current Document's Catalogs:** Inserts the content and applies the paragraph and other tags from the current document. Any matching tags are overwritten, so if the text inset uses a Body tag that's defined with Helvetica, and the container document has a Body tag that's defined with Garamond, all of the Body tags in the text inset will be displayed with the Garamond tag.

- **Reformat as Plain Text:** The entire imported text flow is formatted using the format of the paragraph tag into which you are inserting the text.

- **Retain Source's Formatting:** The inset retains the formatting specified in the inset document.

8. Specify how you want the imported flow updated:

- **Automatic:** The text inset is updated when you open or print the container document and when a book that includes the container document is updated with an update for text insets specified.

- **Manual:** The text inset is updated only when you select **Edit > References** and specify that you want to update text insets set for manual updates or when you update the book and specify that you want to update text insets.

9. Click the Import button. The specified flow is imported into the container file.

Managing text insets

Text insets are updated differently from imported graphics. When you change a graphic file that's imported by reference, FrameMaker immediately picks up the change in the graphic and updates the version displayed in the FrameMaker file. Text insets are handled like cross-references; an update occurs when you open a file, update the book, and the like. However, unlike cross-references, you can set text insets for two different types of updates—manual or automatic. This setting determines when the text insets are updated.

When you first import a text inset, you are prompted to specify whether the inset should be updated manually or automatically (see step 8 in the preceding section).

Managing text insets

To change the update method for a text inset, follow these steps:

1. Double-click the text inset. This displays the Text Inset Properties pod.
2. Click the Settings button. This displays the Import Text Flow by Reference dialog box (see page 522).
3. In the Updating of Imported Flow section, select Automatic or Manual.
4. Click the Import button to update the text inset's settings.

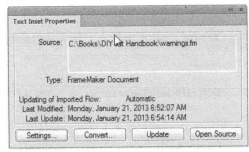

FrameMaker provides several different ways to update text insets. Some methods update only the insets set up for automatic updates; some update both.

Text insets set for manual updates are refreshed only when you explicitly request an update.

Tip To update a single text inset, double-click it to display the Text Inset Properties pod (see page 523), then click the Update Now button. To Update All the text insets in a file, complete the next procedure.

To update text insets, follow these steps:

1. In the container document, select **Edit > Update References** to display the Update References dialog.
2. Check the items you want to update. To Update All text insets, make sure that both the Text Insets Marked for Manual Update and the Text Insets Marked for Automatic Update checkboxes are checked.
3. Click the Update button to update the specified items.

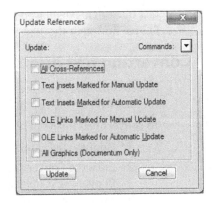

523

There are a number of other ways to update text insets; as shown in the following table.

Action	Automatic Updates	Manual Updates
Open file	Yes	No
Edit > Update References	Yes, if Text Insets Marked for Automatic Update checkbox is checked	Yes, if Text Insets Marked for Manual Update checkbox is checked
Update book	Yes, if Update All Text Insets is selected	Yes, if Update All Text Insets is selected
Print file	Yes	No
Save file	Yes	No
Double-click text inset, then click Update Now button	Yes	Yes

Opening the source file

You can open the text inset source file just as you would any other FrameMaker file. To access a text inset that's displayed in the document window, double-click it, and then click the Open Source button in the Text Inset Properties pod (Figure 1 on page 523).

Converting text insets to text

If necessary, you can convert text insets to regular text. Doing so destroys the link from the container file to the text inset file.

To convert text insets to text, follow these steps:

1. Double-click a text inset to display the Text Inset Properties pod (see page 523).
2. Click the Convert... button. This displays the Convert Text Insets to Text dialog.
3. Do one of the following:
   - To convert only the current text inset, click the Selected Text Inset radio button.
   - To convert all text insets in this file to text, click the All Text Insets radio button.
4. Click the Convert button. The specified text insets are converted to text.

Caution Converting text insets to text destroys the links. The information is no longer linked to the source file, so updating the source file doesn't change anything in the container files.

Chapter 30: Dropbox and Cloud Collaboration

Starting with FrameMaker 12, you can use a Dropbox or other cloud file sharing service to manage content directly from within FrameMaker. Here are a few of the benefits:

- Collaborate with reviewers who may have limited access. Make it easy to deliver files to reviewers and team members who don't have access to content management systems or specific internal networks.
- Access files on the go. Dropbox and other cloud-based services are accessible from any location with an Internet connection. Work with content on any machine, at any time, from anywhere.
- Keep content up-to-date. Dropbox and other cloud-based services automatically synchronize and update files, giving you access to content as soon as it becomes available.
- Work offline when needed. Dropbox and other cloud-based services allow files to be downloaded to a local folder for times you need to work but aren't connected to the cloud.

Here's a look at what's in this chapter:

> Configure Dropbox 525
> Share for review 525
> Open files. 526
> Save files locally and access content offline 526

Configure Dropbox

For this example we'll focus on Dropbox, the service identified by name in the FrameMaker File menu. Follow these steps to start working with Dropbox:

1. Download and install the Dropbox application. To work with Dropbox the Dropbox desktop application must be downloaded and installed. This provides access to Dropbox directly through a folder on your hard drive. Open up a web browser and navigate to www.dropbox.com. Log in or create an account as needed.
2. Select Edit > Preference > Global > Dropbox to configure Dropbox within FrameMaker. See "DropBox" on page 556.
3. Save to repository. Select **File > Dropbox > Save to Dropbox folder**.

Select **Save to Dropbox folder with Dependencies** if the file has links to other content (images and cross-references for example) or if saving a book to Dropbox to ensure all related files are also saved to the repository.

Select the location and complete saving as needed.

Share for review

Share your files for review by providing links via Dropbox. Sharing a link provides access to a specific file. From within Dropbox you can also share entire folders.

To share a link, do the following:

1. Select **File > Dropbox > Share**. The folder that contains the current file opens.
2. Right-click the file to share to display a list of options. Select **Share Dropbox link**. The path and file info from Dropbox are copied to the system clipboard.
3. Share the link with your team.

Open files

You can open and modify files hosted on Dropbox without losing any FrameMaker functionality.

Files are stored locally and synchronized with the Dropbox website. This allows you to work locally with files and have them automatically synchronize with the web-based service. To access Dropbox content, do the following:

1. Select **File > Dropbox > Open** and select the file you wish to open.
2. Click **Open** and work with the file using all the normally available features in FrameMaker.

Save files locally and access content offline

Files can be stored on Dropbox by saving files to a synchronized folder. If you will be offline or disconnected from Dropbox you can save a copy of required files to another location. To save files locally, do the following:

1. Select **Dropbox > Save locally**.
2. Select the location and complete saving as needed.

Select Save locally with Dependencies if the file has links to other content (images and cross-references for example) or if saving a book from Dropbox to ensure all related files are also saved to the local location.

Chapter 31: Using a Content Management System (CMS)

As your content grows, and as you add more people to the editing process, your document reuse needs are likely to grow. Variables, cross-references and text insets are useful, but limited in allowing sophisticated content reuse. A content management system (CMS) extends your content reuse capabilities. Starting with FrameMaker 11 there is a new Connection Manager (**CMS > Connection Manager**) with connections to commercial CMS systems (Documentum and SharePoint) or to define connections to any other WebDAV CMS. FrameMaker 11 also introduced a Repository Manager to manage interactions with defined CMS systems as needed. Adobe provides a specific forum for FrameMaker CMS discussions at forums.adobe.com/community/framemaker/cms.

Here's a look at what's in this chapter:

Default connectors	.527
Creating a custom connection to a CMS	.528
Working with WebDAV	.528
Managing workgroups	.528
Setting up a CMS connection	.529
Using the Repository Manager	.530
Uploading the current document	.530

Default connectors

FrameMaker provides connectors for both the Documentum and SharePoint CMS products. Starting with FrameMaker 12, there is also native support for Adobe Experience Manager. For specific information on versions supported, and for help configuring these connectors, see Working with Content Management Systems at help.adobe.com/en_US/framemaker/using/index.html or refer to your Documentum and SharePoint documentation.

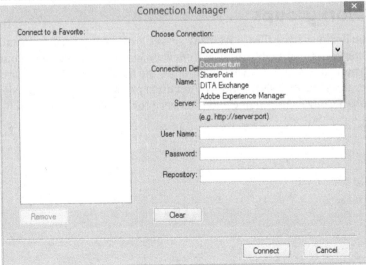

Select **Edit > Preferences... > CMS > (Documentum or SharePoint)** for preference settings related to these products.

Creating a custom connection to a CMS

FrameMaker connects to standard content management systems (CMS), supporting API's and ExtendScript. Adobe has provided enhanced documentation for creating connectors at blogs.adobe.com/tcs/2012/12/uncategorized/your-cms-and-framemaker11.html.

If creating your own connection seems a bit daunting, see the folks listed in "Third-Party Tools and Plug-Ins" on page 534 for assistance.

Once installed, follow the instructions in the "Setting up a CMS connection" section.

Working with WebDAV

Web-based Distributed Authoring and Versioning, or *WebDAV*, is a technology that lets you share and manage documents over the Internet. Using WebDAV, you can take advantage of basic content management features without buying a separate software package. The documents you want FrameMaker to manage are stored on a server running the WebDAV protocol. To access the files, you set up a *workgroup* in FrameMaker that points to the WebDAV server. To edit files, you check out the files, which means working copies of the files are placed in a folder on your hard drive. FrameMaker also locks files to prevent users from modifying the same documents simultaneously. While other users can open the files you checked out and can download the last version saved on the server, they can't modify the files until you have checked the files back into the system.

FrameMaker files can be managed in a workgroup along with non-FrameMaker files, such as imported graphics or PDF files. Applications other than FrameMaker also support WebDAV, so you can open a managed graphic in Adobe Illustrator, for example, or open a managed PDF file in Adobe Acrobat.

Managing workgroups

Before you begin sharing files, a WebDAV server must be installed and configured. Many WebDAV servers are available for free from www.webdav.org. Typically, your system administrator (or an eager coworker) handles the installation. The server can be difficult to configure, and security policies must be set up to provide for secure transfers over the Internet. You'll probably have a user name and password to download and upload files.

Setting up a CMS connection

FrameMaker comes with default CMS connectors to Documentum and SharePoint, and Adobe Experience Manager. To connect to installations of available CMS systems, follow these steps:

1. Select the **CMS > Connection Manager...** menu item to display the Connection Manager dialog.

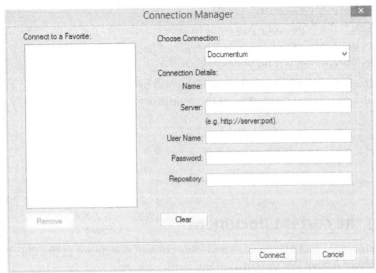

2. Select your CMS, fill in the appropriate connection details, and click the Connect button.
3. For more information on connecting to your CMS, see your system administrator, or your CMS documentation.

Using the Repository Manager

You can control content from any of your connected CMS via the Repository Manager.

To access the Repository Manager, follow these steps:

1. Select the **CMS > Open Repository...** menu item to display the Repository Manager pod.
2. Select the repository you want to access.
3. *(optional)* Use the Search and Advanced Search options to locate content within your CMS.
4. Use the buttons at the bottom of the pod to check in, check out, or remove content from the repository.

Uploading the current document

When connected to a CMS, you can add your current document to the CMS by selecting the **CMS > Upload Active Document...** menu item.

Part VII

Appendixes

Appendix A: Resources

This appendix lists FrameMaker-related resources, including add-on software, web sites, and mailing lists. We have verified all of these sites at the time of printing, but web site addresses can change at any time.

Note These tools are listed as a convenience for you, the reader. Not all of them are part of our everyday workflow, but many FrameMaker users speak highly of them. Also, this appendix is not an exhaustive list of web sites and tools; there are many good tools that are not listed here.

FrameMaker and Technical Communication Web Resources

Adobe Systems. Adobe Systems makes FrameMaker software. Their web site contains product information, the current user guide, new feature videos, the latest reviewer's guide, support information, product updates, printer drivers, and more.

http://www.adobe.com/products/framemaker.html

help.adobe.com/en_US/framemaker/ini/index.html

Tech Comm Tools. Matt Sullivan's blog on tech comm video, FrameMaker, RoboHelp, and other tech comm topics. An excellent resource for those using FrameMaker and/or RoboHelp to produce online output of their FrameMaker files.

Matt has been deeply involved in the development of every version of the Technical Communication Suite. He represents Adobe regularly at industry conferences and is also the author of this book. Matt develops workflow and deployment solutions for technical and mobile content.

www.techcommtools.com

www.mattrsullivan.com

Scriptorium Publishing. Scriptorium Publishing offers expert advice on how to develop, deploy, and manage information. The company provides content strategy analysis and implementation services to transform documentation into business assets.

www.scriptorium.com

Mailing Lists/User Groups

Adobe Community Help. A series of active peer-to-peer user forums. One of your best options for free support and access to experts.

forums.adobe.com/community/framemaker

Friends of FrameMaker. A LinkedIn group devoted to FrameMaker news and developments.

https://www.linkedin.com/grp/home?gid=897507

FrameMaker-DITA Yahoo User Group. For folks interested in editing structured DITA content in FrameMaker.

tech.groups.yahoo.com/group/framemaker-dita/

DITA-FMx-Users Yahoo User Group. For users of Scott Prentice's DITA-FMx product.

tech.groups.yahoo.com/group/dita-fmx-users/

HATT. A high-volume list for help authoring tools and technology.

groups.yahoo.com/group/hatt

TCS-Users Google User Group. For folks interested in the integration available with the Adobe Technical Communication Suite applications. Primarily FrameMaker and RoboHelp linking content.

https://groups.google.com/forum/?fromgroups#!forum/tcs-users

Techwr-l (TechWhirl). A high-volume list for technical writers, run by Al Martine and the folks at INKtopia. Their website includes a jobs board and online magazine.

www.techwr-l.com

Third-Party Tools and Plug-Ins

FrameMaker offers lots of features, but every now and then you run across something you really wish it had. In many cases, a third-party plug-in or add-on software can accomplish what you want. This section lists some sites for software that extends FrameMaker's features.

Carmen Publishing Inc. Rick Quatro of Carmen Publishing Inc. specializes in FrameMaker automation with ExtendScript and FrameScript. He also has a series of FrameMaker plug-ins and scripts to make FrameMaker easier and more productive, including TableCleaner, ImportFormatsSpecial, and PageLabeler. He offers ExtendScript and FrameScript training.

www.frameexpert.com

www.frameautomation.com

CudSpan Tools. CudSpan provides tools for working with FrameMaker documents, including tools to list paragraph format overrides, work with markers, remember page and line breaks in generated files, associate icons with paragraphs, and automatically create MIF copies when saving files.

cudspan.net/plugins/

IXgen. The IXgen tool suite provides a number of tools for creating markers, then editing them. Create markers from the text of selected paragraph tags, from keywords (automatic or interactive), or from character tags. Edit index markers in a single alphabetic list. Spell-check, find/change, character formatting applied from the Character Catalog.

Fully-functional demos at www.fsatools.com

FrameScript. FrameScript from Finite Matters lets you automate tasks inside FrameMaker. FrameScript is a commercial product, however, ExtendScript is similar and installs automatically with FrameMaker and other Adobe products. See Chapter 26, "Automation with ExtendScript."

www.framescript.com

Klaus Daube. Toolbar enhancements for FrameMaker.

www.daube.ch/index.html

Leximation Tool Search. Scott Prentice maintains this site, which has a comprehensive list of many FrameScript scripts and FrameMaker plug-ins (including some of his own). If there is a tool you need for FrameMaker, it is likely you can find it by looking here. His MarkerTools plug-in makes it much easier to edit markers that are stacked on top of each other, lets you use variables within markers, and lets you create templates for marker text (which are handy for all those "see also" index entries).

www.leximation.com/toolsearch

Lin Sims' List of Plug-ins. A fairly extensive list of plug-ins compiled by Lin Sims. Found at either:

www.microtype.com/links.html

MicroType. Shlomo Perets offers tools that let you automate many FrameMaker-to-PDF items (PDF initial view settings, colored & custom bookmarks, custom named destinations, hyperlink highlights that show up online but don't print, and more). The main tool is FrameMaker-to-Acrobat TimeSavers, which has a number of add-ons available for specialized PDF tasks like adding form fields, web-based multimedia, custom navigation and setting additional options for displaying 3D files.

www.microtype.com

Setprint. This plug-in from Sundorne Communications makes the Adobe PDF Printer (or whichever printer you choose) the default printer for FrameMaker while allowing your other applications to use the default printer selected in the Windows Control Pod.

www.sundorne.com/FrameMaker/Freeware/setPrint.htm

Silicon Prairie Software. The Auto-Text plug-in lets you associate keystrokes or menu choices from a custom menu with text or graphics. Your menu can contain graphics, text, formatted tables, paragraphs with pictures beside them, symbols, and other items you use frequently. To change the menu, you update a single FrameMaker file.

Other tools from this site include Alert Tool, Character Tools, Index Tools, Index Tools Professional (which includes instructions for creating a master index for a set of FrameMaker books), LEP Tools for creating Lists of Effective Pages (often required for military documents), Master Page Tools, Outline Tools, Paragraph Tools, Readability Tools, and Table Tools.

www.siliconprairiesoftware.com

SnagIt. This screen capture tool lets you take screen captures, preview and edit them, and then import them by reference without leaving FrameMaker. The FrameMaker add-in is not installed by default, so be sure to select the Custom installation choice to add it.

www.techsmith.com

West Street Consulting. Russ Ward of West Street Consulting specializes in plugins for structured FrameMaker. These plugins add features such as enhanced conditional text and text inset functionality, all built around the usage of structural metadata.

www.weststreetconsulting.com

Database Publishing from FrameMaker

The following tools retrieve information from databases and use FrameMaker to apply complex formatting. For example, these tools can create catalogs of information stored in a database.

Miramo. Miramo lets you bring database content into FrameMaker. They also have a free Miramo Personal Edition (MPE) available from www.miramox.com.

www.miramo.com

PatternStream. PatternStream lets you bring database content into FrameMaker.

www.fml.com

Online Manuals

You can download manuals and tutorials from the Adobe web site:

www.adobe.com/support/documentation/en/framemaker

Reporting Bugs

If you have a FrameMaker problem you want to report to Adobe Systems, use this bug report form:

www.adobe.com/go/wish

Appendix B: Shortcuts

Here it is, a list of shortcuts-the one thing I've been asked about in every class I've ever taught. The thing I thought I couldn't possibly put together, yet here it is!

Unfortunately, a comprehensive list of shortcuts with descriptions might be as long as this book itself. Instead I've compiled a few resources that you might use to create your own shortcut sheet.

Here they are, and I hope you enjoy them.

Most leaned-upon shortcuts while editing this book

Object styles catalog. While technically not a shortcut, Object Styles really speed up my work

esc j j Repeat last paragraph tag

esc c c Repeat last character tag

ctrl+8 Smart Insert character tag

ctrl+9 Smart Insert paragraph tag

ctrl+0 Smart Insert variables

ctrl+f Find/Change, and everything I can do with it

FrameMaker Keyboard Shortcuts

A poster from Adobe, showing keystrokes that will help just about everyone.

http://blogs.adobe.com/tcs/files/2011/02/FM-Keyboard-Shortcuts_A3size.pdf

FrameMaker Help

The last few releases of FrameMaker have help files available on the web. While you can directly search the web for shortcuts, I find it more useful to search the help system for version-specific content, and if necessary, click on the web link to open in a browser. Opening in a browser allows me to save the content and process as needed to make my own shortcut sheets.

Search on **Keyboard Shortcuts** in the FrameMaker help system for a version-specific list of shortcuts.

Search on **Character Sets** in the FrameMaker help system for a version-specific list of characters, shortcuts, and codes for entering characters in dialogs and other locations.

Compiled FrameMaker 11 commands

Many thanks to Klaus Daube for generating a list of all commands available in FrameMaker 11. This list is a few versions old, but excellent for understanding the logic used in all the shortcut assignments. At 37 pages, you can see why I hesitate to list shortcuts, even in line-item format.

http://www.daube.ch/docu/files/etb-fm11-commands.pdf

The list was created using a FrameScript available at i-frame. For more details on free and paid FrameScripts from itl, see

http://www.i-frame.itl.info/en/feature-description.html

Control+1 through Control+9

Starting with FrameMaker 11, the Smart Insert/Edit dialogs have become a mainstay of my daily workflow Here are the functions you have at your disposal:

Table B-1. Control key functions

Keystroke	Smart Insert dialog
Structured Functions	
Ctrl+1	Insert element
Ctrl+2	Wrap element
Ctrl+3	Change element
Ctrl+7	Edit attributes
Conditional Text Functions	
Ctrl+4	Add conditional tag
Ctrl+5	Remove conditional tag
Ctrl+6	Remove all conditional tags
Other functions	
Ctrl+8	Character catalog
Ctrl+9	Paragraph catalog
Ctrl+0	Variable catalog

My favorites are the Ctrl+8 and Ctrl+9 keys. I use them much more often than the actual Character and Paragraph catalogs.

Close behind them, I use the conditional keys (Ctrl+4, 5, 6) far more often than the Conditional Tags pod, which I find cumbersome.

Function keys (F1 through F9)

Here's a list of Function keys, based on Klaus' PDF. While many of the functions introduce overrides to the text, a good many of them are useful even in a tightly controlled template-based workflow.

Table B-2. F key (Function key) functions

Key	Function	Control	Shift	Alt
F1	Help	Align top	Context sensitive help	Set lower case
F2	Plain text	Align middle	Go to previous view	Set initial cap
F3	Underline	Align bottom	Overline	Set uppercase
F4	Bold	Quit	Cascade Windows	Exit
F5	Italic	Show/Hide conditional text	Tile Windows	Anchored frame
F6	Repeat last command*	Show next	Show previous	Toggle between document windows
F7	Smart insert conditional text	Select same condition tags	Set focus back to Document Window	Copy conditional text settings
F8	Smart insert character tag	Select same character format	Set dialog box to "As Is"	Copy character format
F9	Smart insert paragraph tag	Transpose characters	Change settings to match current text	Copy paragraph format
F10	Repeat last element command	Select paragraph	Context menu	Copy table column width

Window (Alt key) shortcuts

While not specific to FrameMaker, using the Alt key to invoke keyboard navigation is still one of my favorite shortcut techniques. Find yourself using the **File > Import > Formats** command regularly? Use the Alt key to quickly divine the ALT F I O shortcut. Using these quick keystroke combos keeps your hands on the keyboard, and saves valuable hand-eye effort and minimizes the mental focus required to perform mundane tasks.

Appendix B: Shortcuts

Appendix C: Building blocks

Building blocks are placeholders in FrameMaker. When FrameMaker encounters a building block, it replaces the building block with a real value. Building blocks are used in cross-reference formats, variables, autonumbered paragraph tags, reference pages, and master pages. For example, the cross-reference format called "See Heading & Page" displays the heading and page of the referenced paragraph tag as:

> See "Managing color definitions" on page 432.

The cross-reference definition is:

> See "<$paratext>" on page <$pagenum>.

When the document is opened or saved, the heading and page number are updated automatically. If the heading changed to "Editing Color Definitions," the cross-reference would include the updated heading.

Building block usage

Building blocks are enclosed by angle brackets and are often preceded by a dollar sign (<$year>, for example). The following table describes FrameMaker building blocks and indicates where each building block is valid. The table is sorted by symbols and then alphabetical entries, so you'll find building blocks beginning with "$" toward the top and building blocks such as "<R>" at the end.

FrameMaker provides Japanese building blocks as well as RTL (Right to Left language) building blocks, but they're beyond the scope of this book. Refer to the FrameMaker help system for details.

Note Some building blocks don't do anything when used alone; instead, they work with other building blocks to display information. For example, <$creationtime> doesn't display the creation time; it causes the building blocks that follow (such as <$hour>:<$minute> <$AMPM>) to display the creation time. Building blocks that don't produce output have "(undisplayed)" in the description.

Building Block	Displays	Autonumber	Cross-Reference	Index Marker	Variable	HTML Reference Page	Other Reference Pages
;	Separates entries in an index marker			•			
:	Separates levels in an index marker			•			
[ ]	Indicates sort order for an index entry			•			
<$alphabetics>	Sort order for alphabetic entries						•
<$ampm>	Lowercase morning or evening designation (am)				•		

541

Appendix C: Building blocks

Used In

Building Block	Displays	Autonumber	Cross-Reference	Index Marker	Variable	HTML Reference Page	Other Reference Pages
<$AMPM>	Uppercase morning or evening designation (AM)				•		
<$autorange>	Automatic page ranges						•
<$chapnum>	Chapter number	•	•		•	•	•
<$condtag[*condtag*]>	Specified condition tag				•		
<$creationtime>	Causes any following time building block to display the creation time (only in running h/f and non-time system variables; undisplayed)				•		
<$curpagenum>	Page number (used only on master pages)						•
<$currenttime>	Causes any following time building block to display the current time (only in running h/f and non-time system variables; undisplayed)				•		
<$dayname>	Name of the day (Monday)				•		
<$daynum>	Number of the day (1)				•		
<$daynum01>	Number of the day with leading 0 (01)				•		
<$defaulttitle>	Text of the first document heading					•	
<$endrange>	End of a page range			•			
<$filename>	Name of the file example: widget.fm		•		•		
<$fullfilename>	Name of the path and file example: c:\Book\widget.fm		•		•	•	
<$hour>	Hours (1)				•		
<$hour01>	Hours with leading 0 (01)				•		
<$hour24>	Hours in 0–24 military format (13)				•		
<$lastpagenum>	Last page number in document				•		
<$marker1> through <$marker8>	Header/Footer $1 marker text through Header/Footer $8, as indicated				•		
<$minute>	Minutes (1)				•		
<$minute00>	Minutes with leading 0 (01)				•		

Building Block	Displays	Autonumber	Cross-Reference	Index Marker	Variable	HTML Reference Page	Other Reference Pages
<$modificationtime>	Causes any following time building block to display the time the file was last opened or saved (only in running h/f and non-time system variables; undisplayed)				•		
<$monthname>	Name of the month (January)				•		
<$monthnum>	Number of the month (1)				•		
<$monthnum01>	Number of the month with leading 0 (01)				•		
<$nextsubdoc>	URL of the next document					•	
<$nopage>	Suppresses the page number			•			
<$numerics>	Sort order for numeric entries						•
<$ObjectId>	Numeric identifier assigned to the linked object						•
<$ObjectType>	Numeric identifier indicating the type of link						•
<$pagenum>	Page number of the linked paragraph		•				•
<$paranum[paratag]>	Autonumber of first matching paragraph tag on page, including text (Chapter 1)	•	•			•	•
<$paranum>	Autonumber of the linked paragraph, including text		•			•	•
<$paranumonly[paratag]>	Autonumber of first matching paragraph tag on page, excluding text (1)	•	•			•	•
<$paratag[paratag]>	Name of first matching paragraph tag on page (ChapterTitle)	•	•			•	•
<$paratext[+,paratag]>	Text of last paragraph on page matching the tag	•	•			•	•
<$paratext[paratag1, paratag2,paratag3]>	Text of first paragraph on page matching the tag, in the order specified	•	•			•	•
<$paratext>	Text from the source paragraph		•			•	•
<$parentdoc>	URL of the first document					•	
<$prevsubdoc>	URL of the previous document					•	
<$relfilename>	Relative path to linked file example: Book\widget.fm					•	
<$second>	Seconds (1)				•		

Appendix C: Building blocks

Building Block	Displays	Autonumber	Cross-Reference	Index Marker	Variable	HTML Reference Page	Other Reference Pages
<$second00>	Seconds with leading 0 (01)				•		
<$sectionnum>	Tracks sections within books	•	•		•	•	•
<$subsectionnum>	Tracks subsections within books	•	•		•	•	•
<$shortdayname>	Name of the day (Mon)				•		
<$shortmonthname>	Name of the month (Jan)				•		
<$shortyear>	Year (30)				•		
<$singlepage>	Designates single page number after <$nopage>		•				
<$startrange>	Beginning of page range		•				
<$symbols>	Sort order for symbolic entries						•
<$tblsheetcount>	Total number of table sheets				•		
<$tblsheetnum>	Number of current table sheet				•		
<$variable[*varname*]>	Text of the variable					•	
<$volnum>	Volume number	•	•		•	•	•
<$year>	Year (2030)				•		
Building blocks used in autonumbers							
\t	Tab	•					•
• or \b	Bullet	•					
< =0>	Resets value to zero or another specified number (undisplayed)	•					
< >	Keeps value of first counter (undisplayed)	•					
<a+>	Lowercase alphabetic numbering; value increased by 1	•					
<A+>	Uppercase alphabetic numbering; value increased by 1	•					
<a=1>	Lowercase alphabetic numbering; value set to 1 (or another number)	•					
<A=1>	Uppercase alphabetic numbering; value set to 1 (or another number)	•					
<a>	Lowercase alphabetic numbering; value unchanged	•					

Building block usage

Building Block	Displays	Autonumber	Cross-Reference	Index Marker	Variable	HTML Reference Page	Other Reference Pages
<A>	Uppercase alphabetic numbering; value unchanged	•					
<char_tag>	Applies the specified character tag to the items that follow in the definition	•	•	•	•	•	•
<Default Para Font> or </>	Removes any character formatting and returns to the regular paragraph formatting of the parent paragraph	•	•	•	•	•	•
<n+>	Numeric numbering; value increased by 1	•					
<n=1>	Numeric numbering; value set to 1 (or another number)	•					
<n>	Numeric numbering; value unchanged	•					
<r+>	Lowercase Roman numeral numbering; value increased by 1	•					
<R+>	Uppercase Roman numeral numbering; value increased by 1	•					
<r=1>	Lowercase Roman numeral numbering; value set to 1 (or another number)	•					
<R=1>	Uppercase Roman numeral numbering; value set to 1 (or another number)	•					
<r>	Lowercase Roman numeral numbering; value unchanged	•					
<R>	Uppercase Roman numeral numbering; value unchanged	•					

Appendix C: Building blocks

Appendix D: Customizing maker.ini

The default settings and paths for FrameMaker are stored in the maker.ini files. You won't need to edit maker.ini often, but there are a few FrameMaker defaults that you may want to change. For example, you can change the clipboard paste order so that Unicode characters are included and not displayed as question marks when you paste from Microsoft Word. Also, some plug-ins for FrameMaker require that you edit maker.ini.

For FrameMaker 12, Adobe released more complete and updated documentation for the maker.ini and the ditafm-output.ini files. Along with expanded information about the .ini files, you'll find information about adding flags that are not in the maker.ini file by default. At time of printing, the new documentation is found at:

help.adobe.com/en_US/framemaker/ini/index.html

The following lists some of the information stored in maker.ini:

- FrameMaker version and license information
- Preferences such as whether to create backup files, remember missing fonts, and so on
- Recently opened files
- Pen thicknesses and patterns
- Directories for templates and samples
- The location of dialog boxes on the screen
- Information needed for plug-ins
- Spelling options
- Default font information
- Filters

Editing maker.ini

To edit maker.ini, do the following:

1. Close FrameMaker.
2. There are two maker.ini files. You will find maker.ini in these locations:
   - **The installation directory.** Your application's copy of maker.ini file might be in C:\Program Files (x86)\Adobe\AdobeFrameMaker11
   - **The user directory.** On Windows 7, your personal maker.ini file might be in C:\Users\(username)\AppData\Roaming\Adobe\FrameMaker\11

   Edit only one of the maker.ini files.

   Note If a setting is displayed in both maker.ini files, the settings in maker.ini in the Documents and Settings folder take precedence over the settings in the FrameMaker folder.

3. Create a backup of maker.ini.
4. Open the original maker.ini file in a text editor such as Notepad or PSPad.
5. Scroll down until you find the setting you want to change. The sections that follow describe some recommended changes.

6. Edit the text, then save maker.ini. Your changes are saved and applied the next time you open FrameMaker.

Setting clipboard pasting order

When you paste into FrameMaker, FrameMaker analyzes the content of the clipboard to determine how to paste the text. For example, if you paste a paragraph from one FrameMaker file into another, FrameMaker maintains the formatting, markers, and so on. However, pasting from some applications may result in an OLE object or text that displays gray boxes or question marks that indicate missing characters. For better results, find this line in maker.ini:

ClipboardFormatsPriorities=FILE, MIFW, MIF, RTF, OLE 2, META, EMF, DIB, BMP, UNICODE TEXT, TEXT

Change the pasting order to move UNICODE TEXT and TEXT closer to the front. This will aid in pasting from other applications, including MS Word. The result could be:

ClipboardFormatsPriorities=FILE, UNICODE TEXT, TEXT, EMF, META, DIB, BMP, MIFW, MIF, RTF, OLE 2

Updating graphics from web addresses

(optional) If you import graphics from the Internet or an intranet site, you can have FrameMaker always update the graphic. Find the following line in maker.ini:

AlwaysDownloadURL flag=Off

Change it to On:

AlwaysDownloadURL flag=**On**

Changing the substitution fonts

The fonts that FrameMaker substitutes for missing fonts are defined in maker.ini. For example, you might open a document that uses the Minion Pro font. If your system doesn't have the font, FrameMaker substitutes the default font, which is Times New Roman or Times Roman. If you know ahead of time that a document uses a font you don't have, you can map the font to a more appropriate font. Find the default font mappings:

; Default used to map unknown fonts:
;
DefaultFamily=Times New Roman, Tms Rmn
DefaultAngle=Regular
DefaultVariation=Regular
DefaultWeight=Regular
DefaultSize=12

Tip When you create a new FrameMaker document using one of the predefined FrameMaker templates, the default fonts defined in maker.ini are selected in paragraph and character tags.

Change the default family setting. You might also want to change the default size. In the following example, the default font is Veljovic and the size is 11:

; Default used to map unknown fonts:
;

DefaultFamily=**Veljovic**
DefaultAngle=Regular
DefaultVariation=Regular
DefaultWeight=Regular
DefaultSize=**11**

Appendix D: Customizing maker.ini

Appendix E: Preference settings

Starting with FrameMaker 11 a number of preference settings are consolidated in a common location under the **Edit > Preferences...** menu item.

Global

General

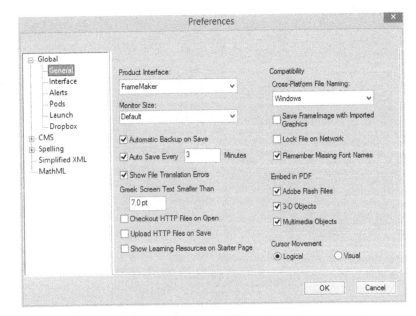

These are more or less the "traditional" FrameMaker preference settings.

Key general preference features

Product Interface. Allows setting of structured or unstructured FrameMaker mode. Requires restart before change will take effect.

Automatic Backup on Save. Creates a backup file in your document directory which you can use in case of file corruption.

Auto Save. Sets an option to create a temporary file (saved in your document directory) and the interval for updating the temp file.

HTTP File options. Tells FrameMaker whether or not to control checkin and checkout options from a content management system.

Show Learning Resources on Starter Page. Removes banner ad below the starter page when starting FrameMaker.

Save FrameImage with Imported Graphics. Creates a low-resolution image of imported graphics to be used if original document cannot be found.

Appendix E: Preference settings

Lock File on Network. Helps prevent colleagues from overwriting files on a network.

Remember Missing Font Names. Ensures that font substitutions made when opening documents are temporary.

Embed in PDF options. Can seriously increase PDF size, so evaluate need for embedding these assets if delivering PDF via network or web.

Interface

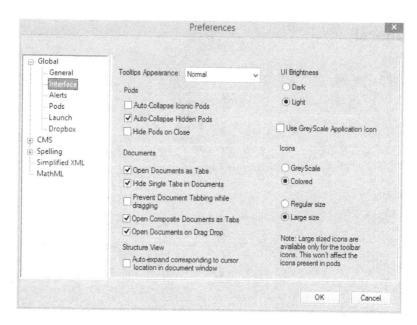

Key interface preference features

Tooltips Appearance. Can be set to Normal, Fast, or Hidden to control tooltip display while hovering over FrameMaker buttons and controls. These settings work on pod buttons, but not the toolbar or graphics tools buttons.

Auto-Collapse Iconic Pods. When displaying iconic pods, this setting allows pods to collapse (or hide) when selecting another pod or clicking in the document.

Auto-Collapse Hidden Pods. Controls behavior of pods when using the UI Visibility feature. Gray bars appear on the sides and bottom of the screen that allow you to access docked toolbars and composite documents (books).

Open Composite Documents as Tabs. Allows books to open into FrameMaker tabs. Deselecting this option forces composite documents to display in floating windows.

Icons. Starting with FrameMaker 12, allows for larger icons with color. This is a huge help to me in both authoring this book, and in teaching or using FrameMaker

Alerts

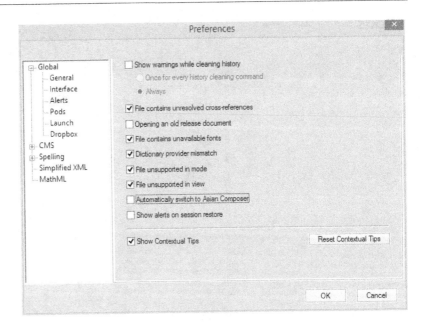

Key alert preference features

Show warnings while cleaning history. While it may be helpful to understand when FrameMaker will be clearing (not "cleaning," as stated in the pod) the undo history, it really doesn't let you change directions. You do stuff, the history gets cleared, you move on... *Recommendation*: Deselect this option to reduce stress and unnecessary alerts.

Other Alert warnings. When opening files, either manually or as part of a book update, FrameMaker analyzes the content. When missing resources or mismatches are detected, they can cause repetitive and time-consuming work to correct. You can suppress some of these messages using the Alerts options. Keep in mind that suppressing the dialog is not the same as resolving the issue, and that if you are part of a workgroup and/or using this content in a single-source workflow, suppressing errors rather than correcting them can have unintended results.

Appendix E: Preference settings

Pods

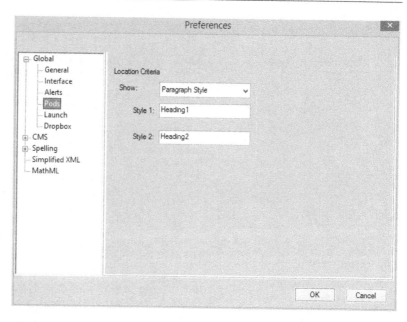

Key pods preference features

The Location Criteria helps you locate markers by displaying information about their location. In structured FrameMaker you can choose to display the text associated with either elements or paragraph styles. In unstructured FrameMaker, you can specify only paragraph styles. The text associated with the location of the marker is displayed as long as it is one of the two styles you specified.

The styles indicated here display in the **Markers** pod (**View > Pods > Markers**). See the following screen shots to see the difference between the Markers pod and the Marker pod.

The **Markers** pod lists the markers contained in the selected document.

The Marker pod lets you insert markers into your content.

Markers Pod

The Markers pod helps locate content containing markers.

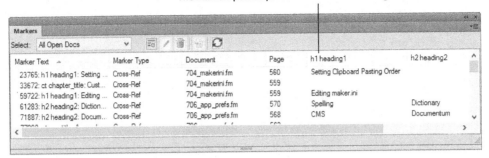

Marker Pod

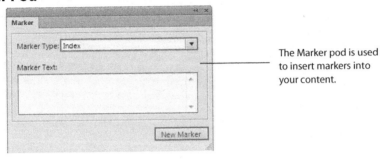

The Marker pod is used to insert markers into your content.

Launch

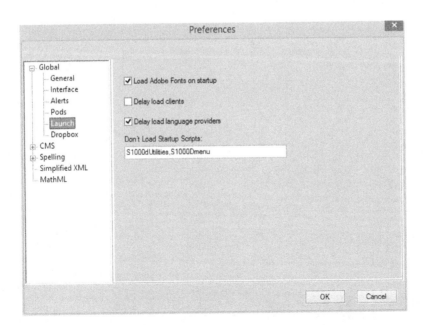

Appendix E: Preference settings

DropBox

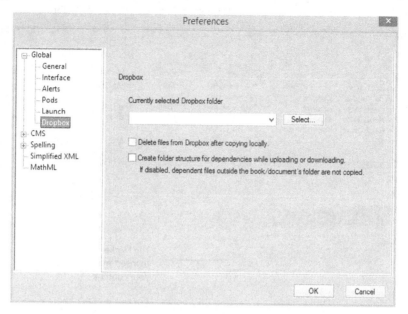

Key dropbox preference features

Works with your existing Dropbox account. You can use the Dropbox folder for shared document reviews. Using the Create folder structure option to see the locations of your referenced files.

CMS

Documentum

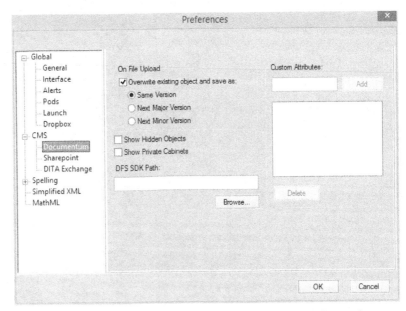

Lets you set Documentum preferences. Refer to your Documentum documentation for guidance on setting options here.

SharePoint

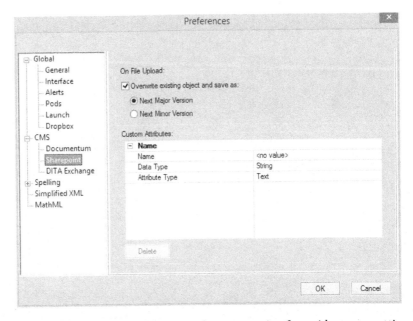

Lets you set SharePoint preferences. Refer to your SharePoint documentation for guidance on setting options here.

DITA Exchange

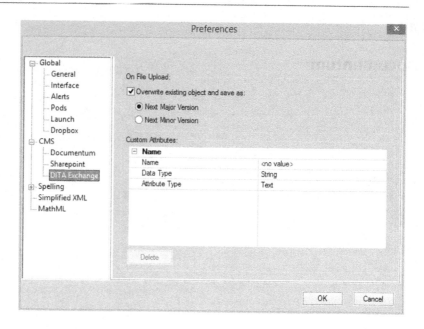

Spelling

Dictionary

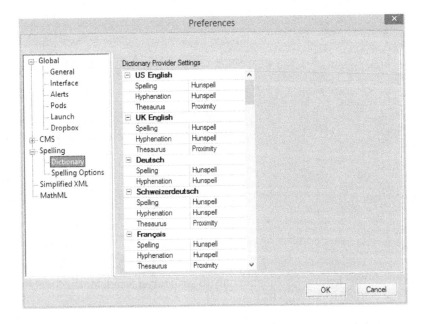

Key dictionary preference features

Starting with FrameMaker 11 there is support for Hunspell dictionaries in addition the Proximity dictionary supported in previous versions. You can customize the FrameMaker spelling checker options for each language by independently setting Spelling, Hyphenation and Thesaurus to either the OpenSource Hunspell dictionary, or the traditional Proximity dictionary.

Spelling options

See "Spelling Checker Options" on page 46 for a description of the options.

Simplified XML

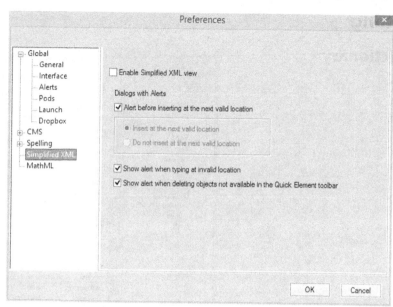

This feature is available only in structured FrameMaker. To access structured FrameMaker, see "Global" on page 551.

MathML

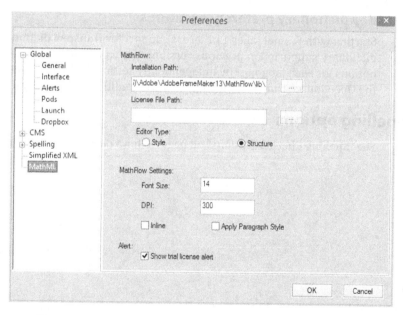

Key MathML Preference Features

Starting with FrameMaker 12, users can access a trial copy of DesignScience MathFlow, and manage MathML content placed in FrameMaker.

Appendix F: Maker Interchange Format

Normally, FrameMaker files are stored in a binary format. In some cases, though, it's useful to have a text version of FrameMaker files. For this, an alternate Maker Interchange Format (MIF) is available. A MIF file contains the same information as a regular, binary FrameMaker file, but it's encoded in a plain text file instead of a binary file.

Editing MIF files directly provides an alternative to working in FrameMaker's graphical interface. For some issues, this is the most efficient way to work. For example, you can use MIF files to perform global search-and-replace operations on fonts. This is much faster than editing each tag in the Paragraph Catalog—even working with the global update options may take more time than a single Find/Change command on a MIF file.

Be careful though, when working "under the hood" in MIF files, small changes in syntax and order of elements can result in files that will not open properly in FrameMaker. Make sure you keep a working copy of your .fm files, rather than relying too heavily on the MIF versions.

Sample MIF code

A MIF file contains all of the information of a .fm file, but in a more verbose format. The following MIF fragment shows one short paragraph of text:

What's displayed in FrameMaker

Cross-stitching is easy to learn, fun, and relaxing. Best of all, it doesn't require a computer!

Corresponding MIF

Paragraph tag Body is assigned to the paragraph.

```
<Para
 <Unique 998321>
 <PgfTag `Body'>
 <ParaLine
  <TextRectID 19>
  <String `Cross-stitching is easy to learn, fun, and relaxing.'>
 > # end of ParaLine
 <ParaLine
  <String `Best of all, it doesn't require a computer!'>
 > # end of ParaLine
> # end of Para
```

Appendix F: Maker Interchange Format

In addition to the text, graphics, markers, and so on, the MIF file also contains a complete listing of your formatting catalogs. The following code excerpt shows the definition for the BodyFirst paragraph tag referenced in the preceding example. Notice that each item available in the Paragraph Designer has its own entry in the list.

```
<Pgf
 <PgfTag `Body'>
 <PgfUseNextTag No>
 <PgfNextTag `'>
 <PgfAlignment Left>
 <PgfFIndent 0.0">
 <PgfLIndent 0.0">
 <PgfRIndent 0.0">
 <PgfFIndentRelative No>
 <PgfFIndentOffset 0.0">
 <PgfTopSeparator `'>
 <PgfTopSepAtIndent No>
 <PgfTopSepOffset 0.0">
 <PgfBotSeparator `'>
 <PgfBotSepAtIndent No>
 <PgfBotSepOffset 0.0">
 <PgfPlacement Anywhere>
 <PgfPlacementStyle Normal>
 <PgfRunInDefaultPunct `'>
 <PgfSpBefore 0.0 pt>
 <PgfSpAfter 0.0 pt>
 <PgfWithPrev No>
 <PgfWithNext No>
 <PgfBlockSize 2>
 <PgfFont
  <FTag `'>
  <FPlatformName `W.Times New Roman.R.400'>
  <FFamily `Times New Roman'>
  <FVar `Regular'>
  <FWeight `Regular'>
  <FAngle `Regular'>
  <FEncoding `FrameRoman'>
  <FSize 12.0 pt>
  <FUnderlining FNoUnderlining>
  <FOverline No>
  <FStrike No>
  <FChangeBar No>
  <FOutline No>
  <FShadow No>
  <FPairKern Yes>
  <FTsume No>
  <FCase FAsTyped>
  <FPosition FNormal>
```

```
<FDX 0.0%>
<FDY 0.0%>
<FDW 0.0%>
<FStretch 100.0%>
<FLanguage USEnglish>
<FLocked No>
<FSeparation 0>
<FColor `Black'>
> # end of PgfFont
<PgfLineSpacing Proportional>
<PgfLeading 2.0 pt>
<PgfAutoNum No>
<PgfNumTabs 0>
<PgfHyphenate Yes>
<HyphenMaxLines 2>
<HyphenMinPrefix 3>
<HyphenMinSuffix 3>
<HyphenMinWord 5>
<PgfLetterSpace No>
<PgfMinWordSpace 90>
<PgfOptWordSpace 100>
<PgfMaxWordSpace 110>
<PgfMinJRomanLetterSpace 0>
<PgfOptJRomanLetterSpace 25>
<PgfMaxJRomanLetterSpace 50>
<PgfMinJLetterSpace 0>
<PgfOptJLetterSpace 0>
<PgfMaxJLetterSpace 10>
<PgfYakumonoType Floating>
<PgfPDFStructureLevel 6>
<PgfLanguage USEnglish>
<PgfCellAlignment Top>
<PgfCellMargins 0.0 pt 0.0 pt 0.0 pt 0.0 pt>
<PgfCellLMarginFixed No>
<PgfCellTMarginFixed No>
<PgfCellRMarginFixed No>
<PgfCellBMarginFixed No>
<PgfLocked No>
> # end of Pgf
```

After reviewing this, it's probably no surprise to find out that MIF files are generally larger than binary .fm files. Expect MIF files to be about 10 times larger than the corresponding .fm file, so saving a 100KB .fm file to MIF might result in a 1MB or larger MIF file.

Many third-party utilities that can read or write FrameMaker files uses MIF as an intermediate format. HTML converters such as ePublisher and MIF2GO save FrameMaker files to MIF, and then process the resulting MIF files to produce HTML or other markup. Many database applications (like MS Access) can generate MIF output, which can then be opened in FrameMaker.

Appendix F: Maker Interchange Format

Note In addition to MIF format, a second markup format, Maker Markup Language (MML), is also available. MML development stopped somewhere around FrameMaker 3, so it's not nearly as rich as MIF. MML, however, is much easier to create than MIF. For this reason, MML is used for some applications because creating a complete MIF file would be too much work. For details on MML, refer to the MML Reference, found online at help.adobe.com/en_US/FrameMaker/8.0/mml_reference.pdf.

Adobe has the following *MIF Reference* available online:

FrameMaker 9 MIF Reference

help.adobe.com/en_US/FrameMaker/9.0/MIF_Reference/MIF_Reference.pdf

FrameMaker 11 MIF Reference

http://help.adobe.com/en_US/FrameMaker/11.0/FrameMaker/11.0/MIFReference/mifref.pdf

FrameMaker 12 MIF Reference

http://help.adobe.com/en_US/framemaker/mifreference/mifref.pdf

The *MIF Reference* describes MIF syntax and options in great detail. We strongly recommend that you consult this manual for detailed information about MIF syntax. In addition, two versions of MIF are available: A version related to your current FrameMaker version, and MIF 7. The MIF 7 version can come in handy if you need to open your document in something more than one version backward of FrameMaker

At the highest level, the MIF file is organized into the following sections:

- **MIFFile:** Identifies the file as a MIF file and provides the FrameMaker version from which the file was created. Every MIF file must start with this statement.
- **Units** and **CharUnits:** Specify measurement units for this file.
- **ColorCatalog:** Lists the color definitions.
- **ConditionCatalog:** Lists the conditional text tag definitions.
- **BoolCondCatalog:** Lists the defined conditional expressions.
- **DefAttrValuesCatalog:** Lists the default attribute values, used for filtering structured documents. Empty for unstructured documents.
- **AttrCondExprCatalog:** Lists defined conditional expressions for filtering structured documents. Empty for unstructured documents.
- **CombinedFontCatalog:** Lists combined fonts (used in double-byte languages such as Japanese).
- **PgfCatalog:** Lists the paragraph tag definitions.
- **FontCatalog:** Lists character tag definitions.
- **Ruling Catalog:** Lists table rule definitions.
- **TblCatalog:** Lists table definitions.
- **KumihanCatalog:** Provides line composition rules for Japanese text.
- **Views:** Lists color view definitions.
- **VariableFormats:** Lists variable definitions.
- **MarkerTypeCatalog:** Lists available custom marker types.

- **XRefFormats:** Lists cross-reference format definitions.
- **Document:** Lists document defaults, including volume, chapter, page, paragraph, and footnote numbering; document window size and location; text options; track text edits settings; change bar properties; and view options.
- **BookComponent:** Lists settings for generated files.
- **InitialAutoNums:** Lists default starting numbers for autonumbered paragraphs.
- **Dictionary:** Lists words not in the dictionary but allowed in this document (added when you click the Allow in Document button in the spell-checker).
- **AFrames:** Lists the anchored frames that occur in this document. The frames' unique identifiers are used to reference the frames in the text flow where they are anchored.
- **Tbls:** Lists the tables that occur in this document. The tables' identifiers are used to reference the tables in the text flow where they are anchored.
- **Page:** Lists the master pages, reference pages, and body pages that occur in this document.
- **Text Flow:** Lists the text flows that occur in this document and their contents.

Performing a MIF wash

If you import text from other applications or work with legacy FrameMaker documents, you may end up with files that are just not quite right. One common technique for sanitizing these files is a process commonly referred to as *MIF wash*. When you export a file to MIF, any problem code is dropped, and thus when re-read into FrameMaker, the offending content no longer exists. You can perform a MIF wash manually on single file, or use any number of scripts or utilities such as MIF2Go to wash books or multiple files at once.

BookMIFWash is included as a DLL in the AdobeFrameMaker11\Samples\ScriptsAndUtilities folder

Remember to use **File > Save As...** after washing to avoid creating a file with a double extension of .mif.fm.

Creating a MIF file

To create a MIF file from a regular FrameMaker file, save the FrameMaker file to MIF format. Follow these steps:

1. Open the file you want to save as MIF.
2. Select **File > Save As** to display the Save Document dialog.
3. In the Save as type drop-down list, click MIF 2015.
4. In the File name field, specify a name for the file. Be sure to change the file extension to MIF; FrameMaker doesn't always do this automatically.
5. Click the Save button.

FrameMaker creates the MIF file.

Opening a MIF file in FrameMaker

To open a MIF file and convert it back to regular FrameMaker format, follow these steps:

1. Select **File > Open** to display the Open dialog.
2. Locate the MIF file on your system and select it.
3. Click the Open button to open the file.

FrameMaker reads the MIF file and displays it as a FrameMaker file. During conversion, a console window may display error messages. FrameMaker will skip any information it cannot process in the MIF file. For example, if you save a FrameMaker 2015 file to MIF and then open that MIF file in version 7 (or earlier), error messages will be displayed.

Viewing a MIF file

You can view a MIF file in any text editor that supports UTF-8 character encoding.

Graham Wideman provides an interesting free utility called FrameMaker MIFBrowse if you don't have a text editor available, or you can use FrameMaker as a text editor.

MIFBrowse is more than a few years old, and can be found at www.grahamwideman.com/gw/tech/framemaker/mifbrowse.htm)

To open a MIF file as text in FrameMaker, follow these steps:

1. Select **File > Open** to display the Open dialog.
2. Locate the MIF file you want to open, hold down the CTRL key, and double-click the file or hold down the SHIFT key and click the Open button.
3. In the Reading Text File dialog box, from the Encoding drop-down list select UTF-8, then click Read.

FrameMaker reads the file as a text file instead of interpreting the MIF commands.

Caution   Do not save the text file as a FrameMaker file. You could inadvertently overwrite your source FrameMaker file with text.

If you see unusual text within the MIF file in your editor, such as the following,

<String `ing. Best of all, it doesnâ€™t require a computer!'>

make sure your editor provides UTF-8 encoding for viewing and working with text files.

Cool stuff you can do with MIF

There are numerous scenarios in which a MIF file can be helpful. Before you begin working in a file, assemble your arsenal; you will need the following items:

- A text editor that can handle MIF files and provides UTF-8 encoding support. I recommend NotePad++. (Notepad and SimpleText cannot open most MIF files because the files are too big.)
- A powerful search-and-replace engine. Some text editors offer advanced search-and-replace options; you can also use a scripting language such as Perl.

- A basic understanding of MIF syntax and the items you need to modify. If, for example, you want to work with index markers and aren't sure how the index markers look in MIF format, create a small sample file, insert an index marker with some unique text, save the file to MIF, and search for your sample text.
- Access to character code references. For MIF files, you can use the Windows Character Map utility to find the Unicode code points, if needed.
- A backup of your file, just in case.

The following sections describe a few common uses for MIF files.

Eliminating file corruption problems

FrameMaker files are usually quite stable. Every now and then, though, a file goes haywire for no apparent reason. Common problems include odd characters at the end of a file, crashes when generating books, tables with odd characters in them, tables that don't display correctly on the page, and others. One technique that can sometimes eliminate the problem is to save the file to MIF, open the MIF in FrameMaker, and save it back to FrameMaker format. This process is known as MIF wash, and is described on page 565. When performing a MIF wash you're not making any changes in the MIF file, so this operation is relatively safe.

There are a couple of problems that occur frequently that saving to MIF solves:

- If Find/Change won't recognize words that you know are present in the file, the problem may be hidden characters that were imported from Microsoft Word files.
- The color catalog becomes bloated with colors you didn't define. If you have many colors with names that start with RGB, you may have imported a PNG graphic that uses an optimized palette. Resave the PNG with the maximum number of colors to keep the colors from showing up in the color palette again.

Making a file available to an older version of FrameMaker

Older versions of FrameMaker cannot open files created in newer versions. For example, a file created in FrameMaker 6 cannot be opened in any release of FrameMaker 5. In many versions, Adobe does include a "save as FrameMaker X-1" option, so that, for example, you can write FrameMaker 2015 files to a format that's compatible with FrameMaker 12. If, however, you need to open files in a version prior to FrameMaker 12, you must go through MIF format.

In FrameMaker 2015, save your file to MIF 7 format, open the older version of FrameMaker, and open the MIF file there. You will see a number of error messages in the console window, but this technique should allow you to open files in all but the earliest versions of FrameMaker.

Keep in mind that any features that aren't available in the earlier version will be stripped from your file, which can cause formatting problems, especially if you are going back to a very old version of FrameMaker.

Creating a character tag for a vertical baseline shift

Some settings are available as MIF file settings but are not accessible through the graphical interface. Vertical baseline shift is one of those items. In the Character Designer, you can set up a character tag that uses a superscript, but you cannot use multiple vertical offsets in a single file—you only have superscripts available, and they always use the same vertical offset.

There is, however, a vertical offset setting available in FrameMaker. In a MIF file, you can assign this offset using the <FDY> token. You can create a character tag definition in a MIF file. You can then use that file as a template and import the tag into a document.

To create a character tag with a vertical baseline shift, follow these steps:

1. Create a text file and insert the following text:

   ```
   <MIFFile 8.00>
   <FontCatalog
   <Font
    <FTag `VerticalOffset'>
    <FDY -60>
    <FLocked No>
   > # end of Font
   > # end of FontCatalog
   ```

   The offset value can range from -100 to 100; negative values move the text up, and positive values move the text down. The quotes surrounding the words VerticalOffset are the backtick character for the opening quote and a straight single quote for the closing quote.

2. Save the file with a .MIF extension.

3. In FrameMaker, open the MIF file and display the document that you want to import the character tag into.

4. Select **File > Import > Formats** to display the Import Formats dialog.

5. Select the MIF file from the Import from Document drop-down list.

6. In the Import and Update section, check the Character Formats checkbox; make sure all other items are unchecked.

7. Click the Import button. The VerticalOffset character tag is displayed in the Character Catalog. When you apply it, a vertical offset results.

Caution If you make changes to the VerticalOffset character tag in the Character Designer and click the Update All button, the baseline offset setting will be deleted.
You cannot access the vertical offset setting from the Character Designer.

Using MIF fragments to update catalog settings

You can use the technique described in the previous section to add information to other catalogs. This is especially useful if you want to import only user variables.

When you import variables from one FrameMaker document to another, both the system variables and the user variables are imported. Often, you may want to update only the user variables. There are probably a few variables—such as book title or copyright date—that you update frequently. Also, you may need to import only the user variable if the system variables conflict across the book. For example, if the templates for the chapter files and the index use the

Cool stuff you can do with MIF

same system variable for different purposes (such as redefining the Running H/F 1 variable on the master pages), then importing the system variables would break the headers in some of the files.

You can update only the user variables by doing the following:

1. In a text editor, create a MIF file that contains definitions for the user variables you want to import:

   ```
   <MIFFile 11.00>
   <VariableFormats
    <VariableFormat
     <VariableName `book_title'>
     <VariableDef `Troubleshooting My Document'>
    > # end of VariableFormat
    <VariableFormat
     <VariableName `version_number'>
     <VariableDef `6.2.5beta'>
    > # end of VariableFormat
   > # end of VariableFormats
   ```

2. Save the MIF file and close the text editor.
3. Open the MIF file in FrameMaker. Do not save the MIF file.

Caution If you save the MIF file, FrameMaker automatically adds pages of default information to the MIF—including default system variables.

4. Import the variables from the MIF file. Only the user variables you defined are imported.

Note After you import the variable definitions, they are available in the Variables pod. You can edit them just like any other variable.

5. Save the FrameMaker files, but close the MIF fragment without saving it.

Performing global search-and-replace operations

Inside FrameMaker, you have a significant number of Find/Change options available, but sometimes, you'll run across a problem that can't be automated in the Find/Change pod.

For example, you might have a situation where you were working on an index. After several hours of work, you realize that you inadvertently inserted markers with a Subject type instead of an Index type.

Here's how to fix this using the find/change feature in FrameMaker:

1. Select **Edit > Find/Change**. The Find/Change pod is displayed:
2. In the Find drop-down list, click Marker of Type; in the field to the right of Marker Type, type in **Subject**.
3. Click the Find button. FrameMaker locates the first Subject marker.
4. Select **Special > Marker** to display the Marker pod.

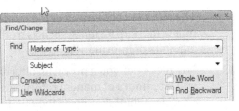

569

5. In the Marker Type drop-down list, click Index.
6. Click the Edit Marker button.
7. Repeat steps 3–6 for each problem marker.

The MIF alternative is much more appealing. Assuming you have not intentionally inserted Subject markers, do the following:

1. Save the problem files to MIF format.
2. Write a script that locates the Subject markers, or just use search and replace in your editor. The items that are the same for each marker are shown in boldface in the following example:

   \<Marker
   \<MType 4**>**
   \<MTypeName `Subject'>
   \<MText `cross-stitching'**>**
   \<MCurrPage `87'>
   \<Unique 1075663>
   > # end of Marker

3. Globally change the \<MType> to 2 and the \<MTypeName> to Index, as shown in the following example:

   \<Marker
   \<MType **2**>
   \<MTypeName `**Index**'>
   \<MText `cross-stitching'>
   \<MCurrPage `87'>
   \<Unique 1075663>
   > # end of Marker

4. Save your modified MIF files.
5. Open the files in FrameMaker, and save them back to regular FrameMaker format.

These types of global search-and-replace operations can save you hours of work.

Writing Your Own Conversion Tools

If you are required to convert information in FrameMaker to other formats and third-party converters are not doing the job, you may want to consider writing your own conversion tool. With some Perl scripting knowledge, you could, for example, write a converter that takes a MIF file as input and creates help pages for a UNIX program.

The MIF file contains all of the information needed to display and print a file, so converting it to online markup formats, such as troff, HTML, or XML, almost always involves throwing away large amounts of information.

MIF files offer you a powerful alternative to working in the FrameMaker interface. The next time you face hours of manual reformatting or other tedious, repetitive work in FrameMaker, consider whether editing the MIF files might be quicker.

Index

< =0> *See Autonumbering* 122

Symbols

< =0> 544
⊥ *See* anchors
< > *See* angle brackets
' *See* apostrophes
* *See* asterisks; stars
` *See* backticks
• *See* bullets
\b *See* bullets
^ *See* carets
: *See* colons
† *See* daggers
$ *See* dollar signs
‡ *See* double daggers
§ *See* end of flow symbols
¶ *See* end of paragraph symbols
< *See* forced returns
? *See* question marks
" *See* quotation marks
; *See* semicolons
_ *See* suppress hyphens
> *See* tabs
\t *See* tabs
</> 97, 351, 545
T *See* Markers

Numerics

3D objects 288–290

A

<a+> 544
Abjad 123
Acrobat Distiller 398
active area, hypertext link 490
ActiveIX 366
actual text property 268
adding
    columns, text flow 245
    columns/rows to table 159–160
    files to book files 323
Adobe
    Acrobat Distiller 398
    FrameMaker. *See* FrameMaker
    Illustrator 78
    PDF Creation Add-On 398
    reporting bugs to 536
    Technical Communication Suite 398
Adobe Experience Manager 527
Adobe PDF Creation Add-On 398
Advanced sheet 124
Alert commands 495
algebra, matrix 497
Alif Ba Ta 123
aligning
    See also positioning
    anchored frames 257–258
    baselines 249
    columns, text 246
    equations 508
    paragraphs 115
    table cells 126
    tables 154
all caps 119
alphabetical
    list of markers 382
    list of paragraphs 382
    numbering 334
<$alphabetics> 368, 542
alternate text 267
<$ampm> or <$AMPM> 195, 542
anchor 150
anchored frames
    See also graphic frames
    attributes 269
    cropping 257
    finding 42

Index

floating 257
inserting 256
object properties for 267
positioning 257–258
shrink-wrapping 265
anchoring graphics 256–269
anchors 18
AND, ANDNOT operators 459
angle brackets 123, 134, 541
apostrophes 52, 180
applying
character tags 140–141
color to text and objects 430
condition tags 446–448
master pages 210–211
paragraph tags 110–112
templates 91–92
arcs, drawing 295
arranging
See also moving; rearranging
graphic objects 310–315
arrowheads 301
articles, threading PDF 405
assigning master pages 210–211
asterisks
See also stars
character overrides 147
in footnotes 67
search wildcards 39
attributes for anchored frames 269
authoring

structured versus unstructured 86
authors
file information 27
index of 385
automated numbering. See autonumbering
automatic
backup and save 28
cross-reference updates, preventing 186
line breaks, in equations 510
automating FrameMaker with ExtendScript 465
automation with ExtendScript 465
autonumbering
See also pagination
building blocks for 122
bullets 131
cautions, notes, warnings 132
chapter and volume numbers 134
counters 122
documenting 107
end of story 137
headings 136
numbered steps 132
paragraphs 121
repeated text or symbols 130
tables 155
tables of contents 354
autonumbering streams 107
<$autorange> 370, 542
Autorun View 468
Auto-Text plug-in 535

B

backslashes
in cross-references 180
escaping special characters 361
backticks 180
backup, automatic 28
backward compatibility 15
balancing columns, text 246
baselines
shifting vertically 567
synchronizing 249
Basic sheet 114
Bezier curves, drawing 295
bitmap images 263, 272
blank
counters 134

documents, creating 13
pages 216, 394
paragraphs 348, 521
bleeding thumb tabs 225–229
BMP files 270
body
pages 209
rows 150, 159
book error log 330
book files
See also books; documents; files
adding files to 323
creating 322
cross-references, updating in 185
non-FrameMaker files in 324

572

Index

numbering in 332–341
printing files from 341
rearranging files in 327
removing files from 327
renaming documents 328
shortcuts 327
spell-checking 342
updating 329–332
book palette. *See* book files
book window 322
bookmark, PDF files 403
books
    comparing versions 60–62
borders
    displaying 17

C

calculus symbols 502
camera-ready. *See* printing
capitalization 105, 119
Captivate 78, 288
carets 39
Carmen Publishing
    TableCleaner 72, 173
catalog settings, updating 568
caution format 132
CDR files 270
Cell fill 163
cell ruling color 164
CGM files 270
change bars
    character tags 148
    color, adding 431
    paragraph tags 118, 137
    using 58–59
changes, tracking 55–62
changing default fonts 548
<$chapnum>
    in book files 333
    in cross-references 183
    overview 542
    in paragraph tags 134
    in tables of contents 351
    in variables 196
Chapter Number variable 196
chapter numbers
    in book files 336–337
    building block for 134

document 243
style 296
boxes. *See* text frames
broken cross-references. *See* unresolved cross-references
bug reports 536
building blocks
    autonumbering 122
    cross-reference 181–183
    reference 541–545
    table of contents 351
    variables 191
bullets 36, 131, 544
Button Matrix command 497

displaying 123
<char_tag> 351, 545
character catalog 140
Character Designer 141
character format, finding 41
Character tags 140
character tags
    applying 140–141
    baseline shift 567
    creating 143–146
    deleting 147
    finding 41
    modifying 141–143
    overrides 143
    removing 146
    renaming 145
    tips 99, 148
    updating globally 145
    vertical baseline shift, adding 567
characters. *See* special characters
checkboxes 144
checking spelling. *See* spell-checking
Chinese 123
circles, drawing 295
clickable text 490
clipboard 43
clone case 39
Close All Hypertext Windows command 498
Close Current Window command 498
closing documents 15, 327
CMYK color model 427–428

colons 359, 363, 541
color
    applying to text and objects 430
    catalog 95
    changing background of 3D objects 290
    CMYK model 427–428
    condition indicators 452
    creating custom 437
    cross-references 431
    definitions 432–434
    deleting 439
    extra colors in catalog 567
    for graphic objects 296
    hiding 438
    HLS model 427
    in PDF files 440
    inconsistent settings in book 332
    knock out 433
    libraries 428, 434
    naming new 434
    for online output 440
    overprint 433
    PANTONE 435
    process 395, 428
    renaming 437
    RGB model 272, 427
    saturation 435
    separations 397
    showing 438
    spot 395, 429
    text 118, 431
    tint 296
    viewing 438–439
color catalog 425
columns, table
    See also rows, table; table cells; tables
    adding 159–161
    aligning 246
    balancing 246
    deleting 159–161
    resizing 151, 160–161
    rulings 156
    selecting 55, 151
    shading 157
    sorting by 168
    width, specifying 161
columns, text
    aligning 246
    gap between 237, 245

columns, text flow
    adding 245
command characters 361
comparing documents and books 60–62
composite reports 60
Computer Graphics Metafile files. *See* CGM files
computing in equations 511
condition tags
    applying 446–448
    condition indicators 449
    creating 452–453
    deleting 453
    displaying 200
    list of 383
    overview 443
    removing 448
    removing conditions from text 449
    Show as per Condition 455
    Show as per Expression 457
    variables 201
conditional text
    alternatives to 451
    choosing method for showing 454–457
    color in 431
    dimensions in 454
    finding 42
    inconsistent settings for 332
    keyboard shortcuts 448
    operators 459
    overview 443
    planning 450–452
    showing/hiding 454
<$condtag[condtag]> 542
<$condtag[hitag,... lotag,nomatch]> 198
Configuring PDF job options 399
configuring text options 51
configuring WebDAV 528
connecting text flows 241
container documents 513
contents, tables of. *See* tables of contents
continuation, table 197
control
    files 107
    points 305
conventions, text xxviii
conversion
    filters 69
converting
    cross-references to text 189

tables to text 172
text from other applications 69
text insets to text 524
text to table 170–171
variables to text 204
copied text and formats, finding 42
copy editing 79
copyfitting
　filling columns 247
　master page overrides 229
　printer driver, effect of changing 397
　production edit 79
　resizing text frames 238
copying
　paragraph tags 126
　special characters from Windows Character Map 37
　text 71
　text frames 239
CorelDraw graphics 270
corners, smoothing in graphics 306
counters 122–123
crash recovery file 28
creating
　See also adding; inserting
　book files 322
　character tags 143–146
　colors, custom 437
　cross-references 177
　documents, new 11–15
　generated lists 380
　graphic objects 294
　hypertext links 487–490
　image maps 491
　index files 358–360
　master pages 209–217
　MIF files 565
　paragraph tags 126–129
　PDF files 401
　table tags 158
　tables 150–152
　templates 93–96
　text insets 521–522
　user variables 201

Creation Date variable 194
<$creationtime> 204, 542
crop marks 396
cropping
　anchored frames 257
　graphic objects 309
cross-platform
　file names 29
　graphics 273
cross-references
　See also hypertext
　building blocks 181–183
　color, adding 431
　converting to text 189
　creating 177
　definitions 179
　deleting formats 184
　examples 184
　external, list of 383
　finding 41
　forcing updates 185
　formats 179
　formatting 180
　overview 175
　paragraph-based 177
　preventing automatic updates 186
　spot references 178
　unresolved 187–189, 332, 385
　updating 185–187
　versus hypertext 487
CudSpan Tools 534
curly quotes 52
<$curpagenum> 196, 542
Current Date variable 194
Current Page Number variable 195
Currently Opened Files 22
<$currenttime> 204, 542
curves, drawing 295
custom markers 382
custom master pages 215
customizing
　hyphenation 49–50
　tables of contents 348
　text flows 240–253

D

daggers 67
daiji 123

dashed lines 296, 302
dashes 36

Index

database publishing 83, 536
Datazone Ltd. Miramo 83, 536
date variables 194
<$dayname> 195, 542
<$daynum> 195, 542
decimal tab stops 116
Default Font sheet 117
default fonts, changing 548
<Default Para Font> 351, 545
<$defaulttitle> 542
deleting
    character tags 147
    colors 439
    condition tags 448
    cross-reference formats 184
    equations 508
    files from book file 327
    graphic objects 299
    index entries 362
    master pages 231
    paragraph tags 129
    table columns, rows 159–160
    table tags 169
    tables 151
    text frames 240
    user variables 205
delimiters, equation 501
deliverables. *See* documents
development (parallel, serial, modular) 83
dialog boxes, special characters in 100
dictionaries
    document 44, 48
    language 118
    personal 44, 47
    preferences 118
    site 44
dictionary-style running header/footer 200
dimensions, conditional text 454
directories, organizing 26
disconnecting text frames 251
discretionary hyphens 18, 50
display
    equations 502
    units 19
displaying
    condition tags 200
    Equations palette 500–502
    grid lines 17

    nonprinting items 17
    rulers 17
distributing graphic objects 312
document
    See also documents
    borders 243
    comparison reports 60
    conventions xxviii
    dictionaries 44, 48
    formats. *See* templates
    properties 51, 92
    window 16
documentation plan 75
documenting templates 107
documents
    See also book files; document; files; managed documents
    all in book 327
    backward compatibility 15
    blank 13
    closing 15
    comparing versions 60–62
    container 513
    creating new 11–15
    editing 78
    formatting 79
    opening 15
    planning 75
    printing 15, 79, 391–397
    read-only 494
    renaming 328
    saving 15
    single- or double-sided 253
    structured versus unstructured 86
    template documentation
    tracking changes in 55–62
    word count 58
Documentum 527
dollar signs 39, 541
dot leaders 353
double daggers 36, 67
double-sided documents 253, 394
DPI settings 261
dragging. *See* moving
drawing tools for graphic objects 294–299
Dropbox 525
duplicating. *See* copying

E

ECMA U3D 3.0 288
editing
    documents 78
    index entries 362
embedded
    indexes 79
    tables of contents 347
EMF files 270
Encapsulated PostScript files. *See* EPS files
end of
    flow symbols 18, 241
    paragraph symbols 18
    story graphic 137
<$endrange> 360, 363, 542
Enhanced Metafile files. *See* EMF files
EPS files 270
equations
    delimiters 501
    display 502
    evaluating 511
    formatting 509–511
    modifying 506–508
    resizing 510
    selecting 504
    spacing 507–508
error log, book 330
evaluating equations 511
Exit Application command 498
exit command 498
expressions, conditional text. *See* Show as per Expression
expressions, mathematical. *See* equations
ExtendScript 465
    Import Formats Special ES 95
ExtendScript Toolkit 469
Extensible Metadata Platform. *See* XMP
Extensible Stylesheet Language transform files. *See* XSLT files

F

facets, image 273
Farsi 123
Favorite View 467
feathering text 247
figures, list of 381
file
    extensions 23
    information, saving 27
    locking 23
    permissions 331
filename variables 196
<$filename> 183, 196, 542
files
    See also book files; documents; managed documents
    generated 323, 487
    grouping. *See* book files
    modifying 342
    organizing 26, 268
    printing 341
    read-only 494
    renaming 187, 328
    view-only 332, 488, 490
fill patterns 296
filters
    conversion 69
    graphic 264
Find/Change 38, 342
    limiting search 40
    search options 40
    troubleshooting problems 567
Finite Matters Ltd. PatternStream 83, 536
flagging changes. *See* change bars
Flash files. ??–284
flipping graphic objects 317
floating anchored frames 257
flow
    definitions, TOC 350
    end of symbols 18, 241
    tags 210, 233, 252
    text. *See* text flows
fonts
    changing defaults 548
    default 117
    in equations 509
    finding renegade 386
    list of 384
    missing 25, 331
    substituting 25, 399
    units 19
footers. *See* running headers/footers

footing rows 150, 159
Footnote formatting 63
Footnote numbering 63
footnotes 62–68
    endnotes 62
    finding 42
    inserting 64
    modifying formatting 66
    numbering 63–65, 67–68, 340
    table footnotes 63, 66–67
forced returns 18
forcing hyphenation 50
foreground pages. See body pages
format catalogs
    character 140
    color 425
    object styles 279
    paragraph 128
    table 152
formats. See character tags; cross-references; paragraph tags
formatting
    characters. See character tags
    cross-references 180
    documents 79
    equations 509–511
    glossaries 137
    indexes 365–366
    overrides 112
    paragraphs. See paragraph tags
    tables of contents 348, 352
Formatting Bar 111
FrameImage facets 273
FrameMaker
    mailing lists 533
    preferences 23
    scripts 534
    starting 10
    strengths 3
    structured versus unstructured 10
    switching between structured and unstructured 29
    toolbar buttons 20
    web sites for 533
frames
    anchored. See anchored frames
    graphic. See graphic frames
    text. See text frames
FrameScript 534
FrameVector files 273
freehand curves, drawing 295
FSA IXgen 362
<$fullfilename> 183, 196, 542
functions (mathematical) 502

G

gap
    columns, text 237, 245
    graphic objects 313
    measurement 15
    tables 155
generated
    files 323, 487
    lists 380
GIF files 270
global
    search-and-replace 569
    updates 145, 158
glossaries, formatting 137
Go to URL command 497
graphic editing software 78
graphic filters 264
graphic frames
    for cropping and masking 309
    drawing 295
    on master pages 210
    on reference pages 275
    renaming 277
Graphic name 286
graphic objects
    arranging 310–315
    control points 305
    creating 294
    cropping 309
    deleting 299
    distributing 312
    flipping 317
    gap settings 313
    gravity 303, 308
    grouping 313
    layering 314
    orientation 315–318
    resizing 303–304
    rotating 315

running text around 313
Graphics
    set graphic name 286
    U3D 288
graphics
    3D objects 288–290
    active areas, creating 490
    alternate text 267
    anchoring 256–269
    automatically updating from web 261
    bitmap images 263, 272
    choosing best 270
    conditional text, applying 448
    CorelDraw 270
    cross-platform 273
    end of story 137
    Flash files 283–284
    formats 270–274
    gray boxes 266, 273, 396
    HTTP file path 262
    image facets 273
    imported, list of 384
    importing 259–260
    on master pages 278
    missing 266
    organizing files 268
    on reference pages 275–278
    resizing imported 264
    scaling 263
    sizing 261
    SnagIt 535
    watermark effect 220
    in workflow 78
Graphics Interchange Format files. See GIF files
gravity 303, 308
gray boxes for graphics 266, 273, 396
grids 17, 297
grouping graphic objects 313
GroupTitlesIX 367, 369

H

handles
    selection 298, 303–304
    table 151
hanging indent 122
hard spaces 18
headers/footers. See running headers/footers
heading rows 150, 159
headings, numbered 136
heads, run-in 120
heads, side 120, 242
Hebrew 123
hidden text. See conditional text
hiding conditional text 454
hira 123
History warnings 35
HLS color model 427
hot spots, creating 490
<$hour> 542
<$hour> variables 195
HTML
    color in 440
    creating 85
hyperbolic functions 502
hyperlinks in PDF files 409
hypertext
    See also cross-references
    commands 495–498
    creating links 487–490
    in generated files 487
    image maps 491
    read-only documents 494
    versus cross-references 487
    web addresses 492
Hypertext Markup Language. See HTML
hyphenation 42, 49–50, 124
hyphens
    discretionary 18, 50
    nonbreaking 50
    suppressing 18, 50

I

Icon size/color 20
IgnoreCharsIX 366–367
illustrations. See graphics
Illustrator 78
image
    facets 273

maps 491
Import Formats Special ES 95
imported graphics, list of 384
importing
    bitmap images 263
    graphics 259–260
    list of graphics 384
    from Microsoft Word 72
    settings from template file 91–92
    SVG files 262
    text from other applications 69
    text. *See* text insets
    user variables only 568
    vector graphics 261
inconsistent
    color settings 332
    conditional text settings 332
indents
    basic 115, 117
    hanging 122
    table 154
indexes
    of authors 385
    creating 358–360
    custom 385–386
    editing and deleting entries 362
    embedded 79
    entry basics 359–365
    file-level 360
    formatting 364–366
    inserting entries 361
    markers 361

page number formatting 370–371
page ranges 360, 363, 370
page separators 367
planning 79
reference pages 366
removing chapter numbers 371
"see also" references 365
"see" references 364
sort order 365, 368
stand-alone 79, 360
subentries 359, 363
tips 362
workflow 79
IndexIX 367, 370
Indic 123
ini. *See* maker.ini
inline equations 502
inserting
    anchored frames 256
    cross-references 177
    equations 502
    footnotes 64
    special characters 35
    tables 150–152
    variables ??–193, 568
insets, text. *See* text insets
interface, switching between structured and unstructured 29
interline and interparagraph padding 247
invisible text 95
IX flow 366
IXgen 362

J

job options for PDF files 399
joining lines 307
JPEG (Joint Photographic Experts Group) files 270

Jump Back command 496
Jump to commands 495
justification, line 115

K

kanji 123
kata 123
keep with settings 120
kerning, pair 119
  506
keyboard shortcuts
    applying conditional text 448
    character tags 140

equations 506
inserting variables 193
moving equations 507
paragraph tags 112
problems with 112, 141
special characters 38
unsupported from previous versions 38
knock out 433

L

landscape master pages 221–225
language
    dictionary 118
    support 87
<$lastpagenum> 196, 542
layering graphic objects 314
layout overrides 229–231
leaders, dot 353
leading. *See* line spacing
Leximation tools 535
libraries, color 428, 434
lighting 3D objects 290
line
    ends 296
    justification 115
    patterns 296
    widths 296, 300
line breaks
    in equations 508, 510
    restricting 53
    in TOC 356
line spacing
    adjusting 508
    negative 137
    setting 115
lines
    dashed 302
    drawing 295
    joining 307
links
    hypertext. *See* hypertext
    web address 492
list of
    figures 381
    markers 382
    paragraphs 381
    references 383–385
    tables 381
literals 361
locking files 23
log 330
logarithmic functions 502
lowercase text 119
low-resolution images, printing 396

M

mailing lists 533
Maker Interchange Format. *See* MIF files
maker.ini
    always download updated graphics from web 261, 548
    customizing 547–549
    setting default substitution fonts 548
    setting pasting order 548
managed documents
    WebDAV overview 528
manual resizing 303
mapping
    tags to master pages 212–214
mapping table 212
margins, table cell 126
marker indicators 18, 361
Marker Markup Language. *See* MML
<$marker1> variables 198, 543
markers
    Cross-Ref. *See* cross-references
    finding 41
    index 361
    index of 386
    Leximation MarkerTools plug-in 535
    list of 382
    text in running header/footer 199
<$markertext> 382
masking graphic objects 309
master pages
    See also body pages
    applying and assigning 210–211
    bleeding thumb tabs 225–229
    creating 209–217
    creating default pages 215
    custom 215
    deleting 231
    displaying 210
    graphics, importing on 278
    landscape 221–225
    mapping table 212
    mapping tags 212–214
    modifying 217–229
    page layout overrides 229–231
    reapplying globally 211

rearranging 217
removing overrides 229–231
renaming 216
running headers/footers 218
text flows 234–240
variables in 192, 197, 568
watermarks 219–221
MasterPageMaps reference page 212–214
math. *See* equations
matrix command 497
merge table cells 165
merging table cells. *See* straddling table cells
Message Client command 497
metadata 27
Microsoft Visio 78
Microsoft Word
  importing content from 72
  tables, converting 173
MicroType, FrameMaker-to-PDF TimeSavers 181, 408, 535
MIF files
  creating 565
  editing 561
  MML 564
  opening 566
  opening in older version 567
  organization of 564
  uses 566–570
  viewing 566
MIF wash 565
<$minute> 195, 543
<$minute00> 543
Miramo 83, 536
missing
  fonts 25, 331

graphics 266
missing text 95
MML 564
Modification Date variable 194
<$modificationtime> 204, 543
modifying
  character tags 141–143
  color definitions 432–434
  condition tags 452–453
  document dictionary 48
  equations 506–508
  files, book-level 342–344
  footnote formatting 66
  graphic objects 299–310
  master pages 217–229
  paragraph tags 113–126
  personal dictionary 47
  system variable definitions 193–201
  table tags 152–158
  text flows 240–253
  user variables 203, 568
Modifying System Variables 193
modular development 83
modular documentation 513, 518–520
  See also text insets
<$monthname> 195, 543
<$monthnum> 195, 543
moving
  equations 506–508
  tables 151
  text frames 239
multichannel publishing 80
multicolumn text frames 237
multifile chapters 337
Multimedia 286

N

<n> building blocks 545
named destinations 409
naming
  colors 434
  graphic frames 277
naming conventions
  templates 104–106
navigating in equations 505
negative line spacing 137
network file locking 23
<$nextsubdoc> 543

nonbreaking hyphen 50, 135
nonbreaking space 135
nonbreaking spaces 18
nonprinting symbols 18
<$nopage> 360, 543
NOT operator 459
note format 132
notification scripts 468
numbering
  See also autonumbering
  chapters 336

Index

footnotes 63–65, 67–68, 340
formats for 334
headings 136
multifile chapters 337
paragraphs 339
steps 132

table footnotes 341
volumes 337
Numbering sheet 121
numerical variables 195
<$numerics> 368, 543

O

object
    linking and embedding 274
    properties, anchored frame 267
Object Model Viewer 473
Object pointer 295, 298
object styles 279
<$ObjectId> 543
<$ObjectType> 543
offset printing 428
offset, text 54
    See also gap; spacing
OLE 274
online documents 440
    See also HTML; PDF files
online help overview 85
Open Document commands 496
opening
    all files in book 327
    documents 15
    file in older version 567
    MIF files 566
    text files 566
openlink command 496
openlinkfitwin command 496
opennew command 497

openpage command 497
operators
    condition 459
    equation 501
optimizing PDF files 407
OR operator 459
organizing
    directories and files 26
    graphic files 268
orientation of graphic objects 315–318
orphan
    lines 120
    rows 155
ovals, drawing 295
overline 118
overprint 296, 433
Overrides 112
overrides
    character tag 143
    formatting 112
    identifying 130
    master page 229–231
    page layout 229–231
    removing 112

P

Package 27
padding, interline and interparagraph 247
page
    layout overrides 229–231
    number formatting, index 370
    numbering. *See* pagination
    ranges, index 360, 363, 370
    scrolling 19
    separators, index 367
page breaks
    avoiding awkward 98
    removing 92

Page Count variable 195
page sizes
    checking 228
    custom 13
    landscape 221
    in PDF files 396
    specifying 14
<$pagenum>
    in book files 333
    in cross-references 183
    overview 543
    in tables of contents 351

Index

pages. *See* blank pages; body pages; master pages; reference pages
pagination
   See also numbering
   managing 332–341
   suppressing in index 360
Pagination sheet 119
pair kern 119
PANTONE colors 435
paper size 14, 393
paragraph
   catalog 128
   symbols 18
Paragraph Designer 113–126
paragraph tags
   applying 110–112
   copying 126
   creating 126–129
   customizing 136–137
   deleting 129
   finding 41
   global changes to 128
   modifying 113–126
   renaming 127
   tips for creating 98
paragraphs
   alignment 115
   blank, removing 348
   color, adding 431
   list of 381
   numbering 339
   positioning on page 119
   space above/below 115
   table cells 126
parallel development 81
<$paranum>
   in cross-references 182
   overview 543
   in tables of contents 351
   in variables 196
<$paranum[paratag]> 198
<$paranumonly[paratag]> 198, 543
<$paranumonly> 183, 196, 351
<$paratag[paratag]> 198, 543
<$paratag> 183
<$paratext[+,paratag]> 198
<$paratext[paratag]> 198
<$paratext> 182, 351, 543
<$parentdoc> 543

pasting
   order, setting in maker.ini 548
   text 71
PatternStream 83, 536
PDF files
   Acrobat Distiller 398
   Adobe PDF Creation Add-On 398
   bookmarks 403
   color in 440
   creating 401
   hyperlinks in 409
   job options files 399
   MicroType FrameMaker-to-PDF TimeSavers 535
   optimizing 407
   page size 396
   registration marks 402
   for review 78
   Save As option 401
   screen readers 405
   setting default printer 397
   settings 96
   tagged 405
   threading articles in 405
   troubleshooting 411
   used as graphics 270
   zoom, default 401
PDF Job Options
   configuring 399
pen pattern 296
personal dictionary 44, 47
Photoshop 78
pipe symbols 39
placeholder counters 123
Placing Captivate output 288
planning
   indexes 79
   projects 75
PNG files 270
pointer in FrameMaker 298
polygons, drawing 295
polylines, drawing 295
Popup Menu command 497
Portable Network Graphics files. *See* PNG files
positioning
   See also aligning
   anchored frames 257–258
   equations 501
preferences
   dictionary 118

general 23
preventing
    automatic updates 186
    hyphenation 50
previouslink command 496
previouslinkfitwin command 496
<$prevsubdoc> 543
printer-resident fonts 397
printers, changing 397
printing
    in black and white 395
    from book files 392
    documents 15, 79, 391–397
    double-sided documents 394
    files 341
    individual document file 392
    low-resolution images 396
    offset 428
    paper size for 393
    registration marks 396

specifying pages 392
    suppressing blank pages 394
    thumbnails 395
    troubleshooting 393
    workflow 79
process color 395, 428
production editing 79
project planning 75
properties
    See also preferences
    anchored frame 267
    document 51, 92
    text frame 235
proportionate resizing 304
pseudo structure 134
publishing
    database 83, 536
    multichannel 80
    workflow 74

Q

QR code 292
question marks
    for equation objects 503
    as search wildcards 39

quit command 498
quitall command 498
quotation marks 52, 180

R

<r> building blocks 545
read/write file permissions 331
read-only documents 494
rearranging
    files in book file 327
    graphic objects 310–315
    master pages 217
rectangles, drawing 295
reference pages
    graphics on 275–278
    IX flow 366
    mapping tables 212
    TOC flow 349
<$referencename> 383
references
    index of 386
    list of 383–385
regex 39
Registered View 468

registration marks 396, 402
Regular Expressions 39
<$relfilename> 544
relinking missing graphics 266
removing
    character tags 146
    condition tags 448
    conditions from text 449
    master page overrides 229–231
    overrides 112
renaming
    character tags 145
    colors 437
    files 187, 328
    graphic frames 277
    master pages 216
    paragraph tags 127
render mode 290
replacing (search and replace) 43

Index

reset counter 122
resetting counters 123
reshape handles 304
reshaping graphic objects 304
resizing
 See also scaling
 columns, table 151, 160–161
 equations 510
 graphic objects 303–304
 graphics 261
 imported graphics 264
 manually 303
 proportionately 304
 SVG files 263
 text frames 238
 zoom settings 16
resolution, image 261, 396
Restore last session 23
reviewing documents 78
RGB color model 272, 427
Rick Quatro 95
right-to-left text variable building blocks See online help

Roman numbering 334
rotating
 graphic objects 315
 pages 221–225
 table cells 167
rounded rectangles, drawing 295
rows, table
 adding and deleting 159–160
 sorting by 168
rubi 42
rulers, displaying 17
ruling
 column 156
 table cell 163–165
run-in heads 120
running headers/footers
 condition tags, displaying 200
 dictionary-style 200
 marker text in 199
 setting up 218
 variables 197, 568

S

saturation, color 435
saving
 all files in book 327
 automatically 28
 documents 15
 file information 27
 as PDF 401
Scalable Vector Graphic files. See SVG files
scaling
 See also resizing
 columns, table 161
 graphics 263
screen readers
 assigning alternate text for graphics 267
 PDF files for 405
Script Library Panel 467
scripting 466
scripts for FrameMaker 534
scrolling, page 19
search and replace 569
 See also Find/Change
search wildcards 39
<$second> 195, 544
secondary entries, index 359, 363

<$sectionnum>
 in book files 334
"see also" references 365
"see" references 364
selecting
 columns, table 55, 151
 equations 504
 graphic objects 298
 table cells 151
 text frames 238
selection handles 298, 303–304
semicolons 359, 363
separations, color 397
SeparatorsIX 366–367
serial development 81
Set graphic name 286
setting
 anchored frame object properties 267
 tab stops 116
shading
 columns 157
 table cell 157, 163–165
SharePoint 527
<$shortdayname> 195, 544

<$shortmonthname> 195, 544
<$shortyear> 195, 544
Show as per Condition
    showing/hiding 457
    tags needed 455
Show as per Expression
    building expressions 458
    layered tags 461
    showing/hiding 458
    single tags 463
    tags needed 457
showing and hiding conditional text 454
shrink-wrapping anchored frames 265
side heads 120, 242
Silicon Prairie Software tools 535
Single author scenario 90
single sourcing
    structured versus unstructured 86
    workflow 85
<$singlepage> 544
single-purpose templates 107
single-sided documents 253
site dictionary 44
sizing
    See also resizing
small caps 54, 119
Smart Insert
    character tags 140
    Paragraph Formats 112
Smart Quotes 52
Smart Select pointer 295, 298
Smart Spaces 53
smoothing graphics 306
SnagIt graphics tool 535
snap grid 240, 297
solid lines 296
sort order, index 365, 368
sorting table data 168
SortOrderIX 366, 368
spaces, nonbreaking 18
spacing
    above and below paragraph 115
    between columns 15
    equations 507–508
    line 115
    preventing extra 53
    Smart Spaces 53
special characters
    ANSI codes 38

finding rubi 42
in dialog boxes 100
inserting 35
keyboard sequences 38
pasted from Windows Character Map 37
symbols toolbar 36
Specify Named Destination command 495
specifying fonts 117
spell-checking 43–50
    book files 342
    dictionaries 44
    document dictionary, modifying 48
    eliminating 137
    hyphenation, customizing 49–50
    options 46
    personal dictionary, modifying 47
splitting text frames 251
spot color 395, 429
spot cross-references 178
spread 118, 507
square brackets
    in building blocks 541
    in indexes 365
    search wildcards 39
squares, drawing 236, 295
stand-alone
    indexes 79, 360
Star the dog 264
stars
    See also asterisks
    character overrides 143
<$startrange> 360, 363, 544
Step Formats 134
steps 132
story. *See* text flows
straddle 165
straddling table cells 165
straight quotes. *See* apostrophes; quotation marks
streams, autonumbering 107
stretch, text 54, 118
strikethrough 118
structured authoring 86
structured documents
    versus unstructured 86
    XML import, export 85
structured to unstructured interfaces 29
style sheets. *See* character tags; paragraph tags
subentries, index 359, 363
subscripts 54, 118

587

<$subsectionnum>
    in book files 334
substituting fonts 25, 399
summary reports 60
superscripts 54, 118
suppress hyphens 18, 50
SVG files
    output 271
    overview 261
SWF 288
symbols. *See* special characters

<$symbols> 368, 544
synchronizing baselines 249
synonyms. *See* Thesaurus
syntax, equation 511
System Variables
    valid location 193
system variables
    importing user variables separate from 568
    modifying definitions 193–201
    overview 191
    updating 201

T

Table
    cell fills 163
table anchor 150
table anchors 150
table catalog 152
Table Cell sheet 126
Table cells
    straddle 165
table cells
    aligning 126
    margins 126
    paragraph properties 126
    paragraph tags in 162
    rotating 167
    ruling and shading 163–165
    selecting 151
    straddling 165
Table Continuation variable 197
Table Designer 152, 154
table footnotes, numbering 341
Table of Contents not updating 349
Table Sheet variable 197
Table Styles
    custom colors 164
table tags
    changing 152
    creating 158
    deleting 169
    finding 42
    globally updating 158
    modifying 152–158
    overview 149
table variables 197
TableCleaner 72, 173
tables
    aligning 154
    autonumbering 155
    color in borders 431
    columns. *See* columns, table
    conditional text in 448
    continuation 197
    converting to text 172
    creating from text 170–171
    deleting 151
    finding 42
    gap 155
    handles 151
    inserting 150–152
    list of 381
    Microsoft Word 173
    moving 151
    rows. *See* rows, table
    ruling 156
    shading 157
    sorting 168
    title 155
tables of contents
    autonumbering 354
    customizing 348
    embedding in a file 347
    file 347
    formatting 348, 352
    line breaks 356
    setting up 345–348
Tabs
    consistent leaders and page numbers 352
tabs 18, 116
tag creep 89
Tagged Image File Format files. *See* TIFF files
tagged PDF files 405

Index

tags. *See* character tags
Tasha the dog 264
<$tblsheetcount> 197, 544
<$tblsheetnum> 197, 544
technical editing 78
TechSmith SnagIt 535
templates
    applying 91–92
    creating 93–96
    default 12
    documenting 107
    documents used as 13
    installed with FrameMaker 91
    naming conventions for 104–106
    single-purpose 107
    structured versus unstructured 86
    tips for designing 98–100
    writers using 76
text
    actual text assignment 268
    alternate for graphics 267
    appearance 117
    block. *See* text frames
    clickable 490
    color, adding 118, 431
    conditional. *See* conditional text
    conditions, removing 449
    configuring default options 51–55
    conventions xxviii
    copying 71
    feathering 247
    finding in marker 41
    FrameMaker format. *See* MIF files 561
    importing. *See* text insets
    lines, drawing 295
    offset. *See* gap; spacing
    pasting 71
    repositioning 508
    running around a graphic 313
    spacing 54
    strange characters in 26
    stretch 54, 118
    symbols 17–18
    tracking edits 55
    watermark effect 219
    width 54
text flows
    adding to master pages 236
    columns 245
    connecting 241
    customizing 240–253
    side heads 242
    understanding 234–240
text frame overflow 235
text frames
    on body pages 237
    copying 239
    deleting 240
    disconnecting 251
    drawing 295
    flow tags 252
    moving 239
    multicolumn 237
    overview 210, 234
    properties 235
    resizing 238
    selecting 238
    splitting 251
text insets
    alternatives to 518
    converting to text 524
    creating 521–522
    finding 41
    formatting 519
    list of 384
    managing 522–524
    overview 513
    planning 518–520
    text behavior 517
Thesaurus 51
threading articles in PDF files 405
thumb tabs 225–229
thumbnails, printing 395
TIFF files 271
time variables 194
TimeSavers tool 535
tint 296
TOC flow 350
TOC reference page 349
Toggle View Only button 60
Tombo registration marks 396
toolbar buttons
    inserting symbols 36
    overview 20
Tools palette 294–296
track text edits. *See* tracking changes
tracking changes 55–62
tracking. *See* spread

589

trigonometric functions 502
troubleshooting
    book updates 330–332
    color catalog bloated 567
    crash recovery file 28
    errors during printing 392
    file corruption 567
    Find/Change problems 567
    finding unresolved cross-references 41
    graphics too large 273
    gray boxes for graphics 273

missing fonts 25
missing graphics 266
PDF files 411
printing 393
reporting bugs 536
RGB color definitions 272
side head area not appearing 245
typing
    equations 504
    special characters 35

U,V

U3D 288
underline 118
Unicode support xxiv, 15, 87, 101
units, font and display 19
unresolved
    cross-references 187–189, 332, 385
    text insets 385
unstructured authoring 85
unstructured documents versus structured 86
unstructured to structured interfaces 29
updating
    books 329–332
    catalog settings 568
    cross-references 185–187
    globally 145, 158
    system variables 201
    user variables only 568
uppercase text 119
user variables
    converting to text 204
    creating 201
    deleting 205
    importing without system variables 568
    modifying 203
    overview 191
Valid Locations for System Variables 193
<$variable[varname]> 544
variables
    building blocks 191
    color, adding 431
    condition tag 201
    converting to text 204
    date 194
    deleting 205

filename 196
finding 41
importing from MIF 568
inserting ??–193, 568
keyboard shortcuts for 193
master page 192, 197, 568
numerical 195
overview 191
running header/footer 197
system definitions, modifying 193–201
table 197
time 194
updating 201
user, creating 201
vector graphics 261, 272
vertical baseline shift 567
viewing
    colors 438–439
    grid lines 297
    MIF files 566
view-only files 332, 488, 490
Visio 78
<$volnum>
    in book files 333, 337
    in cross-references 183
    overview 544
    in paragraph tags 134
    in tables of contents 351
    in variables 196
Volume Number variable 195
volume numbers
    building block for 134
    displaying 123
volumes, numbering 337

W

warning format 132
washing MIF 565
watermarks 219–221
web address links 492
web sites, FrameMaker-related 533
Web-based Distributed Authoring and Versioning. *See* WebDAV
WebDAV 24, 528
    See also managed documents; workgroups
WebDAV server 528
widow lines 120
width
    table column 161
    text 54
Wildcards 39
wildcards 39
Windows Character Map, pasting from 37
Windows Metafile files. *See* WMF files
WMF files 271
word
    count 58
    spacing 124

Word. *See* Microsoft Word
workflow
    database publishing 83
    editing 78
    evaluating options 80
    illustrations 78
    indexing 79
    modular development 83
    overview 74
    parallel development 81
    planning 75
    printing 79
    production editing 79
    serial development 81
    single sourcing 80, 85
    writing 76
workgroups
    overview 75
    WebDAV overview 528
wrapping anchored frames to content 265
writers, role in workflow 76
writing workflow 76

X, Y, Z

XML
    import and export 85
    in structured and unstructured documents 87
XMP 27
xrefs. *See* cross-references

XSLT files 85
<$year> 195, 544
yen character 36
zenkaku 123
zoom 16, 401

Index